Understanding MicroStation/J

A Basic Guide to MicroStation/J®
2D Drawing and 3D Modeling using SmartSolids™

Karen L. Coen-Brown, P.E.

Upper Saddle River, New Jersey
Columbus, Ohio

Never say never! This book is dedicated to my family who knows me best (and at my worst) and still loves me! Thank you, Whitney, Evan, Alex, and Jordan for being supportive when I said yes to another book. You're great kids and I'm extremely lucky to have you in my life. Remember to always do your best and be there for each other. To my parents, the way hasn't always been easy, but you've shown me the value of hard work and determination, for which I am thankful. To my in-laws, thank you for taking over my taxi and kid duties on such short notice. I truly appreciate the support. Last but definitely not least, I am especially grateful to Steve, my soul mate, for <u>still</u> being the other half of the Brown and Coen-Brown team. I'm so glad you love a challenge!

With an even bigger sigh of relief and an infinity (plus one) of love.
Anyone ready for some roller coaster rides?

Library of Congress Cataloging-in-Publication Data

Coen-Brown, Karen L.
  Understanding MicroStation/J: a basic guide to MicroStation/J 2D drawing and 3D modeling using SmartSolids / Karen L. Coen-Brown.
    p. cm.
  ISBN 0-13-025707-9 (pbk.)
  1. Computer graphics. 2. MicroStation. 3. Computer-aided design. I. Title.

T385 .C541366 2001
620'.0042'02855369—dc21   00-058008

Vice President and Publisher: Dave Garza
Editor in Chief: Stephen Helba
Acquisitions Editor: Debbie Yarnell
Production Editor: Louise N. Sette
Design Coordinator: Robin G. Chukes
Cover Designer: Jeff Vanik
Cover art: Top image: © State of Nebraska Department of Roads
                Middle and bottom images: © Charles J. Wood, Jacobs Facilities, Inc.
Production Manager: Brian Fox
Marketing Manager: Jimmy Stephens

This book was set in Arial by Karen L. Coen-Brown. It was printed and bound by Banta Book Group. The cover was printed by Phoenix Color Corp.

The following are registered trademarks of Bentley Systems, Incorporated: AccuDraw®, IntelliTrim®, MicroStation®, MicroStation Geographics®, MicroStation/J®, MicroStation Modeler®, MicroStation 95®, MicroStation SE®, MicroStation Triforma®, QuickVision®GL, SmartLine®.

The following are trademarks of Bentley Systems, Incorporated: CivilPAK™, Java™, MicroStation Schematics™, ProjectBank™, SmartSolid™.

The following is a registered trademark of Microsoft Corporation: Windows®.

The following are registered trademarks of Hewlett-Packard Company: Hewlett-Packard®, LaserJet®, PCL5®.

Copyright © 2001 by Prentice-Hall, Inc., Upper Saddle River, New Jersey 07458. All rights reserved. Printed in the United States of America. This publication is protected by Copyright and permission should be obtained from the publisher prior to any prohibited reproduction, storage in a retrieval system, or transmission in any form or by any means, electronic, mechanical, photocopying, recording, or likewise. For information regarding permission(s), write to: Rights and Permissions Department.

10 9 8 7 6 5 4 3 2 1

ISBN 0-13-025707-9

PREFACE

This book, *Understanding MicroStation/J*, follows in the footsteps of my successful previous text, *Understanding MicroStation 95/SE*. The text and its illustrations focus on the core software called MicroStation/J®. This includes the new SmartSolid™ technology used for 3D modeling that is now incorporated into the core product. This book continues the tradition of being an easy-to-read text covering the basics of creating 2D drawings and 3D models using MicroStation/J.

I have been using MicroStation® in my engineering graphics class since the fall of 1995. Throughout the years, I found that MicroStation can be intimidating, especially to someone new to CAD because of its overwhelming number of options and settings. I set out to create a text that shows you actual screen captures and concise explanations—all on the same page! *Understanding MicroStation/J* cuts through all the bells and whistles of the software and covers the practical essentials of MicroStation/J for the engineering and architectural fields. It is not intended to be an in-depth, extensive technical production reference. Rather, it is for the person wanting to learn the basics of MicroStation/J without having to sort through all the highly technical jargon. This straightforward, concise presentation of a complicated subject is advantageous whether you are teaching yourself or others.

Several aspects of this book set it apart from others:

Extensive use of illustrations. These highlight the various tools' settings and options that appear in the dialog boxes of the software. You, the reader, can see exactly what is being discussed since important points are indicated directly on the dialog box. You see information where and when you need it.

2D and 3D in a single book. You don't need to go out and purchase two large technical references to get started with MicroStation. This book covers the basics in both worlds.

Numerous examples. These show you how the software techniques and concepts covered are applied.

Exercises and questions. Included at the end of the chapters, these can be used in the classroom setting or for hands-on experience and review.

Understanding MicroStation/J is organized so that you learn to walk and then run through the power of MicroStation/J. The first fifteen chapters cover MicroStation/J in 2D. Starting with the graphical interface, the text moves quickly into basic 2D elements such as lines, circles, and text, and their modification and manipulation. Chapter 3, "Nitty Gritty," is a quick step-by-step example of how to create, edit, and print a simple drawing. The benefit of this chapter is that you can become productive early on rather than having to read several chapters before producing and evaluating a rough drawing. This is extremely important in a classroom setting but is also advantageous to any beginner because it exposes the power of the software. Chapters dealing with printing and measurements emphasize the concept that elements are drawn to represent real-size objects. Chapter 11 covers AccuDraw® and SmartLine®, topics that are easier to understand after some experience has been gained. Chapters 12 through 15 deal with the more advanced topics of cells, patterning, dimensioning, and reference files.

After experiencing the two-dimensional world, you are ready to work in three dimensions. Chapters 16–21 discuss several important 3D topics relevant to the beginning MicroStation/J user. Chapter 16 gives you the ability to view (and manipulate the viewing of) a three-dimensional design file. This ability is important even when you are not involved in the creation of the 3D model. The use of auxiliary coordinate systems in both 3D and 2D design files is discussed in Chapter 17 along with the advantage of using AccuDraw in 3D. Chapter 18 on rendering and visualization tools benefits anyone working with a 3D model. The chapter is set up so that existing files that come with MicroStation can be used to experiment with rendering. That way, when you go to the next chapter of creating models, you already know how to render them! Chapter 18 also includes an example showing how to produce a FlyThrough animation.

Chapters 19–20 deal with creating a 3D model. The emphasis here is on using SmartSolids to create and modify three-dimensional solid models. Chapter 19 focuses on the basic solid primitives and Boolean operations. Chapter 20 discusses using a 2D shape to create a 3D model as well as making modifications to the model. Chapter 21 covers Drawing Composition and its use in both 2D and 3D. With Drawing Composition, the 3D model itself is used to develop the standard engineering views, including sectioned views. Chapter 22 discusses Engineering Configurations, and some of the new technology, such as ProjectBank™ and Java™. This discussion will give you some idea of the direction that MicroStation is heading.

Finally, the text includes an appendix that has information about the default settings of MicroStation/J seed files and the directory structure. Tables indicate the default view attributes, working units, active attributes, and other important details of seed files.

This text's contents are a result of what works and what doesn't (after many semesters of trial and error) in the simplification of a vast amount of information and techniques. Because this text covers the **basics**, there are numerous tools and utilities found in the software that intentionally are not covered. For example, this text does not cover tags, points, parabolic elements, grouping, multi-lines, and curves. Some of these are not widely used; others are just specialized extensions of the simpler tools covered in the text.

Understanding MicroStation/J is an easy-to-read guide to using MicroStation/J. It won't make you an expert, but it will give you a great foundation on which to base your knowledge of the software. You'll experience the important aspects of the software without all the pain of digging through extensive information. It is a great text to teach out of, but, most importantly, its format and coverage have a great track record. This approach has proven to be a considerable benefit to learning MicroStation—whether you're a student or not!

ACKNOWLEDGMENTS

Many thanks to all the people involved in this project. First, I want to thank all of my students at the University of Nebraska-Lincoln. They served as an informal advisory board and understood when I needed a coffee break during class! Thanks go to the Mechanical Engineering Department at UN-L for their past and present support and to the State of Nebraska Department of Roads for making material available for my use. The awesome image(s) of Nebula GNX used for this text were provided by Charles J. Wood of Jacobs Facilities Inc. located in Arlington, Virginia. My sincere thanks to Charles. The people at Bentley Systems, Incorporated are also acknowledged for their support of this project. Steve Ethofer of Clarinda deserves kudos for quick and reliable service for production testing. A big thank you to Rande J. Robinson of the North Carolina Department of Transportation, who reviewed the manuscript in its development. Rande's straightforward feedback was greatly appreciated.

Once again, I want to applaud the efforts of the folks at Prentice Hall. I am extremely grateful that they still had the patience to work with me! Thanks go to Debbie Yarnell, she inherited this project (and me) and did a commendable job. Thank you to Louise Sette and the entire production team at Prentice Hall for putting it all together. Special thanks to Steve Helba for remaining available even after his promotion. Many, many thanks to Katie Bradford for her hard work, encouragement, and understanding. She was a great asset to the project and again deserves a round of applause.

BRIEF CONTENTS

Chapter 1: Introduction to MicroStation/J 1

Chapter 2: Starting Up 11

Chapter 3: Nitty-Gritty Drawing 37

Chapter 4: Drawing Aids and Tools 53

Chapter 5: Printing a Hard Copy 93

Chapter 6: Lines, Circles, Arcs, and Polygons 111

Chapter 7: Text 137

Chapter 8: Measurements and Information 159

Chapter 9: Changes and Modifications 173

Chapter 10: Manipulation 203

Chapter 11: AccuDraw and SmartLine 227

Chapter 12: Cells and Cell Libraries 255

Chapter 13: Patterning 281

Chapter 14: Dimensioning 303

Chapter 15: Reference Files 337

Chapter 16: Special Considerations of 3D Space 361

Chapter 17: Auxiliary Coordinate Systems and AccuDraw in 3D 381

Chapter 18: Rendering and Visualization 395

Chapter 19: 3D Modeling with SmartSolids 429

Chapter 20: Advanced Model Construction and Modification 463

Chapter 21: Drawing Composition 491

Chapter 22: Engineering Configurations, ProjectBank, and Java 535

Appendix: Seed File Information 543

Index 547

CONTENTS

Chapter 1: Introduction to MicroStation/J 1

What's New for MicroStation/J 2
Conventions Used in the Text 3
    Warnings 3
    Notes 3
    Key-ins 3
    Keyboard Shortcuts 3
    AccuDraw Keyboard Shortcuts 3
Graphical Interface 4
Mouse Actions 5
Windows and Dialog Boxes 6
Tool Buttons and Boxes 8
Pull-Down Menus 10

Chapter 2: Starting Up 11

MicroStation Manager 12
Workspace 14
    Project 14
    User 16
    Interface 19
    Style 19
Creating a New Design File 20
    Accessing the Create Design File Dialog Box 20
    Create Design File Dialog Box 20
Seed Files 21
Status Bar 23
Key-in 24
Standard Tool Box 25
    Accessing the Standard Tool Box 25
Undo 26
Redo 26
Help 26
Settings 28
    Accessing the Design File Settings Dialog Box 28
    Design File Settings Dialog Box 28
    Save Settings 28
Coordinate System and Working Units 29
    Example of Working Unit Parameters 30
    Changing the Working Units: Unit Names and Resolution 31
    Examples of Different Ways to Enter the Same Values 31
Global Origin 31
    Default Global Origin 32
Enhanced Precision 33
Coordinates 33
    Absolute and Relative 33
    Rectangular and Polar 33
    Key-Ins 34
    AccuDraw and Precision Input 34
Preferences 34
    Accessing the Preferences Dialog Box 34
    View Windows Category 35
    Operation Category 35
    Input Category 36
Saving Your Work 36

Chapter 3: Nitty-Gritty Drawing 37

Creating Your First Drawing 38
Before Starting 39
    Mouse Buttons 39
    Correcting Mistakes 39
    Display of Design File 40
New Design File 41
Tools Needed 41
Elements 42
Body, Screen, and Speaker 42
Knobs 45
Axis Lock Off 47
Diagonal Lines 48
Text 50
Edit 51
Printing 52

Chapter 4: Drawing Aids and Tools 53

Element Attributes 54

Accessing Active Element Attributes 54
Level Attribute 55
Color Attribute 56
Style Attribute 56
Weight Attribute 57
Class Attribute 57

View Controls 58
Update View 59
Zoom In 59
Zoom Out 60
Window Area 61
Fit View 62
Pan View 62

View Attributes 63
Accessing the View Attributes Dialog Box 63

View Levels 64
Accessing the View Levels Dialog Box 64
Display Level Numbers 65
Display Level Names 66

Naming a Level 68
Accessing the Level Names Dialog Box 68
Add Level Name 69

Snaps and Tentative 70
Accessing Snaps 70
Keypoint 71
Intersection 72
Tangent 72
Perpendicular 73
Parallel 74
Snap Mode and Override Snap 74

Grid 75
Grid Parameters 75
Grid Master and Grid Reference 75

Locks 77
Accessing Locks 77
Locks Dialog Box—Full Display 78
Grid Lock 78
Snap Lock 78
Axis Lock 78

Element Selection Tool Box 79
Selection Display 79

Element Selection 81
Selecting More Than One Element 81
Quick Way to Change the Attributes of an Existing Element 81

PowerSelector 82
PowerSelector Methods 83
PowerSelector Modes 85

Fence Tool Box 86

Place Fence 86
Fence Type 87
Fence Mode 88

Delete Fence Contents 89

Delete Element 90

Cursor 91

Questions 92

Chapter 5: Printing a Hard Copy 93

Plot Dialog Box 94
Accessing the Plot Dialog Box 94
Tools and Menus of the Plot Dialog Box 94
Status Fields of the Plot Dialog Box 95
Defaults of the Plot Dialog Box 95

Entity 95
View 95
Fence 95
Configuration 95

Plot 98

Preview Refresh 98

Page/Print Setup 99

Plot Layout 100

Plot Options 101
Plot Border 101
Fence Boundary 101

Plotter Driver 103
Page Setup Dialog Box for Non-Windows Printer 103

Printing to Scale 104
Specific Ratio for Scale to 105

Summary of Printing Sequence 108

Questions 109

Chapter 6: Lines, Circles, Arcs, and Polygons 111

Coordinates 112
Absolute and Relative 112
Key-Ins 112
Coordinate Entry Example 112

Linear Elements Tool Box 113

Place Line 114
Constrained Values 114
Example of Drawing Three Lines 116

Ellipses Tool Box 118

Place Circle 119
Diameter/Radius Size Constraint 119
Center Method 119
Edge Method 120
Diameter Method 121

Place Ellipse 122

x   Contents

Arcs Tool Box   122
Place Arc   123
    Center Method   123
    By Edge Method   125
Polygons Tool Box   126
Place Block   127
Place Regular Polygon   129
    Inscribed/Circumscribed Methods   129
    Edge Method   130
Questions   131
Exercises   132

Chapter 7: Text   137

Text Attributes   138
    Accessing the Text Attributes Settings   138
    Fonts   139
    View Button   139
    Height and Width   140
    Line Spacing   140
    Justification   140
    Underlining Text   141
Text Tool Box   142
Place Text   143
    Method Options   144
Edit Text   145
Match Text Attributes   147
Change Text Attributes   148
Enter Data Field   149
    Data Field Character   149
Fill In Single Enter-Data Field   150
Determining Text Size for Printing to Scale   152
Questions   154
Exercises   155

Chapter 8: Measurements and Information   159

Measure Tool Box   160
Measure Distance   161
    Between Points   161
    Perpendicular   162
Measure Radius   163
Measure Angle   163
Measure Length   164
    Single Element   164
    Shape   164
Measure Area   165
    Element Method   165
    Methods Involving Boolean Operations   165
    Flood Method   167
Analyze Element   168
Questions   169
Exercises   170

Chapter 9: Changes and Modifications   173

Change Attributes Tool Box   174
Change Element Attributes   175
Match Element Attributes   176
SmartMatch   176
Modify Tool Box   177
Modify Element   178
    Modifying a Block   178
    Modifying a Line or a Circle   180
    Modifying Arcs   181
Partial Delete   182
Extend Line   183
Extend Elements to Intersection   184
    Example of Extending Linear Elements   185
    Example of Extend Elements to Intersection on Arc and Line Elements   186
Extend Element to Intersection   186
Trim Elements   187
    Noun/Verb or Verb/Noun Order   188
IntelliTrim   189
    Quick Mode   189
    Advanced Mode   192
Construct Circular Fillet   195
Construct Chamfer   196
Questions   197
Exercises   198

Chapter 10: Manipulation   203

Manipulate Tool Box   204
Copy   205
Move   208
Move Parallel   209
    Example of Move Parallel with Distance Checked ON   209
    Example of Move Parallel with Make Copy Checked ON   209
Scale   210
    Active Scale   211
Rotate   212

Contents xi

    Active Angle 212
    Example of Rotate Using Fence 213
Mirror 214
    Mirror About Vertical 214
    Mirror About Horizontal 215
    Mirror About Line 215
Align Edges 216
Construct Array 218
    Rectangular Array Type 218
    Polar Array Type 219
Questions 221
Exercises 222

Chapter 11: AccuDraw and SmartLine 227

Starting and Quitting AccuDraw 228
    AccuDraw Dialog Box 229
AccuDraw Compass and Coordinate System 229
    Example of the Dynamic AccuDraw Compass and Coordinate System 230
AccuDraw Keyboard Shortcuts 232
AccuDraw Settings 233
    Unit Roundoff 234
    Context Sensitivity 234
Applications of AccuDraw 235
    Place Text 235
    Isometrics with AccuDraw 235
SmartLine 238
Place SmartLine 238
    SmartLIne Placement Settings 239
    Segment Type: Lines 239
    Segment Type: Arcs 243
    Modify Element and SmartLine Modifications 244
Groups Tool Box 245
Drop Element 246
Create Complex Shape 247
Create Region 247
Popup Calculator 248
Table of AccuDraw Keyboard Shortcuts 249
Questions 250
Exercises 251

Chapter 12: Cells and Cell Libraries 255

Cell Types 256
    Graphic Cell Type 256
    Point Cell Type 256
Cell Library 256
    Attaching an Existing Cell Library 257
    Creating a New Cell Library 259
    Detaching a Cell Library 259
Cells Tool Box 260
Creating Cells 261
    Basic Steps for Creating Cells 261
    Things to Remember When Creating Cells 262
Different Uses of Cells 263
Place Active Cell 263
    Active Placement Cell 263
    Active Angle 264
    XScale and YScale Settings 264
    Relative 264
    Interactive 264
    Example of Active Cell Placement 265
Replace Cells 266
    Update Method 266
    Replace Method 267
Select and Place Cell 270
Cell Selector Utility 271
Other Buttons in the Cell Library Dialog Box 272
    Edit Button 272
    Delete Button 272
    Create Button 272
    Share Button 272
Questions 273
Exercises 274

Chapter 13: Patterning 281

Area Type 282
Change Element to Active Area 282
Displaying Patterns 282
Patterns Tool Box 283
    Associative Pattern 283
Delete Pattern 284
Hatch Area 285
Element Method 286
    Hole and Solid Area Types Together 288
    Uses of Element Method 289
Flood Method 289
    Non-Element Methods and Associative Patterns 290
Boolean Methods 291
Crosshatch Area 291
Pattern Area 292
Linear Pattern 294
Show Pattern Attributes 295

Match Pattern Attributes 295
Questions 296
Exercises 297

Chapter 14: Dimensioning 303

Dimensioning Terminology 304
Dimension Settings 304
    Text Height Units 305
    Accessing the Dimension Settings Dialog Box 306
    Common Settings for Each Category 307
    Dimension Lines Category 308
    Extension Lines Category 308
    Placement Category 309
    Terminators Category 310
    Text Category 313
    Tolerance Category 314
    Units Category 315
    Unit Format Category 317
Dimension Tool Box 318
    Associative Dimension 319
    Alignment 320
    Length of Extension Line 320
    Settings Used in Examples 321
Dimension Element 321
Dimension Size with Arrows 322
    Continue to Place Dimensions 323
Dimension Location (Stacked) 323
Dimension Angle Size 324
    Continue to Place Dimensions 324
Dimension Angle Location 325
Dimension Angle Between Lines 326
Dimension Radial 327
    Center Mark 327
    Radius 327
    Radius Extended 328
    Diameter 328
    Diameter Extended 329
Update Dimension 329
Geometric Tolerance 330
    Place Note 330
Questions 331
Exercises 332

Chapter 15: Reference Files 337

Capabilities and Limitations 338
Accessing Reference Files Tools 339
    Reference Files Tool Box 339
    Reference Files Dialog Box 340
Attach Reference File 341
Reference Files Settings 343
    Display 344
    Snap 344
    Locate 344
    Reference Levels Dialog Box 345
    Snapping to a Reference File's Elements 346
Manipulation Tools for Reference Files 347
Move Reference File 347
Scale Reference File 348
Copy Attachment 349
Merge Into Master 349
Detach Reference File 350
Attach URL 351
Reload Reference File 351
Other Related Tools 351
    Fit View 351
    Print 352
    Copy 352
Active Design File Referencing Itself 353
    Drawing Composition 354
Questions 355
Exercises 356

Chapter 16: Special Considerations of 3D Space 361

Positive Right-Hand Coordinate System 362
3D Coordinate System 363
    Drawing Coordinates 363
    Screen Coordinates 364
3D Seed Files 365
3D View Controls 366
    Accessing the 3D View Control Tool Box 366
3D View Control Tool Box 367
Zoom In/Out 368
Change View Perspective 369
Active vs. Display Depth 370
Set Display Depth 371
    Fit View 373
Set Active Depth 373
Show Display Depth 375
Show Active Depth 375
View Rotation 376
Change View Rotation 376
Rotate View 377
    Dynamic Method 378
    3 Points Method 378
Set View Display Mode 379

Questions 380

Chapter 17: Auxiliary Coordinate Systems and AccuDraw in 3D 381

ACS Triad 382
Uses of ACS 382
ACS Key-Ins 383
ACS Tools and Utilities 384
ACS Plane Lock and Snap 385
Define By Element 386
Define By Points 387
Saving an ACS 388
Define By View 388
AccuDraw in 3D 390
    Coordinate System Rotation: View Option 390
    Coordinate System Rotation: Top, Front, and Side Options 391
AccuDraw and ACS 392
    Rotate ACS 392
    Write to ACS 393
    Get ACS 393
Questions 394

Chapter 18: Rendering and Visualization 395

Render 397
    Target 397
    Render Modes 398
    Shading Types 399
Saving Rendered Images 399
    Accessing the Save Image Dialog Box 400
    Format Options 401
Example Design Files 402
Rendering Tools 403
Define Light 404
    Global Lighting 405
    Source Lighting 407
    Shadowing 408
Place Light Source 408
Edit Light Source 410
    Scan 410
    Delete Light 410
Materials 411
    Palette 411
    Material Assignment Table 411
    Bump and Pattern 411
Apply Material 412
    Open Palette File 413
    Select Material Definition 414
    Assign Material 414
    Other Modes Dealing with Materials 415
Transparent Materials 415
Rendering View Attributes 416
Rendering Setup 417
Animation 418
    Keyframe 418
    FlyThrough 418
    Parametric Motion Control 418
Example of FlyThrough Animation 419
    Accessing the FlyThrough Producer Dialog Box 420
    FlyThrough Producer Dialog Box 420
    Define Path 421
    Preview 422
    Record 425
Movies 425
    Accessing the Movies Dialog Box 425
    Load Movie 426
    Play Movie 426
Questions 427

Chapter 19: 3D Modeling with SmartSolids 429

3D Seed File 430
Locating Solids 431
3D Main Tool Box 432
3D Primitives 433
Common Settings 433
    Type 433
    Axis 434
Place Slab 436
Place Sphere 438
Place Cylinder 439
Place Cone 441
Place Torus 442
Place Wedge 443
3D Modify Tool Box 444
    Keep Originals 444
Solids to use for Boolean Operations Example 445
    Place Slab 445
    Place Cylinder 445
    View Display Mode 448
Construct Union 449
    Place Cone 450
Construct Difference 451
Construct Intersection 452

Export 453
    Accessing the Export Visible Edges
       Dialog Box 453
    General 454
    Hidden 454
    Visible 455
    Output 455
Questions 458
Exercises 459

Chapter 20: Advanced Model Construction and Modification 463

3D Construct Tool Box 464
    Closed Shape Profile 464
Extrude 465
Construct Revolution 469
Extrude along Path 471
    Define By Circular 471
    Define By Profile 472
Shell Solid 473
Thicken to Solid 474
3D Modify Tool Box 475
Modify Solid 475
Remove Face and Heal 478
Cut Solid 479
Fillet Edges 482
Chamfer Edges 483
Questions 484
Exercises 485

Chapter 21: Drawing Composition 491

General Sequence 492
    Sequence for 2D Design Files 492
    Sequence for 3D Models 492
Drawing Composition Dialog Box 493
Sheet View 494
    Open Sheet View 494
Attach Border 495
Attaching Views 497
    View Parameters 497
    Hidden Line Removal and Method 498
    Hidden Line Settings Dialog Box 498
    Attachment Parameters 499
Attach Saved View 499
    Saved Views 499
    Accessing Attach Saved View 501
Attach Standard 502
Attach Folded 502

Example of Establishing a Sheet View Layout 503
    Attach Standard View—Front 505
    Attach Folded View—Orthogonal 506
    Attach Standard View—Right Isometric 507
    Editing the Model in the Sheet View 508
    Drawing in the Sheet View 508
Modification of Existing Attached Views 509
    Detach 509
    Move 509
    Modify Hidden Line 510
Example of Modifying a Sheet View Layout 511
    Detach—Single 511
    Attach Folded View—About Line 511
    Move—Single 513
    Modify Hidden Line 514
Dimensions and Text 515
    Levels 515
Printing 516
Sectioned View 516
Generate Section Dialog Box 517
Section by View 518
    Horizontal 518
    Vertical 521
Sectioned View in a Sheet Layout 522
    Reference Levels 524
Questions 528
Exercises 529

Chapter 22: Engineering Configurations, ProjectBank, and Java 535

Engineering Configurations 536
    MicroStation Modeler 536
    MicroStation Triforma 536
    MicroStation Geographics 537
    MicroStation CivilPAK 537
    MicroStation Schematics 537
    Future of Engineering Configurations 537
ProjectBank 538
    Server-Based Usage 538
    Database Functionality 538
    History Information 540
Java 541
Future Direction of MicroStation 541

Appendix: Seed File Information 543

Index 547

Chapter 1:
Introduction to MicroStation/J

SUBJECTS COVERED:

- ➲ What's New for MicroStation/J
- ➲ Conventions Used in the Text
- ➲ Graphical Interface
- ➲ Mouse Actions
- ➲ Windows and Dialog Boxes
- ➲ Tool Buttons and Boxes
- ➲ Pull-Down Menus

MicroStation is, in its simplest form, a computer-aided drafting and design (CAD) software product. Technical fields such as engineering and architecture use drawings to present specific information as well as to aid in visualization. CAD uses lines, arcs, and other geometric elements in an electronic format to produce these technical drawings. **The elements can be positioned at precise coordinates and are created at the ACTUAL SIZE of the object they represent.** The computer and software together do all the work of storing, calculating, manipulating, and displaying these elements. The elements themselves can easily be modified and then printed at various scales, unlike in manual drafting where changes and different scales often require creating a new drawing. However, calling MicroStation a drafting product is like saying baseball is about hitting a ball with a stick and running—the description is valid but doesn't quite cover it all! The software is capable of much more than two-dimensional drawings. MicroStation has the ability to store, retrieve, link, and manipulate vast amounts of graphical and numerical information. Three-dimensional models can be constructed to help you visualize almost any object, from gears to roads. A model can then be used to generate the standard two-dimensional technical graphics. These models can also contain technical data necessary to evaluate and analyze the real-life object they represent. This just scratches the surface of what MicroStation and its related software is capable of—the possibilities seem endless.

Picking up a pencil and drawing a line is quicker and easier than using CAD to draw a line, but it is not nearly as powerful! Because MicroStation is a high-end powerful product, to a beginner without any CAD experience it can seem extremely complicated to use and master. Don't get frustrated. This text will help you cut through all the technical jargon and get to the essentials that you need to create two-dimensional drawings. Once you have mastered those skills, then you'll be ready to move into three-dimensional space and modeling. All of this knowledge will enhance your ability to create technical drawings and models. It will not make you an expert in MicroStation, but you'll have a great foundation to build on.

WHAT'S NEW IN MICROSTATION/J

This section will briefly discuss the new stuff you will find in MicroStation/J. If you are already familiar with MicroStation 95® or MicroStation SE® (perhaps from using my other textbook, *Understanding MicroStation 95/SE*) then this information will help you know which parts of the text have changed drastically. If you are totally new to MicroStation you can at least get a preview of what you're getting into!

Generally speaking, the user will not see too many obvious changes from MicroStation SE. Instead the majority of the new stuff for MicroStation/J is "under the hood" so to speak. The underlying programming has undergone the most drastic changes. This includes its incorporation of the Java Virtual Machine (JVM) that allows MicroStation/J to execute any Java program. Chapter 22 discusses Java and also ProjectBank that utilizes MicroStation/J. The other major change in MicroStation/J has been the introduction of SmartSolids tools that now include true 3D solid modeling in the core MicroStation product. Previously you needed to use MicroStation Modeler® to achieve this. This text will focus on these new solid modeling tools and their use in creating 3D models.

As far as user productivity, here is a list of the some of the new enhancements that you will notice in MicroStation/J and the chapter that discusses each of them.
- New directory structure (Chapter 2)
- New Workspace components (Chapter 2)
- On-Line Help (Chapter 2)
- Enhanced Precision (Chapter 2)
- Reworked PowerSelector (Chapter 4)
- Align Elements By Edge tool (Chapter 10)
- Replace Cells (Chapter 12)
- SmartSolids (Chapters 19-21)
- QuickVision®/GL (Chapter 18)

CONVENTIONS USED IN THE TEXT

Shown below are some special conventions that are used in this text.

Warnings

After instructing numerous students, I have noticed a pattern emerge from the problems they encounter. These warnings are to help you avoid these typical difficulties. Pay close attention to Warnings; they are important.

 A common problem when starting out is remembering that **tentative points must be accepted with a data point.** If you tentatively snap to another element, you get to "see it before you buy it" ... just don't forget to buy it!

Notes

Notes contain information that is nice to know but not necessarily critical. Sometimes Notes serve as a reminder to jog your memory of a previous concept. Other times, they recommend certain settings and values to use.

 With any of the Method settings, the polygon will **not** have a center keypoint. The line segments of its edge do retain their keypoints to snap to even though they are not single elements.

Key-ins

Key-ins refer to entering commands and information by typing in a special Key-in field of the software. Key-ins are preceded by the : symbol. Also, a different font will distinguish key-ins from other keyboard actions. For example, the coordinate of Point D is specified by :DL=4,2. In this case, the DL=4,2 is typed into the Key-in field.

Keyboard Shortcuts

Keyboard shortcuts aren't typed in the Key-in field. These are just keystrokes that are used as a shortcut to quickly open dialog boxes or start commands. Keyboard shortcuts are preceded by the 7 symbol. For example, the keyboard shortcut for Undo is 7CTRL-Z.

AccuDraw Keyboard Shortcuts

AccuDraw keyboard shortcuts are used when the AccuDraw dialog box is active. They are usually one or two letters that are typed. They are noted as <*keystroke*>, and you don't type in the < or the >. For example, AccuDraw can be set so that its coordinate system is based on the View that you are working in. The AccuDraw keyboard shortcut for this is <*V*>. It is also possible that instead of one or two letter characters, the AccuDraw shortcut is a certain key on the keyboard. For example, the <*spacebar*> will toggle you between a polar and a rectangular compass.

GRAPHICAL INTERFACE

Technical drawings are created because a picture is indeed worth a thousand words. Since MicroStation software is in the business of making technical drawings, it only makes sense that the software operates using a graphical interface as shown here. A graphical user interface (also known as GUI) is one that is based on a visual menu system to operate the software. Small pictures called icons may represent actions or tasks of the computer software. Menus and options are listed and accessed using the mouse. MicroStation is based on a graphical interface that has the general look and feel of Windows® as illustrated below. The interface is similar even if you are running the software on a different platform. Other types of graphical interfaces are available, but this text will cover this most common one. As its name implies, Windows uses a visual menu system consisting of windows that can be opened, closed, sized, and moved. These windows contain the information and workings of the software. This text assumes that you already have some experience with Windows, but let's look at some of the components that make up the MicroStation graphical interface.

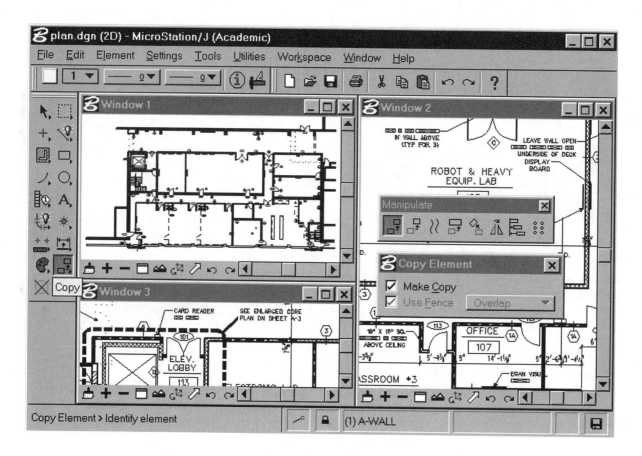

MOUSE ACTIONS

The mouse is one of the most important players in the graphical interface game. Some of the terms that deal with the mouse and its buttons are as follows:

ACTION TERM		WHAT IT MEANS
Windows	**MicroStation**	
Point	Point	Move the mouse so the on-screen pointer is pointing at the correct location (tool, menu, screen location, etc.).
Click in Windows	**Click** on tools and menus **Data Point** when in graphical drawing area	Press and release the **left** mouse button.
Right-Click in Windows	**Reset/Enter** in MicroStation	Press and release the **right** mouse button.
Drag or Hold	**Drag or Hold**	Press and **hold** the **left** mouse button; then move the pointer until it's pointing at the new location and then release.
Double-Click	**Double-Click**	Press and release the **left** mouse button **two** times quickly.
	TENTATIVE in MicroStation	*2-Button Mouse*: Press and release **BOTH** the left and right mouse buttons **at the same time.** *3-Button Mouse:* Press and release the **middle** mouse button.

Here is an illustration of both a 2-button mouse (marked with 2) and a 3-button mouse (marked with 3) and what their actions will do in MicroStation:

Left Mouse Button
- Click on tools
- Data point in graphical drawing area

Right Mouse Button
- Reset
- Accept/Enter

2-button mouse: Left and Right simultaneously
3-button mouse: Middle button
- **TENTATIVE POINT**

WINDOWS AND DIALOG BOXES

A window consists of the basic parts shown in the figure below. You can resize a window by pointing to its borders and waiting for a double arrow to appear. When the double arrow appears, you can drag the window's border to the new size. If more than one window is open, you may also rearrange them on your screen.

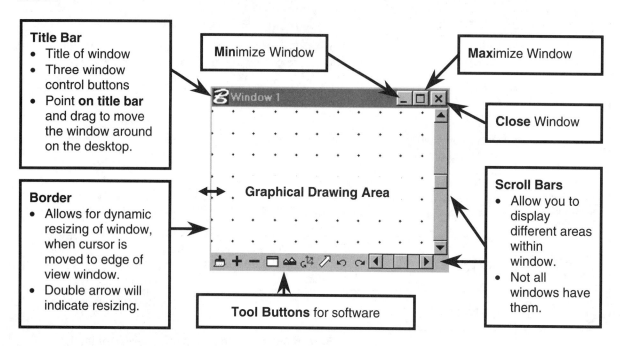

When using Windows, the MicroStation software that will be running will have its own window. There are also windows that open up for MicroStation. The graphical area that is drawn in is called a window. Sometimes this will also be referred to as a **view** window. Lots of windows! You can actually have a design file open, but all the windows to view it can be closed—don't panic. The window running MicroStation has a pull-down menu that allows you to open and close view windows (as well as arrange them in the MicroStation window). It is very similar to opening documents and arranging their windows in a word processing program. We'll look at that in the next chapter.

The figure on the right shows the window running MicroStation and two view windows (Window 1 and Window 2) open that show different views of a drawing of columns.

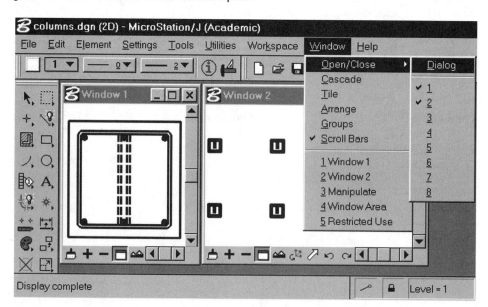

A dialog box allows you to "talk" to the software and tell it what you want. A dialog box contains settings such as fields (to type in), option buttons, check boxes, list boxes, and other items. Dialog boxes usually cannot be resized, but you may close a dialog box by clicking on the ⊠ in its upper right-hand corner. The name of the dialog box shows up in the title bar. The text will refer to the dialog box by name. For instance, illustrated below is the Place Circle dialog box.

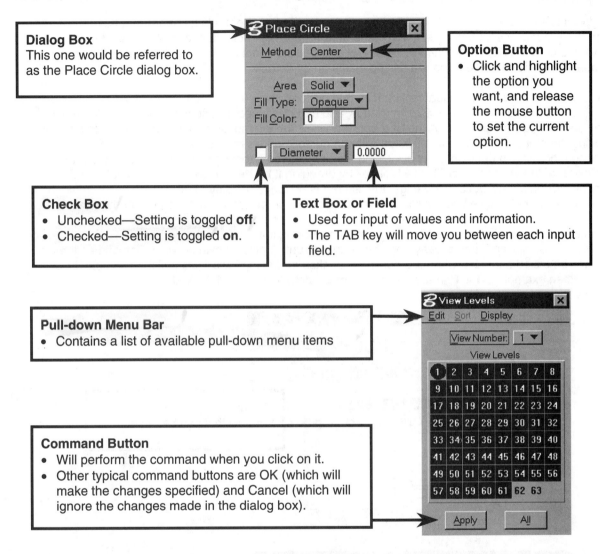

Take a closer look at the command buttons and some of the other labels of the functions—you'll notice that one of the letters is underlined. The underscored letter indicates the ALT-letter keystroke that also activates the function. This keystroke is done by having both keys depressed on the keyboard **at the same time.** The notation for specific keys on the keyboard will be all caps. For instance, ENTER will mean the specific key, not typing in the characters E,N,T,E,R. The ⌨ symbol will indicate these keyboard shortcuts. For example, ⌨**ALT-A** would be the same as clicking on the Apply button and ⌨**ALT-E** would be the same as clicking on the Edit pull-down menu.

 Having the keys depressed at the same time doesn't necessarily mean you have to be quick. An easy way to do this keystroke is to first press and **hold down** the ALT key and then press the other key (while still holding down the ALT key).

You may have also noticed the arrow symbols like ▶ or ▼ as shown in the option buttons. These arrows indicate that there are additional items that can be accessed. An arrow will also appear on some of the tools to indicate that others are available in the tool box. Let's take a closer look at tools and tool boxes.

TOOL BUTTONS AND BOXES

A typical graphical interface for MicroStation utilizes tool buttons (or tools for short) that have icons. An icon is a graphic description (or symbol) of the button's corresponding function. As in Windows, if you move the cursor so it is positioned over the tool and **pause,** a tool tip showing the name of the button will appear. This display of tool tips is a setting that is on by default.

A tool box is a group of individual tool buttons that perform related functions. A tool box can be floating or docked. A floating tool box has its own window and can be moved around in the window. You must move the cursor so you are pointing **at the title bar**; then click the left mouse button **and hold it down** while moving the mouse; the tool box will move with the cursor—this is referred to as dragging. Sometimes a floating tool box can get in the way of viewing the drawing, so a tool box can also be docked to the perimeter of the window. You dock a tool box by dragging it **by its title bar** to the perimeter of the window. When it gets close to the edge, the title bar is no longer displayed, and when you release the left mouse button that you've been holding down, the box is incorporated into the border of the window.

A docked tool box can be moved back to floating by tearing it away from the perimeter. To tear it off, you point on its border and drag it away from the perimeter until it gets its own window. Sometimes the tool box will be called a tool bar when it is docked. Docked or floating—we'll just call it a tool box. Shown below is an example of the Primary Tools tool box as both floating and docked.

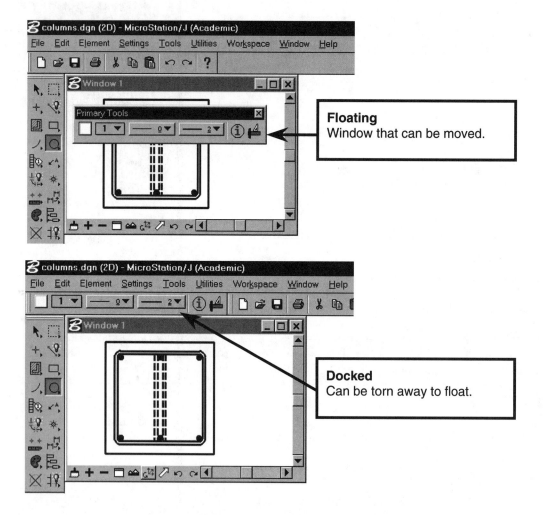

Floating
Window that can be moved.

Docked
Can be torn away to float.

Introduction to MicroStation/J 9

Just as related tools can be grouped into tool boxes, tool boxes can also be grouped together into a larger tool box (sometimes referred to as a tool frame). Such is the case of the very important **Main** tool box that contains various tool boxes.

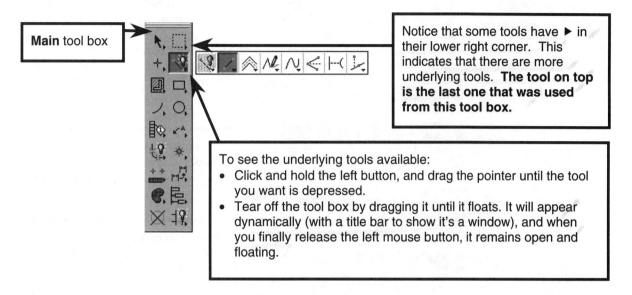

Main tool box

Notice that some tools have ▶ in their lower right corner. This indicates that there are more underlying tools. **The tool on top is the last one that was used from this tool box.**

To see the underlying tools available:
- Click and hold the left button, and drag the pointer until the tool you want is depressed.
- Tear off the tool box by dragging it until it floats. It will appear dynamically (with a title bar to show it's a window), and when you finally release the left mouse button, it remains open and floating.

When you click on a tool, its corresponding dialog box will appear. For example, when you click on the Place Block tool (shown with its tool tip appearing) the Place Block dialog box appears. This is due to an option called Tool Settings, which is on by default.

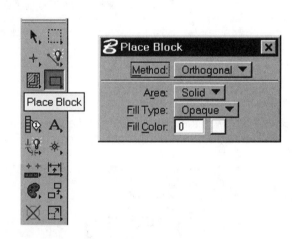

 It is important to become familiar with the tools that are grouped together. What you see is not necessarily what you want. The tool you want may not be on top in the Main tool box but instead be underlying a related tool. The last tool picked will go to the top.

PULL-DOWN MENUS

Pull-down menus are another part of the graphical interface. Most functions can be accessed from both the pull-down menus and a tool—so you have a choice. The text will not cover every possible way to get to a function. However, the pull-down menu path will be noted in some cases when it is easier to get the job done that way. The pull-down menu found at the top of the window running MicroStation will be referred to as the **main** pull-down menu. A pull-down menu found at the top of a specific dialog box will be named after the dialog box, such as the MicroStation Manager pull-down menu. If the pull-down menu isn't specified, assume it refers to the main one.

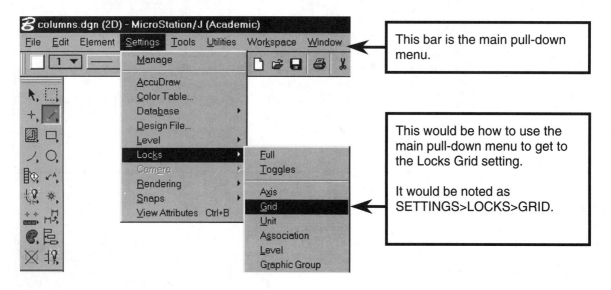

Shown below is the Reference Files dialog box. It has its own pull-down menu, therefore it would be noted as the Reference Files pull-down menu SETTINGS>LEVELS.

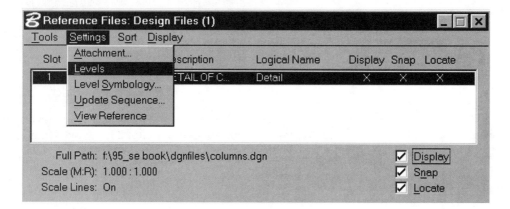

Chapter 2: Starting Up

SUBJECTS COVERED:

- ⊃ MicroStation Manager
- ⊃ Workspace
- ⊃ Create a New Design File
- ⊃ Seed Files
- ⊃ Status Bar
- ⊃ Key-in
- ⊃ Standard Tool Box
- ⊃ Undo
- ⊃ Redo
- ⊃ Help
- ⊃ Settings
- ⊃ Coordinate System and Working Units
- ⊃ Global Origin
- ⊃ Enhanced Precision
- ⊃ Coordinates
- ⊃ Preferences
- ⊃ Saving Your Work

This text assumes that you are using MicroStation straight out of the box, without any customization. Even so, there are many choices still to be made. Depending on the type of work you are doing—whether it is architecture, mechanical engineering, or civil engineering—there are many facets of the software already tailor-made for you. Items such as typical units of measurement, annotation standards, color preference, and even the size of the icons are just a few of the choices to be made. Although this text will cover the basics to build on, it is still highly recommended to do some reading in the software documentation to see if you can find some areas in which the work is already done for you. This chapter will help you become familiar enough to know what you are looking for, recognize it when you see it, or be able to modify it to your needs.

MICROSTATION MANAGER

MicroStation Manager dialog box manages what files you can choose from, where you can find them, and which existing file to open. The Workspace consists of 4 different components: User, Project, Interface, and Style. These allow you to customize the configuration of MicroStation, a must for companies. As you progress, you will realize that MicroStation gives you so many options that it can be overwhelming, especially for a beginner. However, this ability to control down to the smallest detail is what makes this an extremely powerful package with a wide range of applications. If there is an option or setting not covered, then often it is best for beginners to leave it as is (but you can always look it up in the software's documentation)!

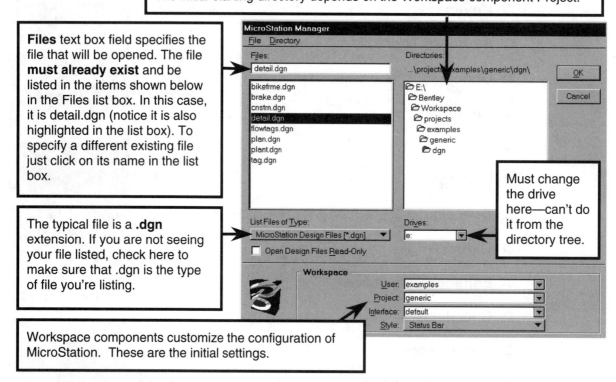

Move through the directory tree here—open various folders as in Windows. The initial starting directory depends on the Workspace component Project.

Files text box field specifies the file that will be opened. The file **must already exist** and be listed in the items shown below in the Files list box. In this case, it is detail.dgn (notice it is also highlighted in the list box). To specify a different existing file just click on its name in the list box.

Must change the drive here—can't do it from the directory tree.

The typical file is a **.dgn** extension. If you are not seeing your file listed, check here to make sure that .dgn is the type of file you're listing.

Workspace components customize the configuration of MicroStation. These are the initial settings.

You must use MicroStation Manager's pull-down menu FILE>NEW to create a new design file. Typing a name into the MicroStation Manager dialog box's "file to be opened" field will **not** make a new drawing of that name (nor will it alert you that you goofed—you will just keep clicking OK and nothing will happen).

Here is what results from opening up the existing detail.dgn file that comes with MicroStation. There are two view windows, Window 1 and Window 2, opened. A Fit View window is open because that is the tool in use. At the bottom of the window running MicroStation is an information area that is called the status bar. We'll discuss the status bar later on in this chapter. If you are running an academic edition of MicroStation/J you will also see the Restricted Use window shown here. You can't close it; rather you just have to drag it out of your way (usually way out to the edge)!

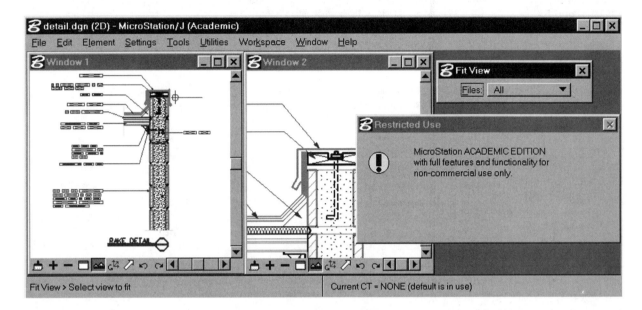

Now that we know how to open an existing design file using MicroStation Manager, let's return to the MicroStation Manager dialog box. When you start MicroStation, the MicroStation Manager dialog box opens so one option is to just close the window running MicroStation and restart MicroStation. Another option is to close this design file: Go under the main pull-down menu FILE>CLOSE (illustrated at the right). This will also get you to the MicroStation Manager dialog box.

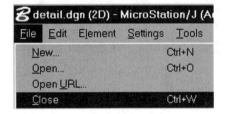

In the MicroStation Manager's pull-down menu FILE you will find these file management items:

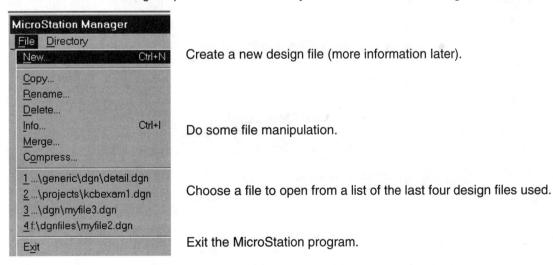

Create a new design file (more information later).

Do some file manipulation.

Choose a file to open from a list of the last four design files used.

Exit the MicroStation program.

WORKSPACE

The Workspace settings of User, Project, Interface, and Style set up the overall running of MicroStation. The User setting specifies a unique set of user configuration and preferences that determines how MicroStation will look and behave. The Workspace is often referred to by its User name. The Project setting helps organize project-related data files. The Interface setting can affect which tools are available for use, and the Style setting is just a minor option with Status Bar being the most used. There are a whole lot of different combinations that are possible, but let's just look at creating a new project and a new user. If you are working on a network or in a large group, these are options that the administrator can modify and control and you may be instructed to change to different Workspace settings.

When you create a new user, there is the opportunity to specify a default project for that user. Therefore, it is easier to **create your new project first** so that it is available to select when you create your user. So we'll look at the Workspace components out of order from how they are listed in the dialog box.

Project

By using projects, you can organize your data files in a reasonable fashion—rather than just putting everything in one directory! Each project can have its own directory structure that can store data specific to that project. **There can be numerous design files located in the dgn folder of a project. You do not need a new project for each new design file.** If the data is needed for numerous projects it can be stored in the Standards directory. If you select a different project, the directory that shows up can change. You can still move around in the directory and choose anything you want, but it will default to looking at this project's directory first. For example, if the Project setting is civil, you will be shown the directories illustrated below.

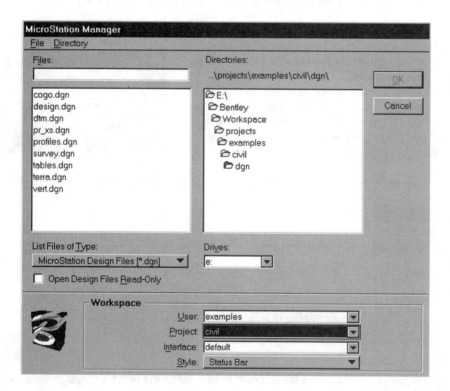

Starting Up 15

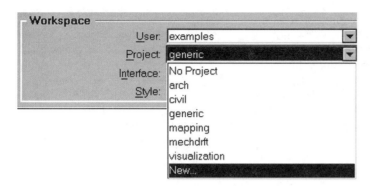

MicroStation comes with some sample projects such as arch, civil, and mapping. These all are located under the Bentley\Workspace\projects\examples directory. Here are the Projects settings that are available with the setting User: examples. These options match up to files with extension .pcf that are found in the Bentley\Workspace\projects\examples directory. These projects match the sample workspaces that in MicroStation95 or SE were located in the wsmod directory.

If you click on the New... option, the Create New Project dialog box will open as illustrated below on the left. After typing in a Name and Desciption, clicking on OK will close the dialog box, the Project will be set to myexample, and the Directories listing will be as shown below on the right.

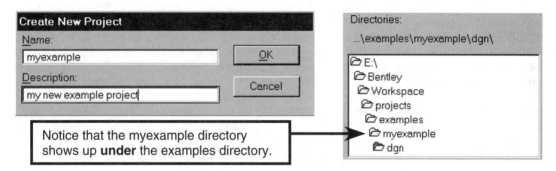

Notice that the myexample directory shows up **under** the examples directory.

If you want the new project to be placed up one level (under the projects directory) then you need to change your User to untitled **before** selecting the New... Project option. Click on the User options and select untitled. The Project options then change to those illustrated below. Select the New... option for the Project setting.

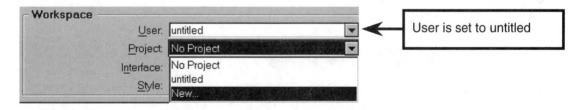

User is set to untitled

The Create New Project dialog box will open. When it is filled out (as shown below on the left) and done by clicking on OK, the new project directory kcb_prj1 will be created and located in the projects directory as illustrated below on the right.

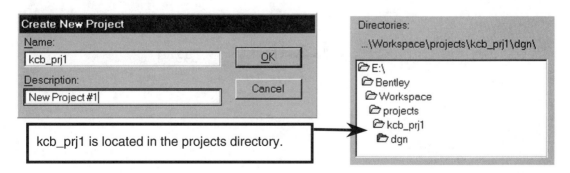

kcb_prj1 is located in the projects directory.

It is important to know that when creating a new project, the dgn sub-directory is not the only one made under the new project directory. There are others as well that form the skeleton directory structure for project-related data files. The dgn folder is the one that shows up because that is where MicroStation Manager wants to look for .dgn files to open! This skeleton directory structure is illustrated on the right. The other sub-directories can contain files needed for this specific project.

Now that we have our project created, let's look at the User Workspace component.

User

Individual users working in a large group may want MicroStation to look and behave differently even if they are using the same project. This is where the User component of the Workspace comes in. There are two files that are named after the User name. One is the user configuration file, which has the extension .ucf. The other is the user preference file, which has the extension .upf. Both of these are linked to the User setting. Later on in this chapter we will look at the Preferences dialog box that allows you to change and save the preferences for the active user.

Shown below are the different options for the User setting. By selecting New..., you can create a new user with the Create User Configuration File dialog box. This is probably the easiest way to create a new user. You will also notice an option called untitled. This shows up because some folks would rather go to the directory, copy the two files, untitled.ucf and untitled.upf, and then rename them both. We'll just use the New... option as shown in the illustration.

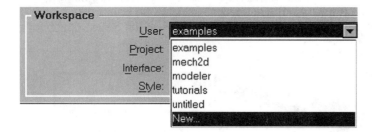

Here is the Create User Configuration File dialog box. The new User will be named kcb. It will also create the new user preference file, which is also named kcb but with the extension .upf rather than .ucf. You need to click on OK to continue with the Create User Configuration File dialog box.

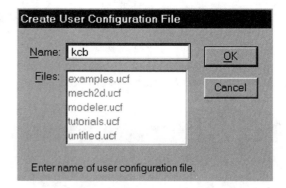

Starting Up   17

Once the user configuration file has been created, the Create User Configuration File dialog box shows its name in brackets in its title bar, and lets you add a description and set the Project and User Interface to use for this User.

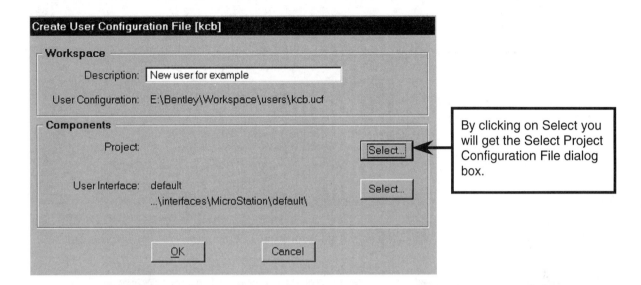

In selecting a project configuration file for this specific user, you will set the default Project setting for MicroStation Manager to use when the specific user is the User setting.

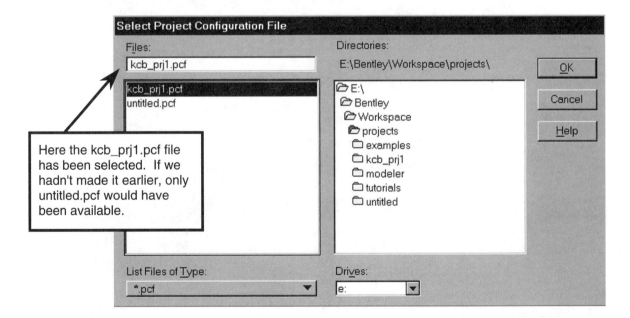

If there is not a project configuration file selected, then the MicroStation Manager Project setting will default to No Project. We'll discuss the ramifications of that later.

Here's what results from completing the selection of the project configuration file.

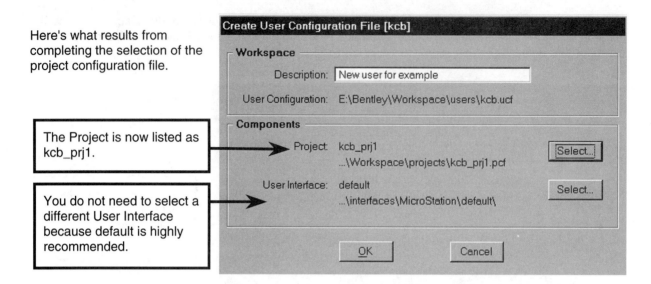

The Project is now listed as kcb_prj1.

You do not need to select a different User Interface because default is highly recommended.

The MicroStation Manager dialog box now shows the Workspace components and directories as illustrated below.

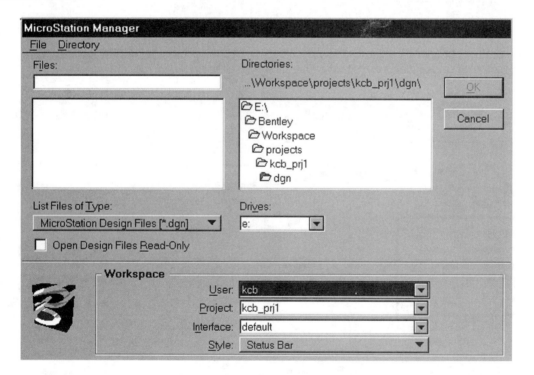

 The Project component just sets up the directory structure and path to start with. You can still move around in these directories and put design files anywhere you want!

If the Project component of the Workspace is set to No Project, then it defaults to the dgn sub-directory of the standards directory as shown below. This would also result from a user configuration file not selecting a project configuration file.

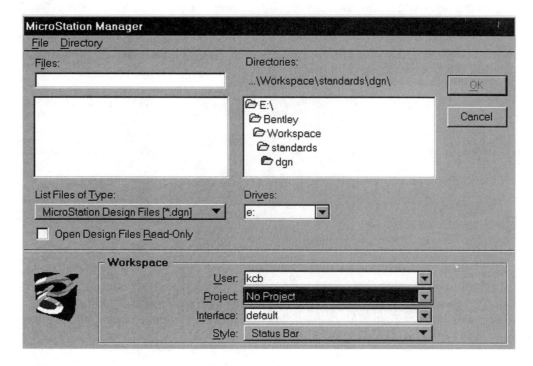

Interface
Leaving the Interface setting at default as shown above is highly recommended. Steer clear of the NewUser option available here—it is missing a lot of tool boxes!

Style
The Status Bar setting for Style is the most common so we won't worry about anything else here.

Whether you create your own projects and user configuration files is up to you. If you are working in a network situation then some of this may have already been done for you and you may only be able to write to certain directories—consult your network administrator. For the purposes of the book, we'll use the examples for user and jbook (a new project created for the book, similar to generic) for the project.
    User: examples
    Project: jbook
    Interface: default
    Style: Status Bar

CREATE A NEW DESIGN FILE

A design file is the electronic version of your CAD drawing. If you need a new design file, you have to use the Create Design File dialog box. There isn't much to creating a new design file other than naming it as well as selecting the seed file to use. More information on seed files in the next section of this chapter.

Accessing the Create Design File Dialog Box

You can open the Create Design File dialog box two different ways, either from MicroStation Manager's pull-down menu or from the main pull-down menu FILE>NEW when a design file is already open. You can also use the keyboard shortcut ⌨ CTRL-N. These are illustrated at the right.

 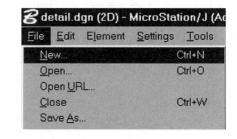

Create Design File Dialog Box

Actually you can create a new file by just giving it a name. Notice that the Create Design File dialog box looks a lot like the MicroStation Manager dialog box. Make sure you're reading the title bar to see what dialog box you are in.

Files text box
Type in what you want to name the file. The .dgn extension (which is an abbreviation for design, as in design file) will be done automatically.

Seed File
Use this default seed2d.dgn for now. The Select button will show the other choices—more on that later in this chapter.

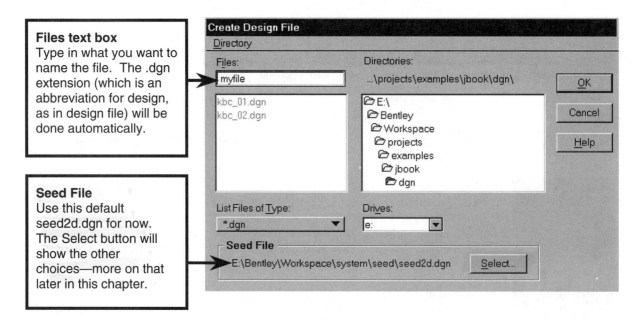

 If you use a different Project setting such as arch, civil, or mapping, there will be a different default seed file. The seed2d.dgn seed file is the default for the New... or generic Project settings.

After creating a new design file, you will be sent back into MicroStation Manager. But this time the new file you just named will already appear and be selected for opening.

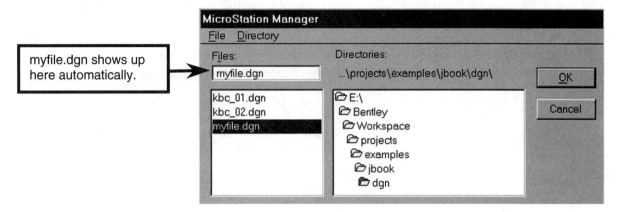

myfile.dgn shows up here automatically.

You can click on the OK button or hit the ENTER key to continue to open the newly created design file. If you click on Cancel, then it won't open the file—but it will still have created the design file.

SEED FILES

When building anything—whether it's a document, a database, or a design file—you usually start from scratch. However, starting from scratch doesn't mean that you start with nothing. A seed file is a design file that appears to be blank, but in fact already has some information stored in it. Seed files are similar to templates that are used in word processing documents, or they can be like a prototype drawing. When creating a new design file, the seed file you specify is actually copied and your design file builds from there. Different seed files exist to allow you choices in your base design file.

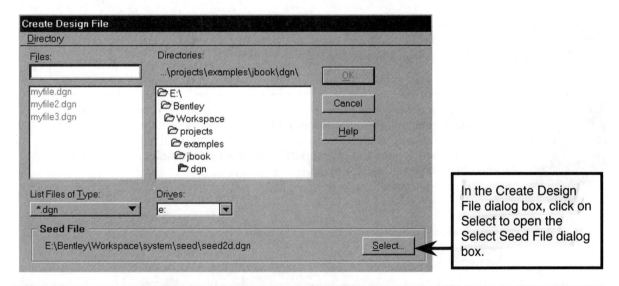

In the Create Design File dialog box, click on Select to open the Select Seed File dialog box.

A common mistake of beginners is actually opening the seed file with MicroStation Manager. When that occurs, the seed file itself is being changed. The result is that any new files created with the selected seed file will reflect any changes that were made to that seed file.

Below is the Select Seed File dialog box. The common seed files are found in the Bentley\Workspace\system\seed directory. Shown highlighted is the seed2d.dgn file, a simple and generic seed file for two-dimensional designs (X and Y coordinates only—like drawing on a flat piece of paper.) Beginners are recommended to use this for a good starting design file. The seed2d.dgn seed file is based on English units. If you want to use metric units, it is recommended that you use 2dm.dgn for the seed file.

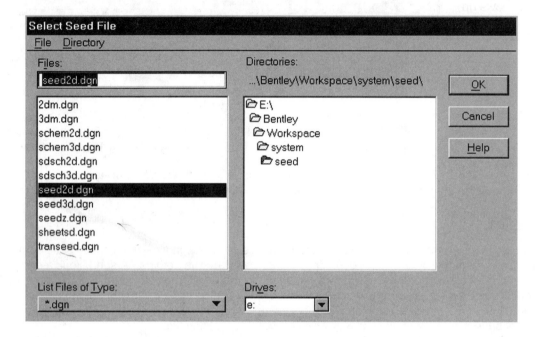

Different seed files are used for a three-dimensional (X,Y,Z coordinates) drawing. Since less information is needed for two-dimensional drawings, having seed files specifically for two-dimensional drawings allows for smaller files. Right now we'll use seed2d.dgn for our seed file.

 The last seed file used will be the default seed file for creating a new design file. It is always a good idea to make sure that you are starting with the correct seed file. Pay attention to the file's directory path too, because there are lots of different seed files in many different directories. See the appendix for specific information about the different seed files that come with MicroStation.

STATUS BAR

The status bar runs across the bottom of the main window running MicroStation. It has a large amount of information such as what function is being done, what the software expects next, basic settings, etc. Beginners should always be aware of the status bar display. **Read what it says—its messages are important.**

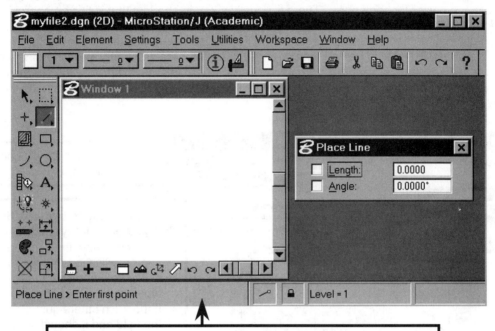

Status Bar
- The tool selected was the Place Line and it is prompting you to Enter first point.
- Items on the right are sections displaying specific settings. These will be discussed when the settings are discussed.

The left portion of the status bar shows the function being done followed by a > and then a prompt for what is needed next. The syntax in the prompt, such as a slash (/) or a comma (,), can actually give you information regarding the mouse actions and what they will accomplish. A slash (/) indicates the difference between what a left mouse button action will do and what a right mouse button action will do. A comma (,) indicates that two things will happen with that single mouse action.

For example, in the middle of measuring the angle between two elements, the prompt reads >Accept, Identify 2nd Element/Reject. The comma indicates that a left mouse button action will do two things: both accept the first identification <u>and</u> identify the second element. The slash indicates that what follows is a right mouse button action. In this case, the right mouse button would reject the first identification.

 Sometimes you can't see the status bar—this can be due to the fact that the bottom of the window running MicroStation is covered up by something. Click on the Maximize button in the upper edge of the main window that is <u>running</u> MicroStation (not a view window in MicroStation) and the entire window, including the status bar, will be visible.

KEY-IN

You may also give instructions to the software by typing the commands and parameters on the keyboard. This is referred to as using a key-in, which doesn't depend on graphical input. **Key-ins require proper syntax**, so you must pay attention to commas (,), equal signs (=), and spaces. **Key-ins are not case-sensitive**, which means it doesn't matter if they are capitalized or not. In most cases, using tools and menus is much easier for beginners. There is usually a key-in shortcut that is like an abbreviation of the key-in (for those of you who hate to type). It is the least number of characters needed to uniquely define the key-in command. For instance, to activate the Place Line command, the key-in is place line and the shortcut is pla lin. The status bar will then prompt you to enter the first point. That coordinate can be specified with a key-in of XY=*xcoordinate,ycoordinate* where the *xcoordinate,ycoordinate* are values giving the precise coordinates of the endpoint of the line. The symbol ⌨ will bring attention to any key-in that may be useful.

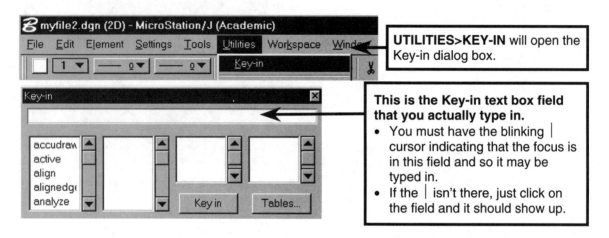

If the Key-in box is docked, you may not see the buttons and other items, but instead only the key-in field is displayed. Look around for the large text box that is the key-in field. A docked key-in is illustrated below. If the window is large enough, the key-in field can be displayed in one line with the Primary and Standard tool boxes. The Primary tool box will be discussed in Chapter 4 in conjunction with Element Attributes. Let's look at the Standard tool box next.

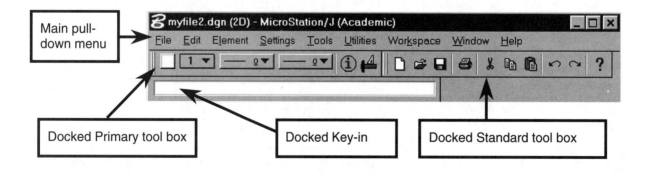

STANDARD TOOL BOX

Some tools common to Windows programs are found in the Standard tool box. These are tools such as New File, Open File, Save Design, Print. The Windows clipboard (this isn't used much with MicroStation) tools of Cut, Copy, and Paste are also here. The Standard tool box is usually docked near the top of the window.

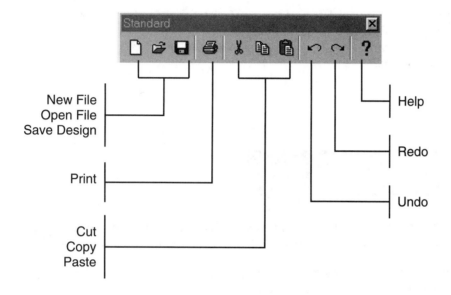

Accessing the Standard Tool Box

If the Standard tool box isn't open then access it by going to the main pull-down menu TOOLS and make sure there is a ✓ beside the Standard.

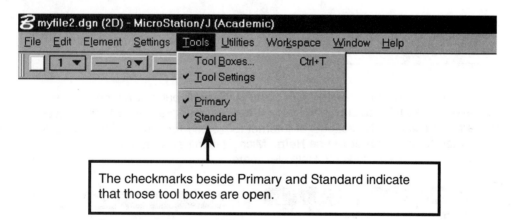

The checkmarks beside Primary and Standard indicate that those tool boxes are open.

UNDO

This will undo the last action taken. It is useful if you catch your mistakes right after you do them.
- You can keep undoing operations (you don't undo the Undo).
- The number of operations you can go back depends on the buffer size.
- ⌨CTRL-Z is the shortcut for Undo.

REDO

The Redo tool is found to the right of the Undo tool. The Redo tool is the one with the arrow going clockwise. It will redo the last action that was eliminated by the undo. Think of it as a way to "undo" the Undo tool action.
- Negates the undo and thereby brings the last operation back.
- Can keep doing and negate more than one undo.
- ⌨CTRL-R is the shortcut for Redo.

HELP

There is a Help menu available. It can be accessed easily through the main pull-down menu HELP. However, that is about the only easy thing about using Help. When you start Help, your External web browser automatically starts (you can use Netscape® or Internet Explorer® for your web browser). **You do not need to be connected to the Internet to use Help.** MicroStation has a program that runs a local host on your machine, and it connects and finds the Help files there.

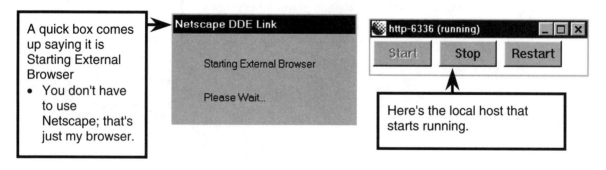

A quick box comes up saying it is Starting External Browser
- You don't have to use Netscape; that's just my browser.

Here's the local host that starts running.

If you're not connected to the Internet, then when your browser starts it may automatically try to connect you. This is where you can run into troubles, or at least I did! If it automatically tries to connect (and you don't want to be connected to the Internet) then you need to cancel the connection. This will allow the browser to locate the local host, otherwise the browser will just keep showing you something similar to "Looking up host:www.localhost" but never get there!

Another problem can be that this Information box illustrated at the right can appear. I have found the easiest thing to do here is just restart MicroStation.

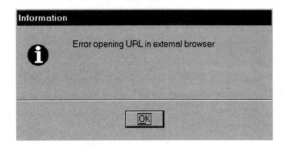

If you get everything just right then the Help stuff should show up as illustrated below. The Help works like a typical web browser with a search engine—you can take it from here!

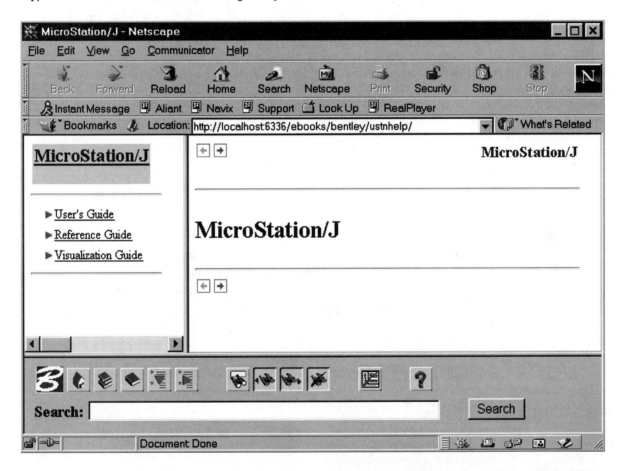

It may be to your advantage to place a bookmark to the Location once you get there. If you run into an error like the one shown at the right, you can just go to the bookmark.

SETTINGS

Each design file will have its own settings. The settings of the seed file will be reflected in the design file; however, setttings of the design file can be changed to suit your needs after you open up the new design file. It is important to remember to either save your settings after you make a change to a setting or to have MicroStation automatically save settings on exit. There are numerous categories of settings, but probably the most important is Working Units. We'll disuss that one in some depth in the next section. The other categories will be considered in various chapters where they come into play.

Accessing the Design File Settings Dialog Box

The Design File Settings dialog box is opened by using the main pull-down menu SETTINGS>DESIGN FILE as illustrated at the right.

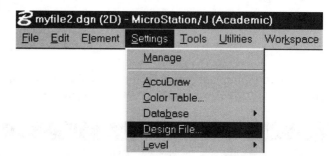

Design File Settings Dialog Box

Illustrated at the right is the Design File Settings. I told you there are a lot of categories! The one selected is Working Units, which is the next topic of discussion. If you make any changes to the Working Units (or any other settings in the other categories) you need to click on OK in order to accept the changes.

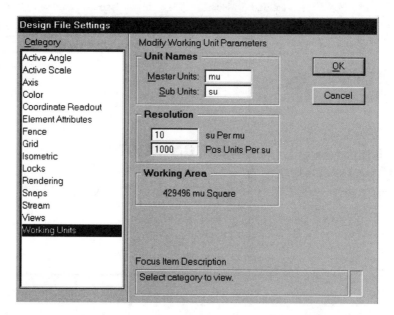

Save Settings

If you want the setting changes to be saved you can do so by using the main pull-down menu FILE>SAVE SETTINGS. You can also use the keyboard shortcut CTRL-F. The pull-down path is illustrated at the right. If the Save Settings is grayed out, it means that your preferences are set to be saved immediately upon exiting the design file. We'll look at specifying that preference later on in this chapter.

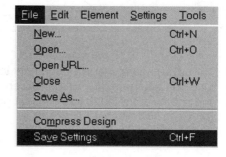

COORDINATE SYSTEM AND WORKING UNITS

Drafting or drawing in the design file requires a coordinate system that allows you to control the size and location of what you draw. MicroStation uses a design plane that is based on positional units. Imagine the design plane looks like a huge grid as shown below.

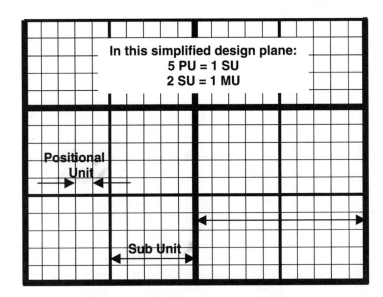

In this simplified design plane:
5 PU = 1 SU
2 SU = 1 MU

To make things easier to work with, the grid is divided into **three** working units. The smallest division of the grid is a Positional Unit (PU). The positional units values are what are stored by the software. However, for CAD to be very precise then it takes a lot of positional units to get a reasonable number of significant digits. The other divisions of working units are Sub Unit (SU) and Master Unit (MU). This is similar to volume being specified in cups, quarts, or gallons where a specific number of smaller units is equivalent to a larger unit (for example, 4 cups to make a quart and 4 quarts to make a gallon). The number of positional units it takes to be equivalent to a sub unit and the number of sub units it takes to be equivalent to a master unit are referred to as resolution. The larger the number, the finer the resolution. Since there is a finite number of positional units to work with, then by varying the resolution you actually vary the number of master units that can be used in one design file. Just as it is easier to express a large volume of water in gallons, MicroStation assumes a single coordinate value to be expressed in master units even though internal values and calculations are done using the positional units.

When entering from the keyboard, you can enter coordinates in *masterunits:subunits:positionalunits*, often stated as MU:SU:PU
Make sure a **COLON** is separating the different values—not a semi-colon.

Even for very experienced users the concept of working units can lead to confusion. Why not just draw in feet? Or meters? You **can** just by labeling the working units. A master unit can be labeled as feet; then if the working units are set as 10 sub units per master unit, a single sub unit would be considered one-tenth of a foot. In any case if you enter a length of 7, then that will be considered 7 master units and you could label them 7 feet, 7 meters, or 7 microns.

IT IS VERY IMPORTANT TO CHANGE THE WORKING UNITS RESOLUTION ONLY AT THE BEGINNING OF DRAWING IN A NEW FILE since by changing in the middle of a design file you will change the actual "size" of all the previous values entered. **YOU ALSO MUST REMEMBER TO SAVE SETTINGS IF YOU MAKE CHANGES TO THE RESOLUTION**—or when you open the design file the next time, the working units will convert back to the original settings, but the elements will be the same number of positional units.

As a beginner you need just a basic understanding of working units in order to enter values into the computer. Otherwise they are not something to be messing around with. Every seed file has its own working units already set and you should be familiar with the ones in the seed file that you are using.

Example of Working Unit Parameters

You can access the settings of the working units after opening a design file. Use the main pull-down menu SETTINGS>DESIGNFILE>WORKING UNITS category and you'll see these working unit parameters.

Shown directly below are the default Working Unit Parameters for the seed2d.dgn seed file.

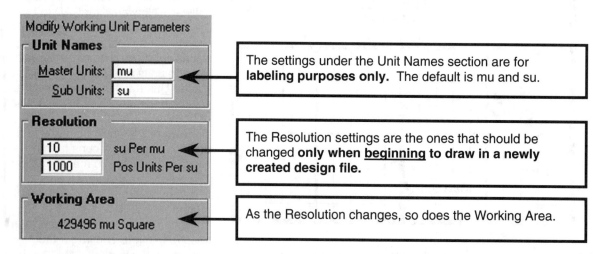

If you wanted to consider basic decimal inches as the values for a design file created using seed2d.dgn, then you would be using these divisions:
- 1 master unit is called an inch
- 1 sub unit = 1/10 of an inch
- 1 positional unit = 1/1000 of a sub unit or 1/10000 of an inch

These are the default Working Unit Parameters for the 2dm.dgn seed file.

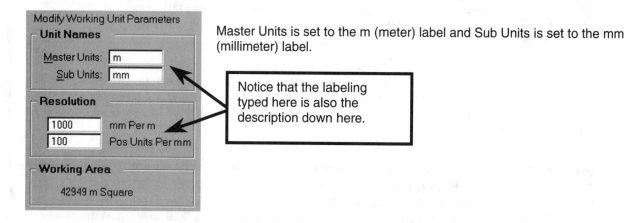

Changing the Working Units: Unit Names and Resolution

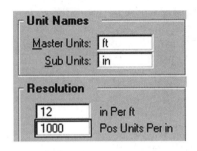

In this case the labeling has been modified and the resolution has also been changed.

1 master unit is called a foot
1 sub unit = 1/12 of a foot
1 positional unit = 1/1000 of an inch

It would take 1000 positional units to make up 1 inch. It would take 12 inches to make up 1 foot. A value of 20 would be considered feet.

Examples of Different Ways to Enter the Same Values

If MU:SU:PU is set at (inches:10 sub units:1000 positional units),
then to enter the value of $3\frac{1}{4}$ inch you could do one of the following:

3.25 *all as master units (this is easiest way to do it)*

3:2.5 *because the 2.5 is actually sub units, so it equals 2.5/10 master units = 0.25 master units*

3:2:500 *because the 500 is actually positional units, so it equals 500/1000 = 0.5 sub units*

3:0:2500 *because the 2500 positional units equal 2.5 sub units, which equals 0.25 master units (in)*

GLOBAL ORIGIN

The working units control the precision of your drawing and consequently the maximum size of your drawing. That is because there is a limit to the number of positional units available. The design plane has 2^{32} positional units on each edge. That's a lot. The "global origin" is the coordinate assigned to the lower left-hand corner of this design plane. The term "global origin" doesn't necessarily refer to where the X=0 and Y=0 coordinate is.

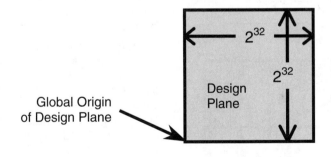

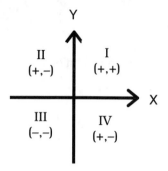

Generally a 2D coordinate system can be thought of as being divided into **four** quadrants. In each quadrant, the X and Y values are either positive or negative as shown on the left. The position where the coordinate is X = 0 and Y = 0 is in the middle, which is where the four quadrants coincide. Now this coordinate system can be positioned anywhere on the design plane. Beginners are not encouraged to change the global origin, but advanced users may want to do so.

If only positive X and Y values are needed then only quadrant I is necessary, so the global origin can be given the coordinate 0,0. This would allow 2^{32} positional units for positive X and 2^{32} positional units for positive Y, but negative coordinates wouldn't be allowed.

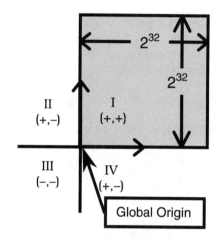

Default Global Origin

Each seed file comes with its own global origin—you should be aware of what it is and how it affects the coordinates. The seed files available with the default workspace have the 0,0 coordinate in the exact middle of the design plane. This is just fine for beginners.

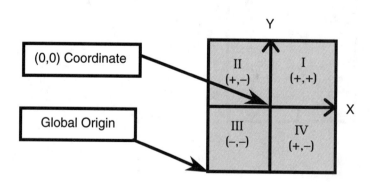

On the left is a schematic of when the 0,0 coordinate is in the middle of the design plane. All quadrants are available, so you can use positive and negative coordinates. In this case the global origin would actually have a negative X and negative Y value of 2^{31} (which is 2^{32} divided by 2) positional units.

ENHANCED PRECISION

Enhanced precision allows coordinate data to be stored at floating point accuracy. This means that instead of rounding to the nearest unit of resolution (uor), it stores the fractional part of the uor as attribute data. This more precise number is used in all geometric computations. Using enhanced precision increases the size of .dgn files but is important when higher geometric precision is required. By default, enhanced precision is enabled for the design file. It is a system setting applied to each design file individually and can be found in the Operation category of the Workspace Configuration.

COORDINATES

Working Units and its settings deal with setting up the coordinate system to use for the design file. The use of coordinates allows you to specify an exact size and location of elements on this coordinate system. A fundamental CAD principle is to **"Draw all objects actual size and with exact precision"** and the use of coordinates allows you to do that. To just eyeball to "close enough" is to waste the power of CAD. Coordinates can be entered by key-in, which means to use the keyboard to type in the values. Coordinates can also be picked graphically using tentative points and snaps to existing elements to maintain the accuracy. Each different graphical element may require its own type of input, whether it is an actual X,Y coordinate; just a magnitude, or a combination. Examples of using coordinates will be discussed in more detail in later chapters that deal with specific elements. For right now let's look at some general concepts.

Absolute and Relative

Coordinates can be specified as relative or absolute. Relative coordinates mean they are calculated from a coordinate that has already been specified. Absolute coordinates means they are calculated from the drawing's (0,0) coordinate, which will be referred to as the origin. Relative coordinates are more useful since you usually know sizes in relation to other elements rather than to the origin.

Rectangular and Polar

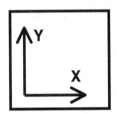

For two-dimensional design files, there are two basic formats for key-in coordinates. These are based on rectangular or polar coordinates. The working units were set up to reflect a rectangular grid. The rectangular coordinate system coincides with this grid, so that you specify an X and Y value for the coordinates. The right-hand rule convention is followed so that a positive X is to the right and a positive Y is straight up. A rectangular system is shown at the left.

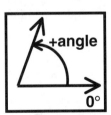

The polar system allows you to specify the distance (or magnitude) and then the angle to measure. A design file is set up so that 0° is to the right and a positive angle is measured counterclockwise, as shown in the example on the left. You can specify negative angles (in degrees) but when you do they're changed to reflect a positive angle. So a –30° angle would end up displayed as 330°. When you type an angle, you don't need the degree symbol, such as °, but for clarity it will be included in the text.

Key-Ins

This table shows the key-ins generally used for specifying coordinates or distances. As stated previously, there is a drawing aid called AccuDraw that is intended to be less cumbersome than key-ins. However, AccuDraw can be tough for beginners to understand, so these key-ins are generally used until you get some experience. AccuDraw will be discussed later on in the text.

ABSOLUTE	
Rectangular system	🖥 XY=*xcoordinate,ycoordinate*

RELATIVE	(the △ means change in—don't type it)
Rectangular system	🖥 DL=△*xcoordinate*,△*ycoordinate*
Polar system	🖥 DI=△*distance,angle*

AccuDraw and Precision Input

There is an advanced drawing aid called AccuDraw. The basic use for AccuDraw is to allow for dynamic precision input rather than using key-ins for coordinates. Shown below on the left is the AccuDraw dialog box and on the right is how it appears when docked above the status bar.

AccuDraw also has other capabilities, such as quick AccuDraw keyboard shortcuts that allow you to run the mouse with one hand and toggle settings by typing with other hand. AccuDraw also behaves somewhat intuitively, that is, it tries to anticipate the next action. This can be a disadvantage for beginners who don't have an understanding of, or experience with the underlying commands. Key-ins, even though they can seem a bit more cumbersome, are recommended for beginners until they have a better understanding of the software. Once a beginner understands how the key-ins work, they will want to investigate AccuDraw more thoroughly. Chapter 11 covers the use of AccuDraw in detail, if you're interested in using it immediately.

PREFERENCES

The preferences do not depend on the design file; rather the preferences come from the file that has an extension .upf, and its file name corresponds to the User component of the Workspace. We won't begin to cover all of the categories found in the Preferences dialog box, since there are a lot of them. There are three different ones that you may want to change from what is obtained by creating your own New... user:
- Background color of the view window
- Saving the Settings changes automatically upon exit
- Having a selection set be highlighted rather than displaying handles on the elements

Accessing the Preferences Dialog Box

The Preferences dialog box is obtained by the main pull-down menu WORKSPACE>PREFERENCES as illustrated on the right. You can also see the other choices listed under Workspace here. If you click on About Workspace you can see which files are being used for the user configuration and the project configuration, as well as the preferences file—but you can't change them!

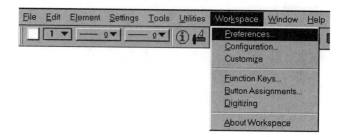

View Windows Category

This category has the setting to change the background color of the view windows.

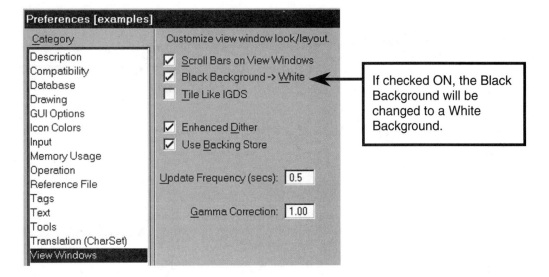

If checked ON, the Black Background will be changed to a White Background.

Operation Category

This category has the setting that will automatically save settings on exit. It is highly recommended that you have it checked ON.

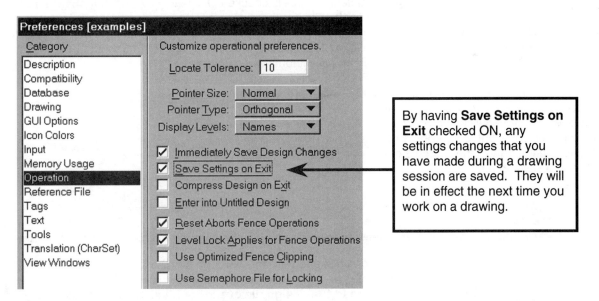

By having **Save Settings on Exit** checked ON, any settings changes that you have made during a drawing session are saved. They will be in effect the next time you work on a drawing.

 The Immediately Save Design Changes is usually checked ON for the preference files that come with MicroStation. However, it is prudent to make sure that it is checked ON while you're looking at this category anyway.

Input Category

The Input category lets you highlight elements rather than display handles on them when they've been selected. This may be preferred because with handles it is very easy to mistakenly change the size and coordinates of an element.

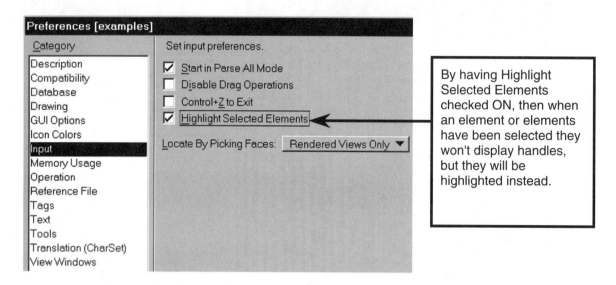

By having Highlight Selected Elements checked ON, then when an element or elements have been selected they won't display handles, but they will be highlighted instead.

SAVING YOUR WORK

Under WORKSPACE>PREFERENCES and under the Operation category, there is also the setting to Immediately Save Design Changes. The default for this is checked ON, which means that **changes to the design file are immediately written to the file**. So there is no need to save your work since it is being saved automatically after each change. The advantage of leaving this default ON is that if you are booted out of the software in the middle of a session, the file will still contain your changes. However, it also means that you do not have an option to exit without saving changes.

If Immediately Save Design Changes is unchecked (OFF) then upon closing the file or exiting the application, you will be prompted if you want to save the changes. Shown on the right is the example of the Alert that will appear.

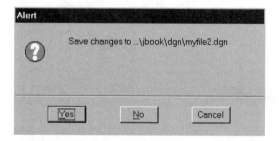

If there is a red icon of a diskette with an X through it in the lower right corner of the status bar (illustrated below), then you are working in a READ ONLY file. **Nothing is being done to the design file even though it will appear as if elements are being placed.** The properties of the file will need to be changed in order for you to actually make changes to the design file.

Chapter 3: Nitty-Gritty Drawing

A BASIC INTRODUCTION TO:

- Creating a New Design File
- Accessing the Basic Tools
- Placing Elements
- Snaps, Locks, and View Controls
- Editing Elements
- Printing

CREATING YOUR FIRST DRAWING

One of the most frustrating things about textbooks is that you usually have to read four or five chapters to be able to get a simple drawing done. That's because it takes about four or five detailed chapters for you to know all about drawing, editing, printing, and saving techniques. In order to avoid this frustration, let's jump right into doing a drawing. What follows are the basic steps and bare bones tools that you need to do a very simple drawing. Granted, some of the time (okay, probably most of the time) you will be doing things without a thorough understanding of why you're doing them. That's the idea! Jump in and get your feet wet! Then in the subsequent chapters, you'll be more comfortable because some things will be familiar. You'll know what to look for while developing a better understanding and more in-depth knowledge of the software.

The purpose of this chapter is to give you a general feel for the software—-you will have gone from a blank drawing to a piece of paper with a CAD drawing that you can hang on your 'fridge.
The end result will be a print similar to the one shown below. The drawing will be made in a stepwise fashion. The general sizes will be given, but most of it will be drawn without specific sizes. That way you don't have to worry about the minute details but instead get to experiment a little on your own.

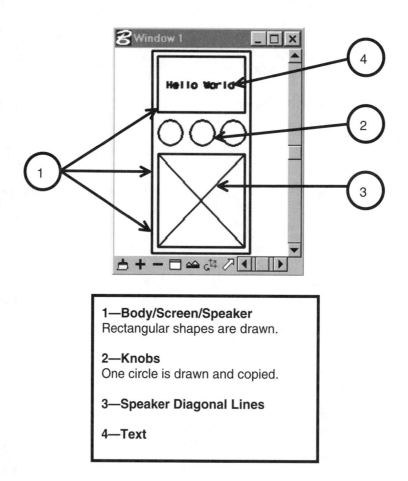

1—Body/Screen/Speaker
Rectangular shapes are drawn.

2—Knobs
One circle is drawn and copied.

3—Speaker Diagonal Lines

4—Text

Nitty-Gritty Drawing

BEFORE STARTING

Mouse Buttons

Using a mouse is important. For this chapter there will be a rebus (little picture) to help you with the mouse buttons that need to be used. Both a 2-button mouse and a 3-button mouse will be shown along with which button or buttons need to be used. The following figures show what they look like and what the mouse action is used for:

Left Mouse Button
- Click on tools
- Data point in drawing area

Right Mouse Button
- Reset
- Accept/Enter

2-button mouse: Left and Right **simultaneously**
3-button mouse: Middle Button
- **Tentative Point**

Correcting Mistakes

Since most of this is new, you will make many mistakes, even if you follow along very closely. That's okay. It is important to know how to either undo your mistake or delete it. Undo and Redo were discussed in Chapter 2 but let's look at a quick review because you're going to need them!

Undo

This tool is found with the standard tools that are usually docked near the top of the window. It may be familiar to you since it is a common tool for Windows programs.

If you make a mistake and catch it immediately, then click on the Undo tool. This will undo the last action taken. You can keep using this tool by clicking on it to keep undoing actions.

Redo

The Redo tool is found to the right of the Undo tool. This is what you use to "undo" the Undo tool action. It will redo the last action that was eliminated by the undo.

Chapter 3

Delete

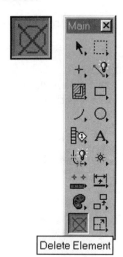

Sometimes you won't catch your mistake early enough to use an undo, and instead you will need to delete an existing element. The Delete tool is found in the lower left corner of the Main tool box.

To delete an element involves three mouse actions.

First start the Delete Element by clicking 🖱 on its tool in the lower left corner of the Main tool box. Next, move the cursor so that it touches the element you want to delete and data point 🖱 on it. The element will change to the highlight color and if it's the one you want to get rid of, then do another data point 🖱 to accept the deletion. If it isn't the one you want, use the Reset 🖱 button to start over with the delete sequence.

Display of Design File

Where's the graphic drawing area? When you first enter a design file, by default there is a window open that is displaying an area of the design file. However, this window can accidentally be minimized or, even more confusing, closed. This causes massive confusion from the start. Shown below is how to open a window that displays the elements in a design file. Make sure that Window 1 is open and Windows 2 through 8 are closed as shown below. It is also a good idea to maximize Window 1 so that it fills the working area.

This main pull-down menu deals with the windows of the design file. A ✓ indicates it is open (such as Window 1 is). There will not be a checkmark if it is closed.

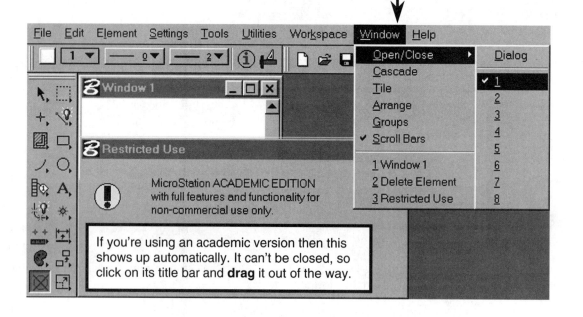

If you're using an academic version then this shows up automatically. It can't be closed, so click on its title bar and **drag** it out of the way.

NEW DESIGN FILE

First, create a new design file; go back over Chapter 2 if you need help. The seed file will be **seed2d.dgn** and its default working units can be left alone. Name it whatever you like—the one shown here is named gadget. Your directory probably won't look exactly like this. It is important to realize that throughout these steps, some of the illustrations will not be exactly what you will see. Items such as tool box placement and directory structures will vary. Be flexible—go exploring.

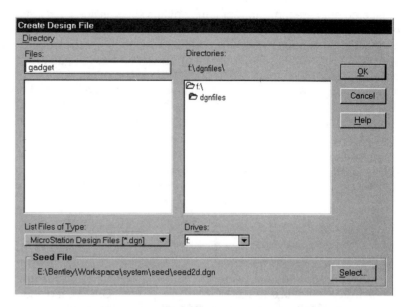

 There are some assumed default settings such as: 1) Snap Lock is checked ON and set to Keypoint, 2) Dynamics is checked ON in the View Attributes, and 3) Tool Settings (under the Tools pull-down menu) has a checkmark by it. These are typical default settings found for seed2d.dgn, but they are not set in stone.

TOOLS NEEDED

To get the bare-necessity tools, use the Tools pull-down menu and make sure there are checkmarks ✓ by Tool Settings, Primary, Standard, and Main as shown below. Dock the resulting tool boxes to have more screen area to see the design file window. Also dock the key-in as illustrated below.

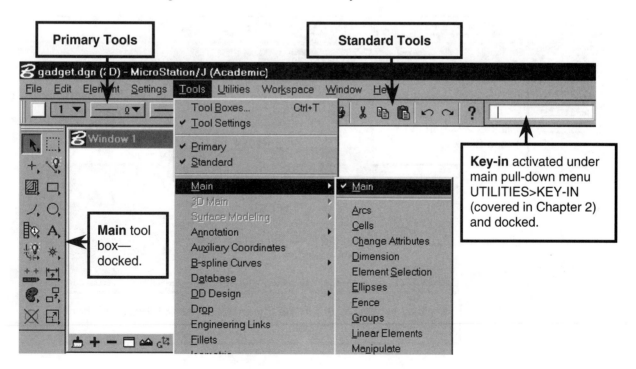

ELEMENTS

Most elements are familiar geometric entities such as lines, circles, arcs, and polygons. An element can be considered open or closed. A prime example of an open element is a single line. It has distinct endpoints. Single open elements can be joined together to form a chain or string. Once joined together, they will then be considered a single **open** element as long as the start and end of the chain or string are distinct. A **closed** element is one entity that can be made up of a combination of single elements that share endpoints and enclose a specific area. A polygon is an example of a closed element. A more detailed discussion of these concepts is found in Chapter 11 dealing with AccuDraw and SmartLine. For a simple example, consider what appears to be a rectangle. It can actually be the following:

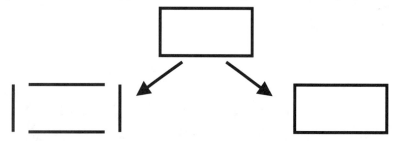

4 separate line elements
Separate elements arranged into a rectangle. Each element is independent of the others. For instance, you would have to delete each element independently.

A single closed element
One closed shape called a block. It is **one** element made up of four segments that share endpoints. If you data point on any of the segments, the entire shape is affected. For instance, one delete element would take care of the entire shape.

BODY, SCREEN, AND SPEAKER

First let's draw three closed rectangular elements to represent the body, screen, and speaker of our gadget with the Place Block tool. The outside large rectangle will use a key-in. The other two rectangles will be done just graphically.

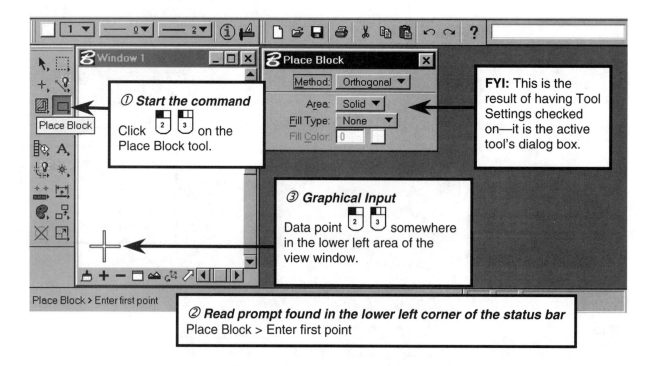

Nitty-Gritty Drawing 43

Now move the cursor around (don't hold down any mouse buttons) and a rectangle will appear and dynamically move with the cursor. It needs the opposite corner of the block—the prompt says so.

④ *Key-in Input*
Click 🖱² 🖱³ in the key-in field so that you get a blinking I cursor. Now you can type in dl=4,8 as shown here. Hit the ENTER key (on the keyboard) to finish. This will specify the opposite corner of the rectangle as being 4 master units in the X-direction and 8 master units in the Y-direction from the previous point specified.

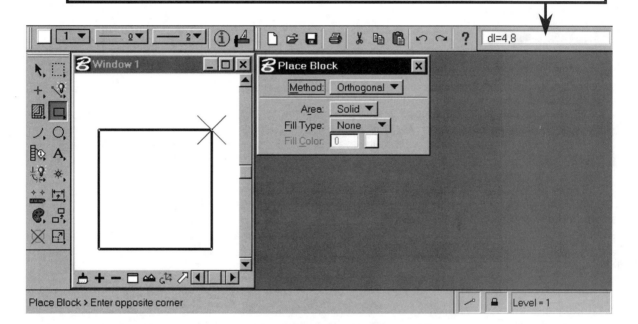

The rectangular shape will be drawn, but you might see only a portion of it as shown here. It also may show up but be very, very small. In either case a quick Fit View will help.

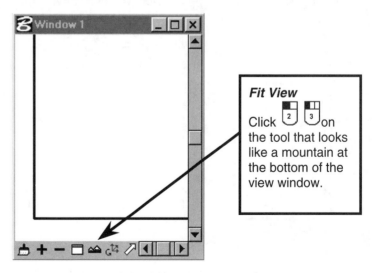

Fit View
Click 🖱² 🖱³ on the tool that looks like a mountain at the bottom of the view window.

Now you should be able to see the larger outer rectangular closed shape that you made with Place Block as illustrated to the right.

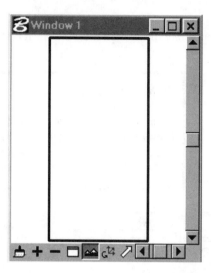

> **NOTE:** This would be a good time to experiment with the Delete Element and Undo tools discussed earlier. Try deleting the rectangle; this will reinforce the concept that it is one element (not 4 separate lines). Immediately after deletion, undo to get it back.

Use the Place Block tool again to do the screen and speaker rectangles. For this quick drawing, their size isn't important. However, normally the size and location of the elements would be precise. For now, just leave room for the knobs. Enter each corner (first and opposite) of each block with a graphical data point when the cursor is where you want it.

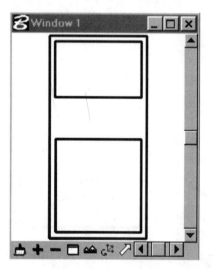

Stop when you have something that looks similar to what is shown here. If you made it through without any mistakes, that's great, but you might want to delete a block and try to place it again—just for practice.

By now you should be very good at using the left mouse button for clicking on tools or a data point in the **drawing area. Therefore, the** rebus for that mouse action will no longer be shown. You can handle it.

KNOBS

First a circle will be placed, and then copying it two times will complete drawing the knobs. The first circle's size will be specified in the dialog box, and then its center will be specified graphically. In order to get all the circles in a straight horizontal line, the drawing axis will be locked on before making the copies.

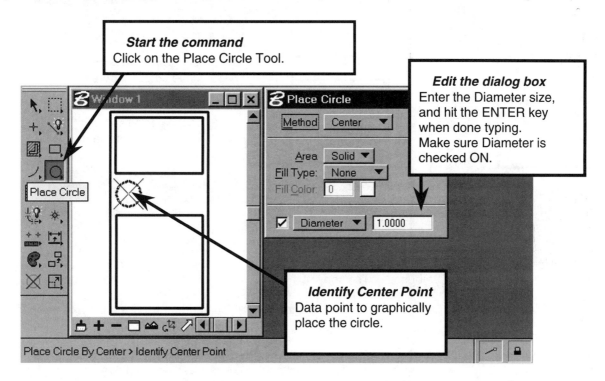

Now to lock the axis so that you can draw only in a horizontal or a vertical direction.

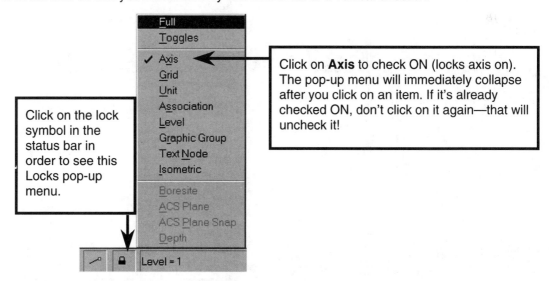

If the lock symbol doesn't show up in the status bar—click on the spot where it should be. The list should still appear. There is also a key-in that will toggle Axis Lock on. Go to the key-in field and type **lock axis on** then hit the ENTER key on the keyboard when you're done typing. This will accomplish the same thing.

The first circle has been placed and the axis has been locked, so on to making a copy of the circle.

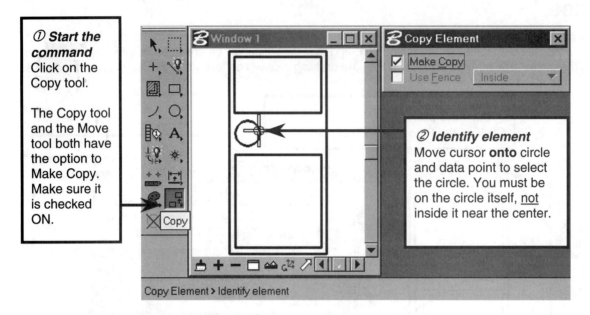

① Start the command
Click on the Copy tool.

The Copy tool and the Move tool both have the option to Make Copy. Make sure it is checked ON.

② Identify element
Move cursor **onto** circle and data point to select the circle. You must be on the circle itself, not inside it near the center.

After you identify the circle as the element that is to be copied, move the cursor (again, do this without holding down any of the mouse buttons). The copy of the circle will appear and move in the general direction of the cursor. The circle should be restricted to going either horizontally or vertically, whereas the cursor can go anywhere. If the new circle isn't restricted then you goofed on locking the axis (go back and try again).

If you do have Axis Lock ON, then your screen will look something like the display shown below. The copied circle appears dashed and the prompt is to Enter point to define distance and direction. A circle will be placed with each data point that you do. Don't worry if the circle is grayed. When you're done, use the right mouse button to reject any more copies and the last circle will go back to the color of the rest.

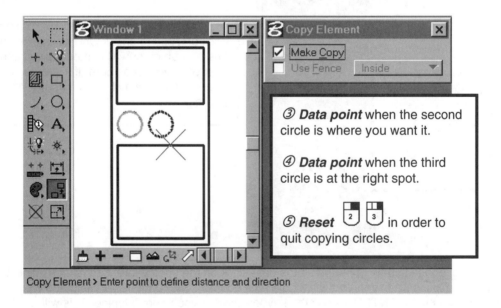

③ Data point when the second circle is where you want it.

④ Data point when the third circle is at the right spot.

⑤ Reset in order to quit copying circles.

Nitty-Gritty Drawing 47

At the end of this sequence your screen should look something like the display shown below. If you made some mistakes then the displayed picture may need to be updated, especially if you see little weird pixels here and there. Update the view by clicking on the Update View tool (paintbrush icon) in the lower left corner of the view window.

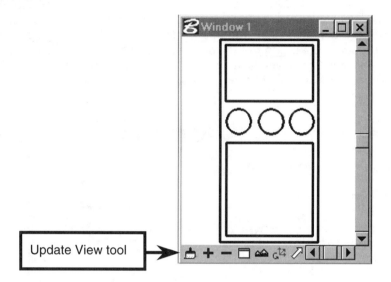

Update View tool

AXIS LOCK OFF

Even if you don't update your view, take the time now to **TOGGLE OFF THE AXIS LOCK**. Go back and click on the lock symbol of the status bar. If the lock symbol isn't there, click where it should be and the Locks pop-up menu should show up. Click on **Axis** so that it is **unchecked (OFF)**.

There is also a key-in that will toggle Axis Lock off just like there was one to toggle Axis Lock on. Go to the key-in field and type `lock axis off` then hit the ENTER key on the keyboard when you're done typing. This is illustrated below.

No matter how you get it done, please make sure that you have Axis Lock OFF.

DIAGONAL LINES

Placing the lines that go from one corner of the speaker rectangle to the other requires the use of snaps.

Snaps are done with the Tentative button on the mouse. This is one of the toughest maneuvers for beginners (at least those with a 2-button mouse). The trick is to be coordinated and <u>get both buttons depressed at the **same** time</u>. You will depress <u>and release</u> both buttons—don't hold them down. You have to remember to accept the tentative point you have just snapped to (if it is where you want it). You accept a tentative point by doing a data point. Using a tentative point is definitely a case of learning by doing—reading about it doesn't cut it.

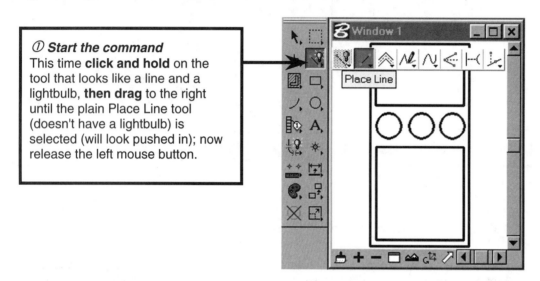

① **Start the command**
This time **click and hold** on the tool that looks like a line and a lightbulb, **then drag** to the right until the plain Place Line tool (doesn't have a lightbulb) is selected (will look pushed in); now release the left mouse button.

The Place Line dialog box will appear, but we won't use it. Instead the existing block will be used to help specify the first point of the line. Now it's time to use a tentative point to snap to an existing element.

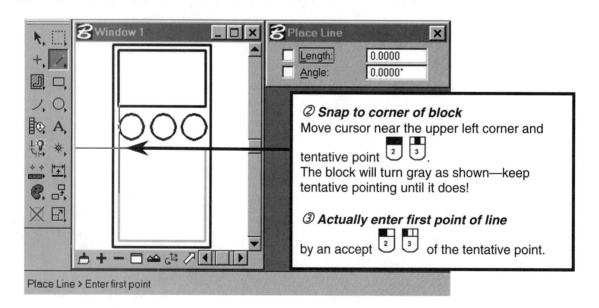

② **Snap to corner of block**
Move cursor near the upper left corner and tentative point.
The block will turn gray as shown—keep tentative pointing until it does!

③ **Actually enter first point of line**
by an accept of the tentative point.

Nitty-Gritty Drawing 49

If everything went well, you should see a line starting at the upper left corner and dynamically moving about as illustrated below. If the line goes only across or up and down, you forgot to turn off axis lock (better do it and then get back to this spot). The prompt indicates that an endpoint is needed. Again "snap to it."

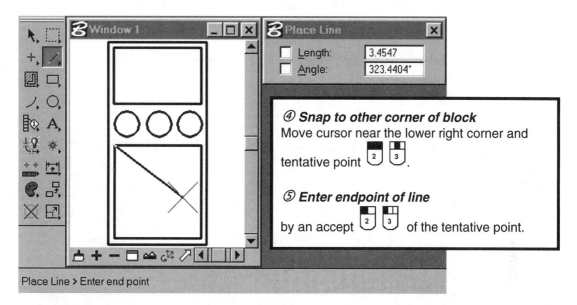

④ **Snap to other corner of block**
Move cursor near the lower right corner and tentative point 🖱️.

⑤ **Enter endpoint of line**
by an accept 🖱️ of the tentative point.

After entering the endpoint of the first line, another line will be started using that endpoint as its first point. Shown below is that next line automatically showing up. We don't want to do that—so the command needs to be reset to start again.

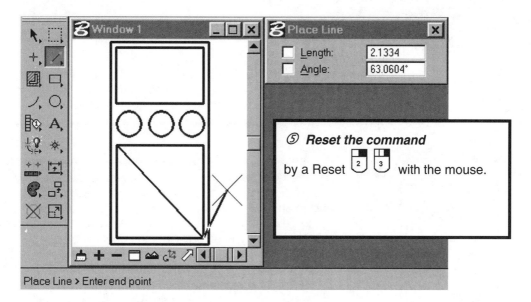

⑤ **Reset the command**
by a Reset 🖱️ with the mouse.

Place Line > Enter first point

The status bar should then return to the prompt of "Enter first point" of a new line. Do the other diagonal line on your own. **Be sure to practice using tentative to snap to the corners—don't eyeball.** If you are using the tentative correctly, the element being snapped to should be highlighted—look for this visual cue.

TEXT

It is important to sign your work, so adding text is part of the drawing. There is a Text Editor dialog box where you type the necessary text—**don't use the key-in field.** You can use Hello World or your own name. Either way, the text size may need to be adjusted before the text is actually placed, but that is fairly easy to do.

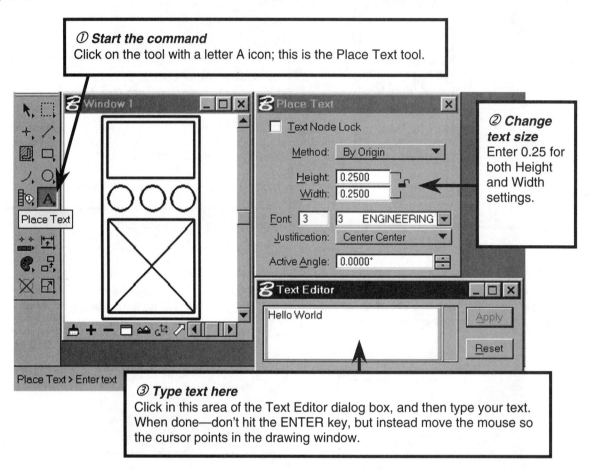

① **Start the command**
Click on the tool with a letter A icon; this is the Place Text tool.

② **Change text size**
Enter 0.25 for both Height and Width settings.

③ **Type text here**
Click in this area of the Text Editor dialog box, and then type your text. When done—don't hit the ENTER key, but instead move the mouse so the cursor points in the drawing window.

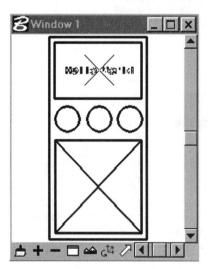

Once the cursor moves back into the drawing area, you will see the text appear. It can still be moved around since you haven't actually placed it. It should look similar to what is shown on the left. If your text looks too big, move back to the Place Text dialog box and re-enter new sizes. These will be reflected when you go back into the drawing area.

④ **Position text and place it**
Once you have the size adjusted so that your text fits, use a data point to place the text.

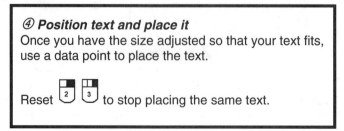

Reset to stop placing the same text.

Now you're done with what we set out to draw, but one of the benefits of CAD is that it is easy to change a drawing. So let's manipulate our gadget drawing a little bit before we print it out.

EDIT

The drawing can be edited by manipulating an existing element, such as moving or rotating it. This manipulation can also result in a new element. Let's change the look of the speaker by using Move Parallel to make smaller blocks from the original block.

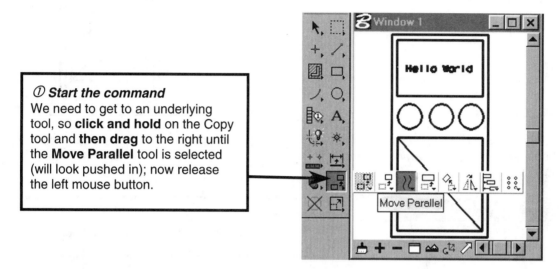

① *Start the command*
We need to get to an underlying tool, so **click and hold** on the Copy tool and **then drag** to the right until the **Move Parallel** tool is selected (will look pushed in); now release the left mouse button.

The title bar of the dialog box will indicate that Move Parallel is the selected tool. There are also two settings in the dialog box (Distance and Make Copy) that need to be changed so that they appear as shown below.

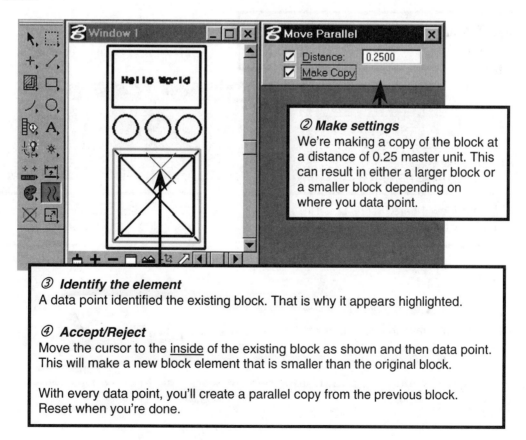

② *Make settings*
We're making a copy of the block at a distance of 0.25 master unit. This can result in either a larger block or a smaller block depending on where you data point.

③ *Identify the element*
A data point identified the existing block. That is why it appears highlighted.

④ *Accept/Reject*
Move the cursor to the inside of the existing block as shown and then data point. This will make a new block element that is smaller than the original block.

With every data point, you'll create a parallel copy from the previous block. Reset when you're done.

Now the drawing should look similar to what is shown here on the right. Make sure that you see the entire gadget. If not, go back and review how to fit the view (earlier in this chapter).

If you didn't have any trouble, that is amazing. Making mistakes is a great way to learn, and so if you goofed up along the way—think of all the new knowledge you gained!

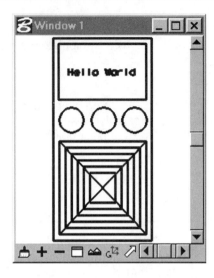

PRINTING

Getting the drawing to print on the paper is easy only if your printer has already been set up to work. If it hasn't been, then go to the software documentation or follow your instructor's directions. Otherwise, keep on going.

① **Start the command**
Click on the Print tool in the Standard tool box.

The following Plot dialog box will appear. Don't worry about the print/plot terms, they're interchangeable.

② **Print**
Click on this printer icon to submit the print.

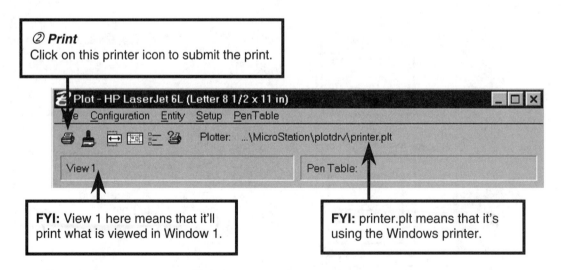

FYI: View 1 here means that it'll print what is viewed in Window 1.

FYI: printer.plt means that it's using the Windows printer.

That's all it takes. The status bar will state when it has "Finished Creating Plot".   ‹Finished Creating Plot›

☝ **Go get the sheet of paper, admire your work and head to the 'fridge!** ☝

Chapter 4: Drawing Aids and Tools

SUBJECTS COVERED:

- Element Attributes
- View Controls
- View Attributes
- View Levels
- Naming a Level
- Snaps and Tentative
- Grid
- Locks
- Element Selection Tool Box
- Element Selection
- PowerSelector
- Fence Tool Box
- Place Fence
- Delete Fence Contents
- Delete Element
- Cursor

ELEMENT ATTRIBUTES

Element refers to items such as lines, arcs, circles, and even text. Attributes determine how an element appears and behaves. Each element that you place in a drawing has at least five common attributes: Level, Color, Style, Weight, and Class. An element's attributes are determined by the settings that are active when the element is created. Attributes can also be modified later, but it is easier to set the attributes that you want before placing an element.

Accessing Active Element Attributes

You can get to the Element Attributes dialog box by the main pull-down menu ELEMENT>ATTRIBUTES.

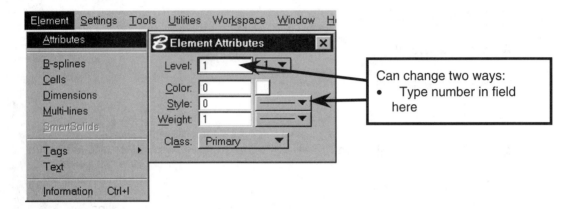

The Primary Tools tool box also gives you access to option buttons for the active attribute settings. It is common to dock this tool box near the top of the MicroStation window. When it's docked, you won't see the window's title bar.

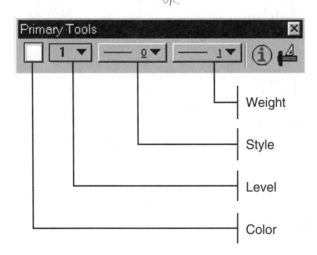

Drawing Aids and Tools 55

If the Primary Tools tool box is not open then you go under the main pull-down menu TOOLS. Any tool box with a checkmark next to it is open somewhere (docked or floating). If there isn't a check by the name of the tool box then click on it, and this will open up the tool box. If there is a check and you click on the name, then the tool box that was open will close. Quite often, one tool box will accidentally get closed while you're trying to open a different one!

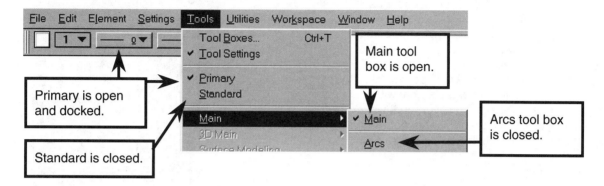

Level Attribute

Using different levels for elements is a powerful organizational tool. A level doesn't mean at a different elevation. All the levels' coordinates coincide. Instead think of a level as a transparent sheet that is being drawn on. If all the levels are on, then all the transparent sheets are stacked on top of each other and you're working with all the elements. However, if a drawing is organized by placing elements on different levels, then you can manipulate the display, colors, printing, etc., of elements by controlling the levels. Turning the level off is similar to not laying that transparent sheet down. You haven't deleted the elements; you just aren't going to work with them.

In this drawing, the elements representing the intersection were placed with Level 1 being active. The arcs for the trees were placed on Level 2. In Window 1, illustrated below on the left, both levels are turned on. In Window 2, illustrated below on the right, Level 2 was turned off. The trees have not been deleted; they just aren't being worked with right now.

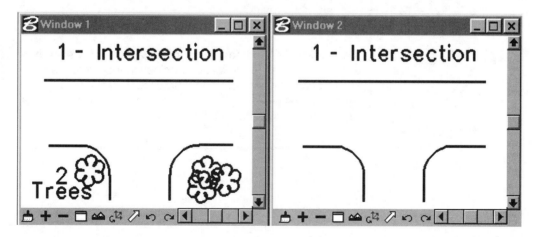

For beginners, it is suggested that the Active Level be left at the default of 1. Just remember that the Active Level must be on in all the views—this is so you can't turn off a level and still be drawing on it.

Color Attribute

Elements can be different colors. Shown here is the default color table. Colors are actually stored by their number, with the color table mapping out what color will be used for each number. The button shows an example of current color. Color 0 looks white. It is actually black on paper but displays as the "opposite" of black/white background. For a white background, the element will appear black. For a black background, the element will appear white. It will not, however, print white on a black piece of paper! If you're not using a color printer but one that's set to do gray scale, then colors will appear as different shades of gray.

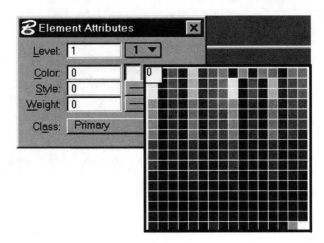

 Usually most users will use one color per level for good drawing organization.

Style Attribute

There are 0–7 standard Styles available. They are usually referred to as line styles even though they are used for elements such as circles too. Advanced users can make custom styles and edit them too.

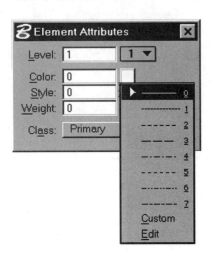

Suggestions for Style settings that work best for standard graphic line styles are:
0 for visible lines
2 or 3 for hidden lines
6 for phantom lines

7 works well for a centerline used in mapping. However, it doesn't work well for typical centerlines in technical drawings. These can be accomplished by drawing 3 segments of a 0 style line ——— — ——— like this, or advanced users can try their skill at custom line styles.

Weight Attribute

The Weight corresponds to the hand-drafting terminology for pens. The higher the number, the heavier or "thicker" the weight of the element.

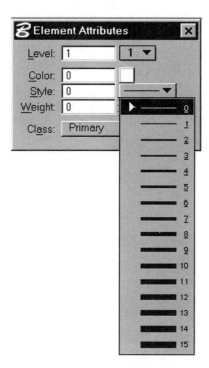

Standard graphic convention usually uses different line weights for different line styles.

These different weights will print out differently depending on the output device (printer/plotter) being used. Use weights that work for you or as specified by your instructor.

Advanced users can adjust the printer/plotter settings rather than make changes to the attributes. This is useful in setting a company standard when multiple types of output devices will be used.

Suggested line weights to use for a **300 dpi inkjet**:
　0 for thinner centerlines
　1 for hidden lines and section lines
　2 for visible lines
　4 for cutting planes

Suggested line weights to use for a **600 dpi LaserJet®**:
　0 for thinner centerlines
　2 for hidden lines and section lines
　3 for visible lines
　5 for cutting planes

Class Attribute

This also is an organizational tool. There are two classes to choose from: Primary and Construction. Primary is the class for the elements that make up your drawing so Class is usually left set to Primary. As beginners, sometimes you'll place elements just to help simplify the drawing process and then delete them when they're no longer needed. That is just fine when starting out (and even encouraged). However, with advanced drawings and users, elements that are not to be part of the final drawing but just aid in its construction are assigned a Construction class. It can be specified that construction class elements aren't printed out, but they don't have to be deleted. This leaves them available to aid in placing primary elements at a later time too. **For the purposes of this text, all elements will be given the class of Primary.**

VIEW CONTROLS

View refers to a window that is open and displays what has been drawn in the design file. There are eight view windows that you may open. Each individual view has its own controls, which manipulate the actual display much like a camera with a zoom, pan, etc. **These controls do not change any sizes of the elements.** Each view has its own controls, so each view can be set independently. Shown below are the view controls found at the bottom of each view window. Each control is labeled, but we'll discuss some specific ones in more detail.

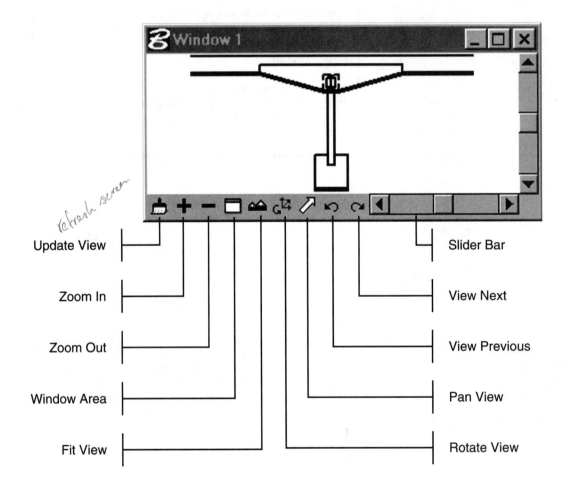

Drawing Aids and Tools 59

Update View

The Update View tool refreshes the display. It may be necessary to use this tool after deleting elements or making modifications. Sometimes what looks like lines or pieces are actually not there—the display just didn't clean itself up.

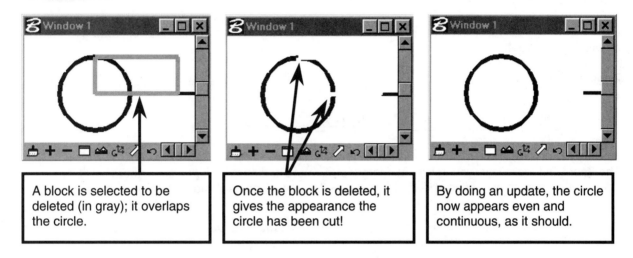

A block is selected to be deleted (in gray); it overlaps the circle.

Once the block is deleted, it gives the appearance the circle has been cut!

By doing an update, the circle now appears even and continuous, as it should.

Zoom In

Zoom In allows you to look closer at part of the display. It uses a zoom ratio that acts as a multiplication factor for the display.

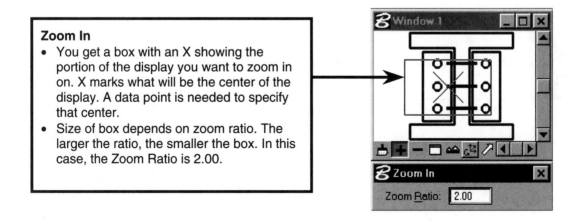

Zoom In
- You get a box with an X showing the portion of the display you want to zoom in on. X marks what will be the center of the display. A data point is needed to specify that center.
- Size of box depends on zoom ratio. The larger the ratio, the smaller the box. In this case, the Zoom Ratio is 2.00.

The result of the Zoom In by a Ratio of 2.00 operation is shown at the right. We can see the circles much better.

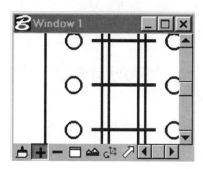

Now that we can take a closer look, what about looking at a larger portion of the drawing area? That will require zooming out.

Zoom Out

The Zoom Out tool allows you to look at a larger portion of the drawing. In this case, there isn't any additional data point needed to specify the center of the view since it will stay the same.

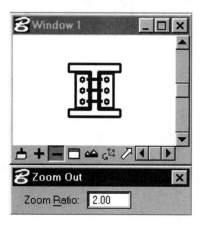

Zoom Out
- Takes the current display and zooms out by the zoom ratio. In this case the Zoom Ratio is 2.00.
- Center of display stays centered.

 Remember that zooming in or out does not change the actual size of the elements.

Window Area

The Window Area view control doesn't use a flat multiplication ratio like zooming does. Instead you graphically pick the area of the current display that you want to take a closer look at. Shown below are the steps taken to window an area of the drawing.

First Step
X marks the spot of the corner of the window area you want to display. A data point picks it.

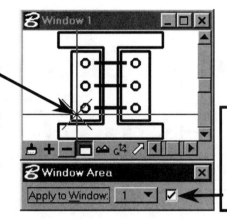

You can define the Window Area in one view and the results can be displayed in a different window.

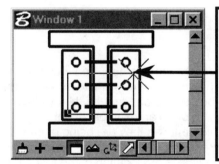

Second Step
Specify the opposing corner with another data point. The actual window area shape is determined by the size of the existing view window. You can specify only an area of the same proportions as the view window.

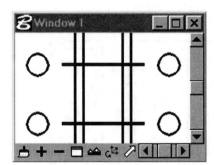

Here's the display that results from the Window Area operation.

Fit View

The view control that looks like mountains is called Fit View. It allows you to fit the entire active design into the window. This is especially useful if you are entering elements by keying in their coordinates, since you can be drawing in an area that is not being displayed. This view control quickly shows all the elements in the design file that the view attributes are displaying including those that were placed off-screen. This tool was also discussed in Chapter 3 when we were working on the nitty-gritty drawing.

Shown below is an example of using Fit View. The view on the left is looking at just a portion of the drawing. By fitting the view, you can see all of the elements in the drawing.

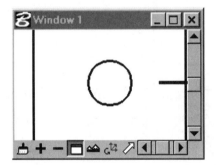
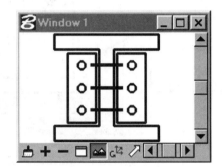

Pan View

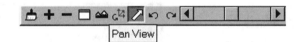

Panning is another way to move the display around on the screen. You data point where you want to grab onto the drawing and then move the cursor in the direction you want to move the display. The slider tools on the right and left sides of the view can also be used to pan around the display. Pan is useful when you just want to scoot the display a little bit. Shown below is an example of panning in a view. With Dynamic Display checked ON you can actually see the elements as they are being panned around in the window. If you have it unchecked (Off) then a pan arrow will show up to indicate direction instead of seeing the actual elements. The picture below on the left indicates the location of the first data point. The picture below on the right shows the second data point and the result of sliding the display down and to the right. Notice that the display size does not change.

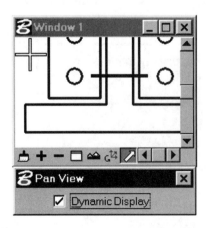
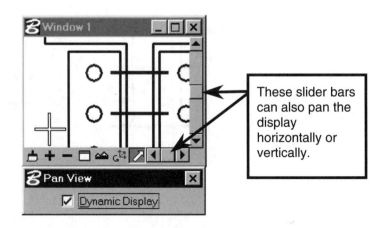

These slider bars can also pan the display horizontally or vertically.

Drawing Aids and Tools 63

VIEW ATTRIBUTES

View controls change how you looked at the display, from close up to far away and lots of points in between. The View Attributes settings control what elements and aids are displayed or not displayed in a view. The attributes of each view can be controlled individually or as applied to all of the views. It is important to realize that you can create an item, but unless the view attributes are set to display it you won't see it. For example, if the view attributes are not set to display text, text can be placed but it doesn't appear on the drawing. However, if the View Attributes Text setting is checked ON, the text will then be displayed. If you find yourself saying "Where did it go?", do a fit view to see if it shows up, but then also check the view window's View Attributes settings.

Accessing the View Attributes Dialog Box

Under the main pull-down menu SETTINGS>VIEW ATTRIBUTES you can access the View Attributes dialog box, but CTRL-B is the shortcut that is often used to accomplish the same thing. One more way to get to the View Attributes dialog box is by clicking on the "B" icon in the upper left corner of a view window.

Shown below is the View Attributes dialog box displaying the default settings for a drawing started with the **seed2d.dgn** seed file. Each seed file will contain its own default View Attributes settings. You should be familiar with the settings for the seed file you are using. A check box with a checkmark means the setting is toggled on. For example, as shown below, Grid checked ON means that the grid would be displayed in View Number 1.

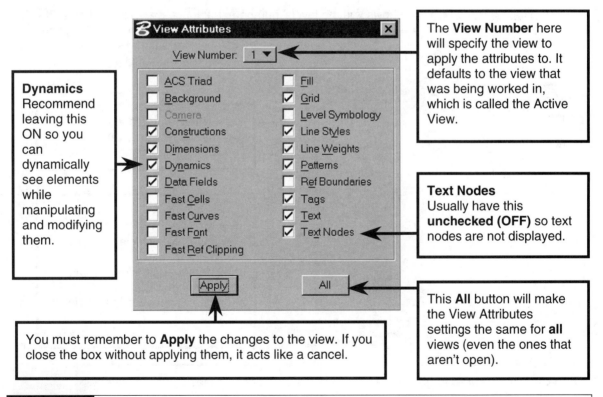

Dynamics
Recommend leaving this ON so you can dynamically see elements while manipulating and modifying them.

The **View Number** here will specify the view to apply the attributes to. It defaults to the view that was being worked in, which is called the Active View.

Text Nodes
Usually have this **unchecked (OFF)** so text nodes are not displayed.

You must remember to **Apply** the changes to the view. If you close the box without applying them, it acts like a cancel.

This **All** button will make the View Attributes settings the same for **all** views (even the ones that aren't open).

The View Attributes settings shown here are for guidance only. You may want to make modifications to suit your own needs.

VIEW LEVELS

Earlier in this chapter, the element attribute called level was discussed. Now we can look at controlling which levels are displayed and in which views. This can be easily accomplished by using the View Levels dialog box. This dialog box can indicate the levels according to their numbers, or there is an option to use their level names. There is also a Level Manager dialog box that puts the settings that deal with levels in one dialog box with various tabs to let you access the different dialog boxes. Level Manager can be complicated to the beginner, so we'll stick to the View Levels dialog box.

Accessing the View Levels Dialog Box

The display of levels can be accessed through the main pull-down menu SETTINGS>LEVEL>DISPLAY or with the keyboard shortcut ⌨CTRL-E.

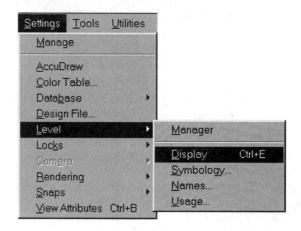

The default View Levels dialog box shows you the levels' numbers, but there is also the option to display the levels by name. Shown below is the View Levels dialog box (Display Level Numbers) and some settings to illustrate the use of View Levels to control the display of levels for each view.

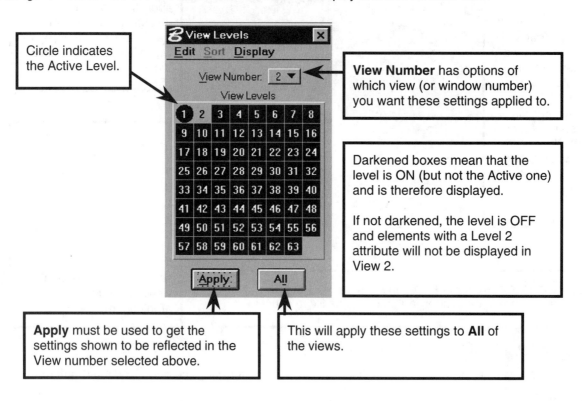

Circle indicates the Active Level.

View Number has options of which view (or window number) you want these settings applied to.

Darkened boxes mean that the level is ON (but not the Active one) and is therefore displayed.

If not darkened, the level is OFF and elements with a Level 2 attribute will not be displayed in View 2.

Apply must be used to get the settings shown to be reflected in the View number selected above.

This will apply these settings to **All** of the views.

Drawing Aids and Tools

Display Level Numbers

Let's look at a simple design file that has elements existing on four different levels. Level 10 has the intersection. Level 11 has utilities such as water and lights. Level 12 has features like trees and shrubs. Level 13 has notes. Shown below are View Window 1 and View Window 2 along with the View Levels dialog box that indicates which levels are on or off in each view. There can be only one Active Level; in this case it is Level 10.

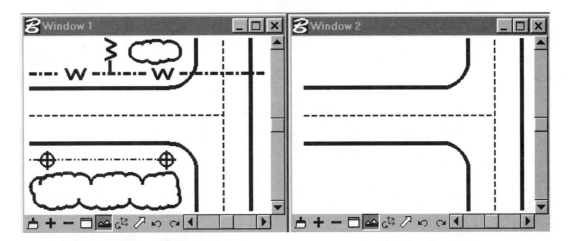

View Window 1
- Level 13 is off.
- All other levels are on.

View Window 2
- Levels 11, 12, and 13 are off.
- All other levels are on.

Display Level Names

With a complicated design file, it is often necessary to name the levels. It is easier to keep track of numerous levels and their elements if the levels have been named. By using the View Levels pull-down menu DISPLAY>LEVEL NAMES as illustrated on the right, you can see the levels displayed by their names rather than their numbers. This display is a bit tougher to look at and determine which levels are on or off in the different views.

Shown below is the View Levels dialog box using the Display Level Names option. The level display for View Window 1 and View Window 2 is repeated here to illustrate how the View Levels settings match the views' displays.

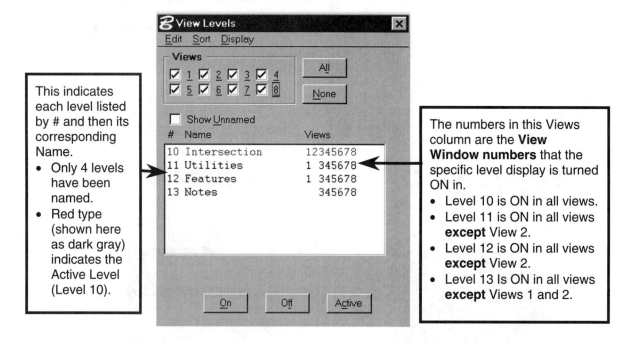

This indicates each level listed by # and then its corresponding Name.
- Only 4 levels have been named.
- Red type (shown here as dark gray) indicates the Active Level (Level 10).

The numbers in this Views column are the **View Window numbers** that the specific level display is turned ON in.
- Level 10 is ON in all views.
- Level 11 is ON in all views **except** View 2.
- Level 12 is ON in all views **except** View 2.
- Level 13 Is ON in all views **except** Views 1 and 2.

View Window 1 shows the Intersection level (#10), the Utilities level (#11), and the Features level (#12), but not the Notes level (#13). View Window 2 shows only the Intersection level (#10).

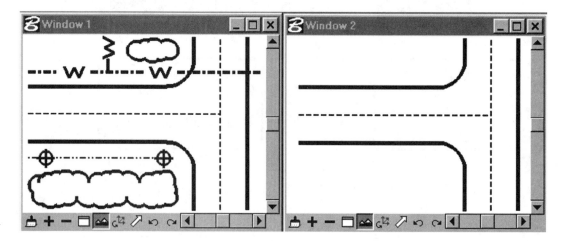

Turning the display of a level on or off with the View Levels displaying the level numbers requires a bit of finesse. However, it is useful because it can be applied to any number of Views, not just one or all. The Views that you want to affect are selected, the Level that you want to control is highlighted, and then you click the On or Off button.

Views 1 and 2 will be affected when the On or Off buttons are clicked.

The Notes level (#13) display will be turned On or Off.
- You can highlight more than one level at a time.

Clicking the On button will turn On the Notes level (#13) in Views 1 and 2.

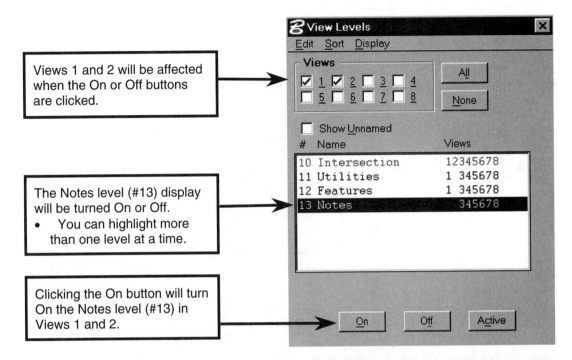

As a result of clicking the On button, the listing will change as shown to the right and Window 1 and Window 2 will now display the Notes level (#13) as illustrated below.

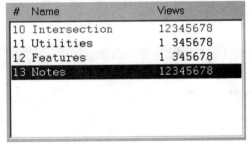

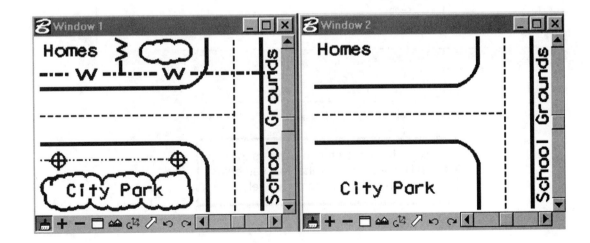

NAMING A LEVEL

If you are going to use named levels, you'd better be able to name them. The names are assigned by using the Level Names dialog box. It also allows you to add, edit, or delete level names. There do not need to be elements on the level in order to name it.

Accessing the Level Names Dialog Box

The Level Names dialog box can be accessed from the View Levels pull-down menu EDIT>DEFINE NAMES as shown below on the left. You can also access it from the main pull-down menu SETTINGS>LEVEL>NAMES as shown below on the right.

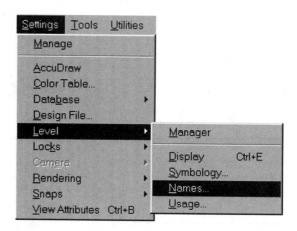

Whichever way you get there, you can see which levels (if any) have been named. You also can add, edit, or delete level names. Shown below is an illustration of the Level Names dialog box.

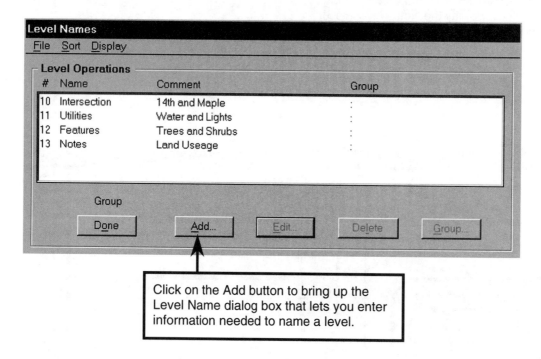

Click on the Add button to bring up the Level Name dialog box that lets you enter information needed to name a level.

Drawing Aids and Tools   69

Add Level Name

Clicking on the Add button will bring up the Level Name dialog box as illustrated on the right. Here you can fill in the Number, Name, and Comment fields. Once you're done, click OK to complete the process and the Level Names dialog box will show the newly named level as illustrated below.

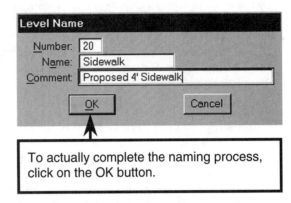

To actually complete the naming process, click on the OK button.

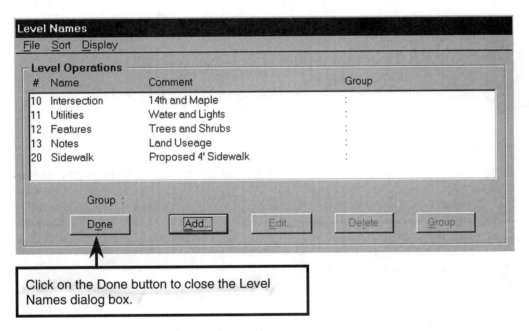

Click on the Done button to close the Level Names dialog box.

The View Levels dialog box will now show the Sidewalk level (#20) in its listing of named levels as illustrated below.

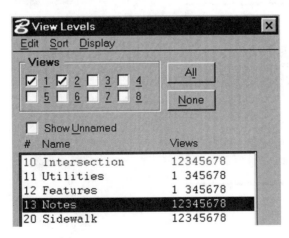

SNAPS AND TENTATIVE

Probably the most useful drawing aids are the snaps that you can use. A snap allows you to use an existing graphical element to establish a coordinate for the new element. You actually "snap" or jump to that specific coordinate of the element. Snaps use key geometrical points of the elements such as quarter points of a circle or endpoints of lines. There are 14 snaps but not all are always available for use. Which snaps can be used depends on what elements already exist and are being placed. The most widely used snaps are Keypoint, Intersection, Tangent, Perpendicular, and Parallel. Keypoint is the default of choice.

Tentative is used in conjunction with snaps. Remember that the Tentative button for a 2-button mouse is done by **pressing both the right and left buttons simultaneously**. The Tentative button for a 3-button mouse is the middle button. By using a tentative, you temporarily jump to the snap point of the element you're pointing at. The element will highlight and + will indicate the snap point that is tentatively selected. This allows you to "see it before you buy it." To accept the tentative point, you data point with the left mouse button. To reject it, you do a Reset with the right mouse button. In the case of more than one element or snap location in the same area, you can keep using the tentative until the snap location and selection you <u>do</u> want are indicated.

 The strong point of CAD is its ability to be precise. With CAD you don't eyeball; instead everything is drawn real size and precise. This precision is often made possible by the ability to snap as well as by key-in coordinate entry.

Accessing Snaps

Snaps are found as a shortcut in the status bar at the bottom of MicroStation's window. You may also set up the snaps as a button bar.

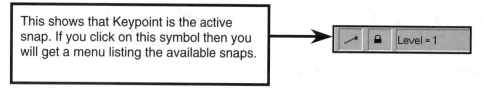

This shows that Keypoint is the active snap. If you click on this symbol then you will get a menu listing the available snaps.

When you click on this symbol, the Snap Modes pop-up menu will appear.

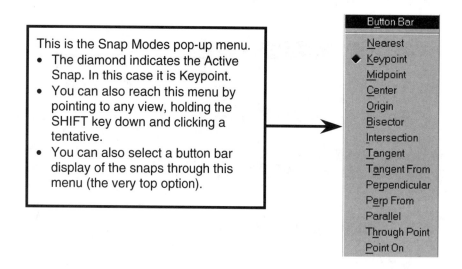

This is the Snap Modes pop-up menu.
- The diamond indicates the Active Snap. In this case it is Keypoint.
- You can also reach this menu by pointing to any view, holding the SHIFT key down and clicking a tentative.
- You can also select a button bar display of the snaps through this menu (the very top option).

Drawing Aids and Tools 71

The Snap Mode button bar is also available for display and use. It can be docked or resized. The buttons follow the order of the listed menu. Most of the icons make sense. Some of these snaps are related, such as Tangent and Tangent From—in these cases, an explanation of the simple one will be given and you can experiment on your own with the other.

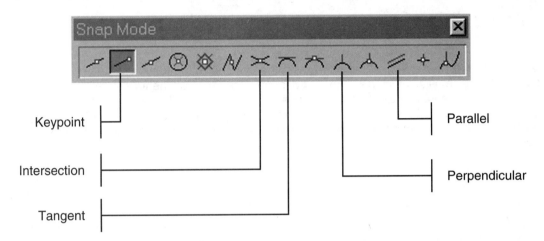

Since most of the snaps are self-explanatory, they won't all be discussed in detail. The basic ones shown will be discussed and the use of the other snaps will be left for you to experiment with on your own.

Keypoint

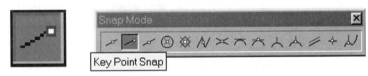

This is the most used snap because it is an "all-in-one" snap. It uses the fact that each geometric element has specific "key" geometric points. A line has its endpoints. A circle has its center, and quarterpoints at 90°. Text has its text justification point that is its insertion point. Polygons have their center, and then each side's endpoints. The points on an element that the Keypoint snap recognizes can be adjusted using the snap divisor value found under the Locks dialog box (covered later in the chapter). The snap divisor allows you to change the division of an element that you can snap to. A snap divisor of 2 would give additional keypoints such as midpoint of a line; a circle would then also have keypoints at 45°.

A keypoint of the circle is shown here (its center). To snap to this one, you do not point to the circle itself, but instead you position the cursor near the center of the circle and then do a tentative (both left and right mouse buttons at same time).

 If the snap divisor were set to 3 then the snap would still go to each endpoint of a line. However, it could also snap to two points that divide the line into thirds—in this case you wouldn't get to the midpoint of the line.

Intersection

This snap goes to the intersection of two elements. When you place the tentative snap point, you will notice that the two elements selected will appear dashed. One of the elements will be highlighted. The two elements dashed represent which intersection is involved. The highlighted element will be the one selected for the command. For example, if you want to delete part of Line A starting at the intersection of Line B then you will want to snap to the intersection but keep using a tentative until element A is the one highlighted.

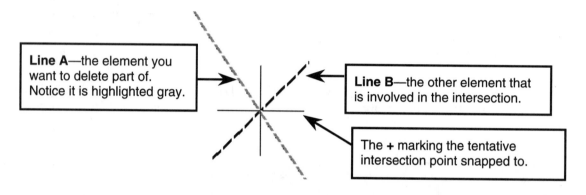

Line A—the element you want to delete part of. Notice it is highlighted gray.

Line B—the other element that is involved in the intersection.

The **+** marking the tentative intersection point snapped to.

Tangent

This snap involves at least one curved element such as a circle. A snap can be specified at the beginning and/or the end of a command. In each of these cases the snaps will work a bit differently.

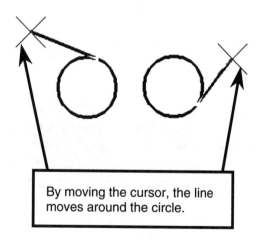

By moving the cursor, the line moves around the circle.

For instance, when specifying the first endpoint of a line by a Tangent snap to a circle, the actual coordinate isn't set because it needs a direction for the line specified. Therefore, the first endpoint can be anywhere on the circle. The location of the line depends on where the second endpoint is placed. So while you move the cursor, you will actually see the line move around the circle and remain tangent to it.

In the case of the line's first endpoint having already been placed, snapping to the tangent of the circle (in the general vicinity) will snap to the **actual point** that would make the line tangent to the circle. By accepting this tentative, the endpoint of the line will be placed at that tangent point and a new line will start at that same point. A Reset cancels the new line.

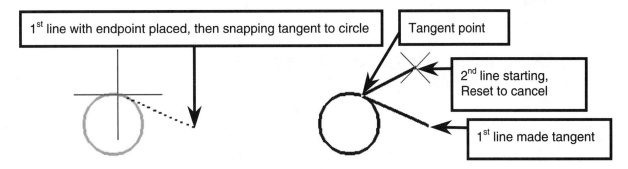

Perpendicular

This snap allows you to make one element perpendicular to another. The elements may or may not end up intersecting. Again, the results differ if you snap the first point perpendicular or if you snap the second point perpendicular.

If you specify the first endpoint as perpendicular, then the orientation of the element has been determined, but you can still move the first endpoint around while placing the second endpoint. For example, shown here is how the second endpoint of the line (marked by X) can be moved to below or above Line A (which Line B is to be perpendicular to) and the first endpoint adjusts also. The only thing constant is the orientation of Line B.

If you place the first endpoint and then use the Perpendicular Snap for the second endpoint, the second endpoint is automatically determined. Here the dotted line shows the line's first endpoint already specified, and its tentative second endpoint has been determined by snapping perpendicular to the gray highlighted line. By accepting this tentative, you place the second endpoint and start a new line.

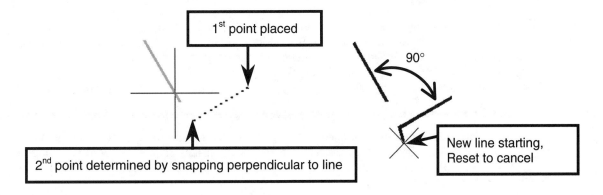

> **NOTE:** Both the Tangent From and Perpendicular From snaps allow the line to continue beyond, instead of stopping at the tangent/perpendicular point. Their icons show a circle that indicates that you can continue from that point.

Parallel

With this snap, one element is placed parallel to another element. It determines the orientation only, and both endpoints can be placed, so it doesn't make a big difference when the snap is applied.

Snap Mode and Override Snap

The snap that is currently being used is called the Active Snap. To help you keep track of which snap is active, its icon shows up in the status bar. The Snap Mode setting specifies the snap that is active until it is overridden. You can override the Snap Mode just by clicking on a different snap. This new snap becomes active for <u>one operation and only one operation</u>; then the Active Snap goes back to the Snap Mode setting. Usually the Snap Mode setting is left as Keypoint. Both the menu and the button bar indicate what the Active Snap is and what the current Snap Mode is.

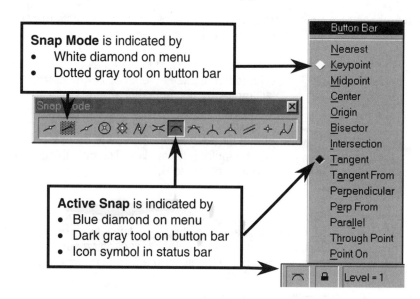

Snap Mode is indicated by
- White diamond on menu
- Dotted gray tool on button bar

Active Snap is indicated by
- Blue diamond on menu
- Dark gray tool on button bar
- Icon symbol in status bar

If you find yourself always overriding the current snap and getting frustrated that the override is good only for one operation, then you may want to change the Snap Mode setting. This allows you to control which snap will be the defaulted to as the Active Snap. The two easiest ways to do this are:
- Double-clicking on the desired Snap's button in the button bar.
- Holding the SHIFT key down while selecting the desired Snap in the pop-up menu.

> **WARNING:** Not only does the availability of the snaps change depending on the elements that exist, but SOME DON'T WORK WITH SMARTLINES (subject covered in Chapter 11). Therefore, you may want to use the regular Place Line tool when you are just beginning.

GRID

When drawing, it is often useful to have a visual guide in the form of a grid in the drawing area. The grid visually divides the drawing area into even increments that you can control. The grid does not actually print out and you can control its display in each view independently. You can also place elements at precise increments by setting Grid Lock ON, which requires each data point to lie on a grid point.

Grid Parameters

By using the main pull-down menu SETTINGS>DESIGN FILE and going to the Grid category you will reach the Grid Parameters that you can control as shown here.

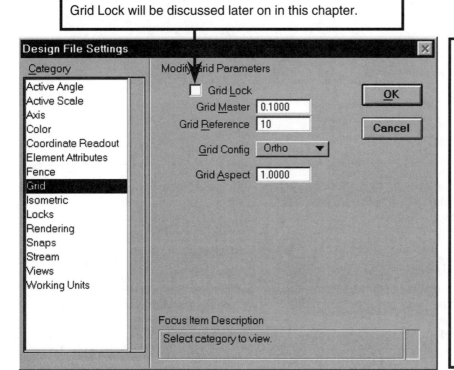

Grid Lock will be discussed later on in this chapter.

Grid Master will be the distance (in working units) between each of the grid points.

Grid Reference indicates the number of grid points to use for a larger reference + grid.

Grid Config has options of Ortho(gonal) or Isometric grid.

Grid Aspect is used when you want different spacing for the X coordinates and Y coordinates.

Grid Master and Grid Reference

The Grid Master setting specifies the distance between the points of the grid. Calling it "Master" does **not** mean that it changes the Master working units for the design file. This distance is just in working units, it doesn't affect working units. It is important to note that the Grid Reference setting is **not a distance**; rather it is just an integer number that indicates how many grid points make up the reference grid. To determine the distance between the + symbols of the reference grid, multiply the Grid Master distance by the Grid Reference value. For the default grid parameters of Grid Master = 0.1 and Grid Reference = 10 the basic master grid would show up every 0.1 master units and the reference grid +'s would show up at every 1.0 master units. This distance is calculated from 0.1 master units x 10 = 1.0 master units.

Example of Grid

Shown below is how the grid would appear if the Grid Master distance was set at 0.2500 master units and the Grid Reference value was set at 4. If one master unit was considered as an inch, then the grid points would show up every 0.25 inches. The distance between the +'s of the reference grid would be 0.25 inches x 4, which would be 1 inch.

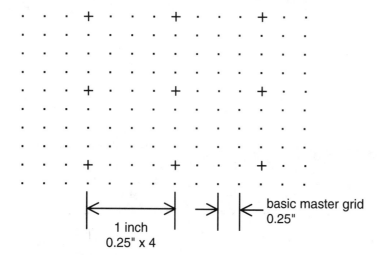

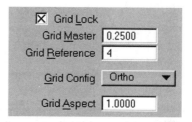

Beginners may want to set the Grid Reference to 1 so that a + will show up for every master grid. This means you'll only see the +'s, but it may eliminate some confusion.

Zooming's Effect on the Grid Seen

With the view's attributes set to display the grid, you still may or may not see the grid in your view. Being able to see the displayed grid depends on how close or far you are looking at the drawing area.

- Zoomed very close: No grid is displayed because you're between master grid points.
- Zoomed close: Only master grid points are displayed because you're between reference grid +'s.
- Zoomed average: Both master grid points and reference grid +'s are shown.
- Zoomed out: See only reference grid +'s (be careful to not mistake these for master grid points).
- Zoomed way out: No grid is displayed because you are too far away to see even the reference grid +'s.

A common mistake is to set Grid Master to 0.25 but then forget to change Grid Reference from 10 to 4. Then when you're counting + symbols, the distance is really 2.5 master units (0.25 master units x 10) but you think it is 1 master unit. Oops—two and a half times too big! **Another mistake is to think you're counting the smaller master grid when instead you are zoomed out and counting the reference grid instead.**

Drawing Aids and Tools 77

LOCKS

There are certain drawing aids that you may want to lock on or off. These aids and their settings are grouped together in a dialog box called Locks. Some of the more widely used drawing aids are Grid, Snap, and Axis. These locks can be viewed as just a check box (where they are toggled on or off) or a full view showing some of the settings of the drawing aids.

Accessing Locks

The full Locks dialog box is obtained by the main pull-down menu SETTINGS>LOCKS>FULL. You can also access the Locks pop-up menu by clicking on the lock symbol in the status bar. The Locks pop-up menu is illustrated below.

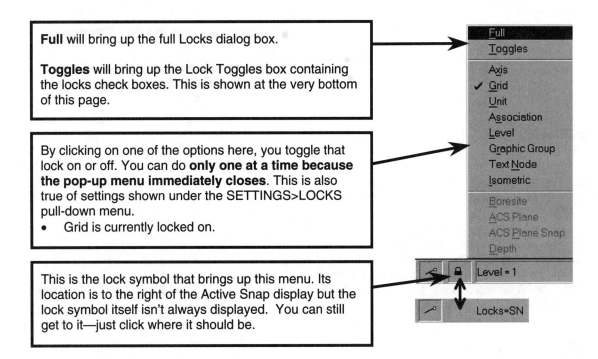

Full will bring up the full Locks dialog box.

Toggles will bring up the Lock Toggles box containing the locks check boxes. This is shown at the very bottom of this page.

By clicking on one of the options here, you toggle that lock on or off. You can do **only one at a time because the pop-up menu immediately closes**. This is also true of settings shown under the SETTINGS>LOCKS pull-down menu.
- Grid is currently locked on.

This is the lock symbol that brings up this menu. Its location is to the right of the Active Snap display but the lock symbol itself isn't always displayed. You can still get to it—just click where it should be.

Shown below is the Lock Toggles box that was mentioned above. It is handy to use for toggling on and off locks because the box **stays open** until you close it. This means that you can toggle more than one lock setting at a time. It can also be accessed by the main pull-down menu SETTINGS>LOCKS>TOGGLES.

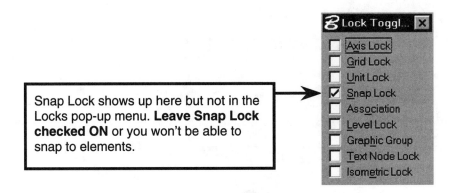

Snap Lock shows up here but not in the Locks pop-up menu. **Leave Snap Lock checked ON** or you won't be able to snap to elements.

Locks Dialog Box—Full Display

The full Locks dialog box allows you not only to turn the locks on or off but also to change some of the settings of the associated drawing aids. Shown below is the Locks dialog box in its full display. The main locks to be discussed are Grid, Snap, and Axis.

Grid Lock

In a view's attributes you can choose to display a grid that serves as a reference when drawing. The Grid Lock checked ON actually allows you to graphically jump from grid point to grid point when drawing elements. This is a very useful feature when drawing in certain increments of length. There are two major things that you need to be aware of:

- Snaps and key-in coordinates override the Grid Lock setting.
- Grid Lock can still be on even if the grid itself is not being displayed in the view.

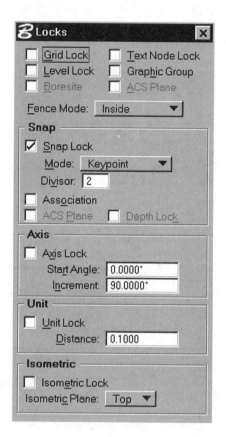

Snap Lock

This lock controls whether or not you can actually use snaps. **It should be checked ON.** When placing a tentative and nothing can be snapped to, check the Snap Lock setting.

Settings that you can control in the Snap section are as follows:

- **Mode** is another route to changing the Active Snap Mode.
- **Divisor** allows you to change the number of integer divisions that you can divide an element by. A setting of 2 will get you to the midpoint and endpoints of a line. A setting of 1 will allow you to snap only to the endpoints of the line. A setting of 4 will go to the midpoint, endpoints, and quarterpoints of a line.

Axis Lock

Axis Lock checked ON will restrict graphical entries to the angle increments specified. The default Start Angle and Increment settings of 0° and 90°, respectively, will restrict you to drawing horizontal or vertical lines. Again snaps and key-in coordinates will override this Axis Lock. Shown below is the result of having Axis Lock ON. The left endpoint of the line was specified and now the second endpoint is being established on the right. The line is locked to staying horizontal even though the cursor isn't positioned that way.

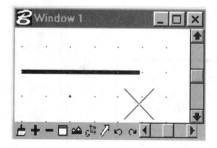

ELEMENT SELECTION TOOL BOX

The element selection tools are used to identify elements before a command. Then when the command is launched, you won't be prompted to select elements because they have already been selected. The Element Selection tool box is found in the upper left corner of the Main tool box. The Element Selection tool allows you to data point on the element to select it. The PowerSelector tool (has the light bulb) is used when you want more control of which elements make up your selection set. It allows you to quickly add or subtract elements from the selection set.

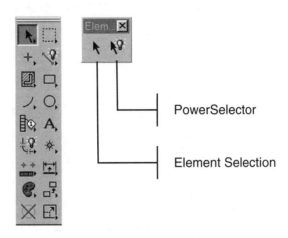

Selection Display

When an element has been selected, it will display handles that are similar to the keypoints of the element. Using these handles, you can interactively change the shape and size of the element. However, this quick method of modifying an element can give unwanted results because it is difficult to specify exact changes—you can lose the precision associated with the size and location of the graphical element. There is a preference setting so the element will highlight instead of displaying handles to indicate that it has been selected. If the handles are not displayed, then you cannot interactively change the shape and size of the element by grabbing on to them.

Here are two lines and one circle that have been selected and are displaying their handles:

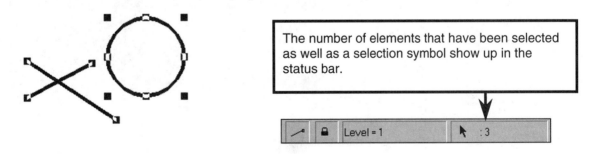

The number of elements that have been selected as well as a selection symbol show up in the status bar.

The Highlight Selected Elements option is found in the Input category of the Preferences dialog box. It can be accessed by the Main pull-down menu WORKSPACE>PREFERENCES. The illustration below shows that Highlight Selected Elements has been checked ON.

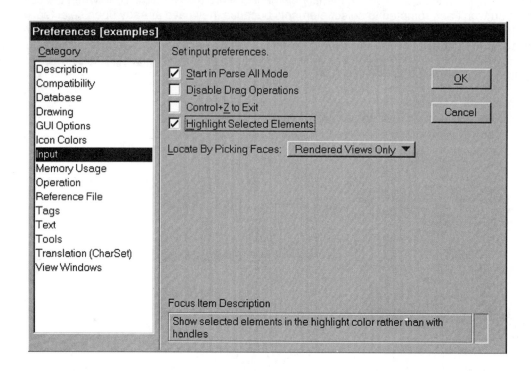

The lines and circle are now highlighted with a color instead of having handles, as illustrated below on the left. This highlight color default is pink. It can be changed in the Color category of the Design Files Settings dialog box as shown below on the right. You can get to this dialog box by the Main pull-down menu SETTINGS>DESIGN FILE.

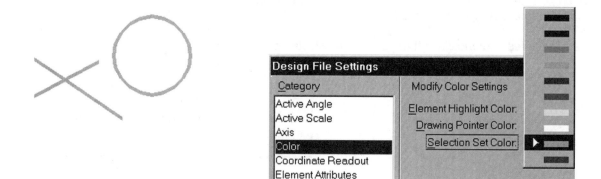

ELEMENT SELECTION

The use of the Element Selection tool requires a data point on the element to select it (a tentative point is not necessary). Once the element's handles are displayed then it can be moved or modified. This requires some mouse action that in simple terminology could be called grab and drag. You "grab" onto the element by positioning the cursor over a point on the element, then press the **left** mouse button **and keep it depressed**. You can then "drag" the element around by moving the cursor (as long as you keep the left mouse button depressed) until the left mouse button is released.

When you want to move the element without resizing it, position the cursor to a point on the element that is <u>not</u> a handle. Now press and hold down the left mouse button and drag the element to its new location by moving the cursor (still holding down the left mouse button). When the element is at its new position, release the left mouse button. This will move the element but not resize it. In order to resize or modify an element, you need to grab onto its handle. Drag the handle and the element's size (and possibly shape) will change.

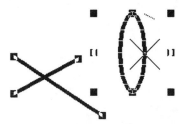

In this case, the selected circle's right middle handle was grabbed and it is being dragged into an elliptical shape.

Selecting More Than One Element

You can use the Element Selection tool to specify more than one element in two different ways:
- Hold down CTRL while you data point on the elements to select. To deselect an element that has already been selected, just data point on the element again.
- As in Windows, you data point **and drag** (remember, this means to keep the left mouse button depressed) the other corner of a box to surround the elements wanted.

There is also a Select All setting that is found under the EDIT pull-down menu. This is sometimes used in order to be able to delete all the elements in a design file quickly.

Quick Way to Change the Attributes of an Existing Element

Using the Element Selection tool and the Primary Tools tool box you can quickly change the basic attributes of an existing element.
- Select the element(s).
- Go to option buttons in the Primary Tools tool box and select the new attribute setting.

This also works to update an element to the existing attributes but you must be sure to <u>click on the specific attribute in the actual list of options, not just the option button</u>.

When reading, please note that the **terminology of "select" an element will mean to use a data point**—<u>not</u> the Element Selection tool.

POWERSELECTOR

The PowerSelector tool allows you to easily specify a selection set using four different methods. Elements can be added to or subtracted from an existing selection set using those same methods. Once a selection set has been established, then using another tool will act on that selection set.

Below is the PowerSelector box with the different Method and Mode settings available. The Mode deals with whether the elements will be added to or subtracted from the selection set, or if the selection set is to be inverted or cleared. The Method determines how the elements are identified, individually, by a rectangular block, an irregular shape, or a line. The button that looks pushed in is the active Method or Mode, which can be changed at any time by clicking on it.

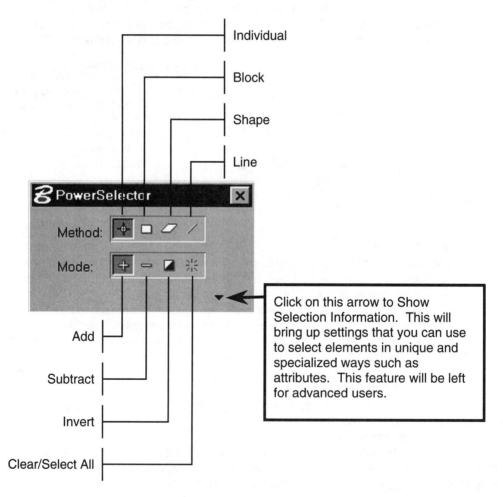

Click on this arrow to Show Selection Information. This will bring up settings that you can use to select elements in unique and specialized ways such as attributes. This feature will be left for advanced users.

Drawing Aids and Tools 83

You may notice that tool tips come up for each of the methods. These are the quick keys for the dexterous power users who run a mouse in one hand and type with the other. They allow a quick switch between the different methods with a touch of a key. Why two letters? They are on opposite sides of the keyboard, so that both left- and right-handed people can take advantage of the feature.

Illustrated at the right is the tool tip that shows the quick keys for changing the Mode to Add. Just by pressing the A key or the J key on the keyboard, you can switch to the mode that will add elements to the selection set.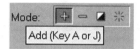

PowerSelector Methods

The four different methods for selecting elements are very straightforward. You aren't constrained to using only one method. You can switch between them if you want. Both the Block and Shape methods use an area shape to select the elements. Then the decision needs to be made if you want to include elements that don't fall totally within the area but instead just overlap the boundary, so these two methods have an Inside and an Overlap setting. All you need to do is click on the button again to toggle it to the other setting.

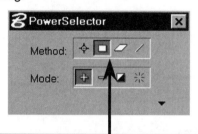
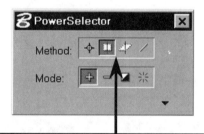

Inside: The area has a solid border (without a line running through it)—this indicates that it is set to Inside. Elements that are only totally within the block or shape specified will be selected.

Overlap: The area has a dashed border and a vertical slash—this indicates that it is set to Overlap. Elements that are within or overlap the block or shape specified will be selected.

Individual
This method lets you select individual elements with a data point. It is very simple and straightforward but a bit tedious.

Block

The block method uses a rectangular-shaped block for the area. It requires two data points. The first one specifies one corner of the rectangular-shaped block and then the second data point specifies the opposite corner.

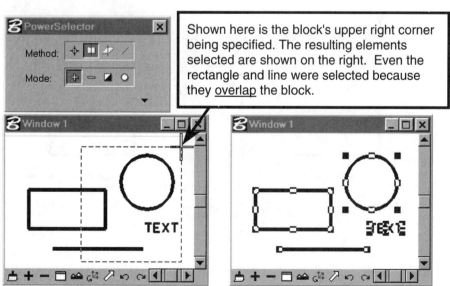

Shown here is the block's upper right corner being specified. The resulting elements selected are shown on the right. Even the rectangle and line were selected because they <u>overlap</u> the block.

Shape

The Shape method uses data points to specify vertices of an irregular shape. The last data point will need to be at the same place as the first data point in order to close the shape area. You can snap to the first vertex or, if you move the cursor close to the first vertex, it will "jump" to it.

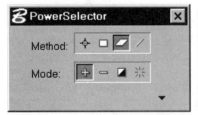

Shown here is the shape's area being defined. The cursor is approaching the first vertex specified. The resulting elements selected are shown on the right. Only the rectangle and line that are <u>inside</u> the shape are selected.

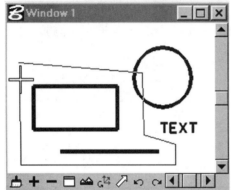 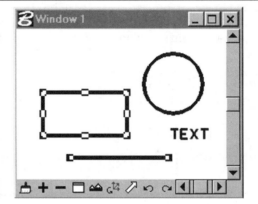

Line

The Line Method uses the two data points to establish a line. Any element that intersects the line will be selected. Illustrated below is using the Line Method to select the circle, text, and line. The illustration on the left shows the line being established. The one on the right shows the handles indicating the selected elements.

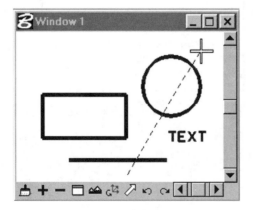 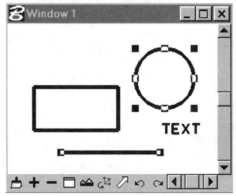

Drawing Aids and Tools 85

PowerSelector Modes

Once a selection set is started, you may want to add elements to or subtract elements from it. The Mode setting controls whether the elements selected with the active method will be added or subtracted. This is pretty easy to understand. If you wanted to add the rectangle to the selection set just established, you could use the Add mode and the Individual Method and just data point on the rectangle.

The Invert Mode just inverts the selection of the elements. If an element is selected, it will be unselected, and vice versa. Read the prompts because the elements to invert will need to be identified using a selection Method—in this case, a block shape will be used. The Invert mode is illustrated below.

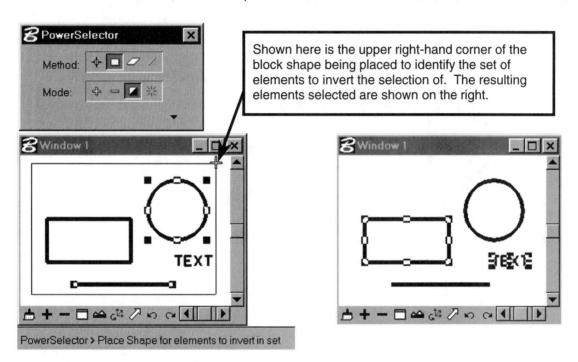

The last mode to discuss is the Clear/Select All. It is the far right button in the Mode choices. It looks like an exploding circle if it is Clear and like a solid circle if it is Select All. Read the prompts in the status bar if you get confused or wait for the tool tips (as illustrated below). The Clear Mode does just that—it clears the selection set. This does not mean it deletes the elements selected; rather any elements that were selected will no longer be selected. The Select All Mode selects all of the elements.

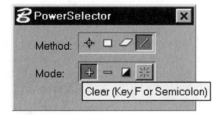 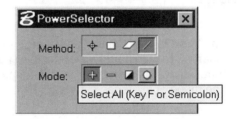

FENCE TOOL BOX

A fence is another way of selecting more than one element at a time. It is not necessarily an element but a closed boundary specified around and/or through elements. There can be only one fence existing at a time. A fence differs from using the selection tool in that it does not activate the handles of the elements. Many of the command dialog boxes such as Copy, Mirror, and Array have a setting that uses the fence to specify the elements to manipulate. Printing also allows you to print the contents of a fence. There are tools that are applied only to fences, such as Stretch Fence Contents, but they will not be covered here. The Fence tool box is found in the upper right corner of the Main tool box. It can be torn off and appears as shown below. The Place Fence is the default tool that starts out on top.

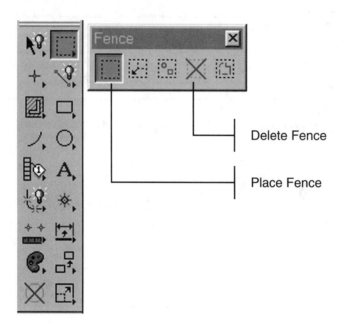

PLACE FENCE

 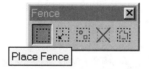

When you go to place a fence, you can specify the type of fence, which deals with what constitutes the fence boundary. Each different type of fence will be placed in its own manner. The mode of the fence can also be controlled and that deals with what elements will be affected by the fence and how they will be affected. The mode of the fence can easily be changed, even for an existing fence and in the middle of a command.

Drawing Aids and Tools

Fence Type

There are six different Fence Type options, but only Block and Shape will be covered since they are the commonly used ones.

Shown here is the Place Fence dialog box and the options for different fence types. To change the Fence Type setting, just select a different option from the list by clicking on it.

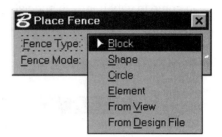

Block

With a Block Fence Type, you specify a rectangular-shaped block by a data point for one corner of the block and then another data point at the opposite corner.

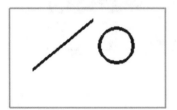

Shown here is a Block fence that surrounds the line and the circle elements.

The fence itself will show up in the same color selected for the element highlight.

Shape

With a Shape Fence Type, you specify the points that the fence will run between, and then, since it has to be a closed boundary, you must end up at the same point where you started from. This fence type is useful when you want an odd-shaped boundary.

The gray Shape fence goes around the line.

The Close Fence button will finish the fence by going in a straight line from the last data point to the first data point.

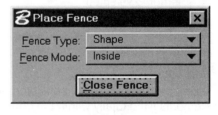

Once a fence is placed, it remains until another fence is placed. To eliminate having a fence, just select the Place Fence tool again but don't specify the corner; instead go on to selecting the next tool that you want to use.

Fence Mode

The Fence Mode actually controls which elements the fence will select. The basic ones covered are Inside and Overlap. The others are useful to more advanced users, so they won't be covered here. However, experimenting on your own can be fun since fences involving Clip and Void give interesting results that can come in very handy in special situations.

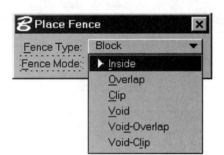

You may change the Fence Mode before placing the fence. Just use the option button for Fence Mode to see your different choices.

This Fence Mode would be Inside.

Remember that the Fence Mode setting can be changed "on the fly" so to speak. For example, the Copy command has the Use Fence setting checked ON. The existing fence will be used to select the elements to copy. It also has an option button allowing you to change the Fence Mode while in the middle of the task.

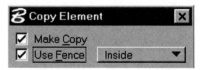

Inside

As the name implies, any element lying totally within the fence boundary will be selected. However, if any part of the element lies outside of the boundary then it will not be selected.

For example:
Fence Type: Block (shown in gray)
Fence Mode: **Inside**

Elements selected:
- Rectangular Block element
- Text

Overlap

The Overlap Fence Mode will select the elements that overlap the fence boundary in addition to selecting any elements lying totally within the boundary. Using this mode allows more flexibility when placing a fence since you can just touch the element with the fence and you don't have to worry about enclosing the entire element.

For example:
Fence Type: Block (shown in gray)
Fence Mode: **Overlap**

Elements selected:
- Triangular Polygon element
- Line
- Circle

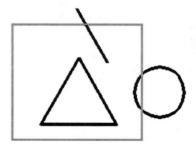

Drawing Aids and Tools

DELETE FENCE CONTENTS

This tool allows you to delete a large number of elements at a single stroke rather than having to select each element individually. Click on the Delete Fence Contents tool and then you'll be prompted to accept (data point) or reject (reset) the fence contents. That way the deletion isn't done until you accept it. It is important to realize that while the prompt refers to fence "contents," that doesn't mean that it is limited to those elements <u>inside</u> the fence. <u>The fence **mode** still applies</u>. Contents refer to those elements selected by the fence mode in conjunction with the existing fence.

 Using a Void-Clip Fence Mode with the Delete Fence Contents tool is a quick way to leave all the elements within the fence but delete everything that lies outside of the fence boundary. If an element overlaps the boundary, it will be clipped off and the outside part will be deleted.

 If you use the Delete Fence Contents tool and then change your mind and use the Undo tool, be aware of the fact that there are a limited number of element deletions that can be undone. The undo buffer sets up this limit. This means that you may not get back all of the elements deleted by this method.

Is "On the Line" Considered In?

The question always arises whether or not an element that is actually in the fence boundary is considered inside the fence. The answer is "No, but....". If an element is actually lying coincident to the fence boundary then it will **not** be selected by an **inside** fence mode for tasks such as Delete Fence Contents or manipulation. <u>But</u> if you are **printing** the contents of the inside mode fence, then those coincident elements will be printed.

Example of Element Inside the Fence Boundary

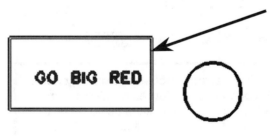

Fence Type: Block **that coincides with the rectangular Block element.**

Fence Mode: **Inside**

Elements selected for Delete Fence Contents:
- Text only

Elements selected for Print Fence:
- Text
- Rectangular Block element

DELETE ELEMENT

The Delete Element tool is found in the lower left corner of the Main tool box. Its icon looks a lot like the Delete Fence Contents tool—don't get them confused. This tool allows you to delete individual elements. You select the element to delete with a data point. The element will highlight; another data point is needed to accept its deletion. The data point that accepts the deletion can also identify the next element to delete.

The use of the Delete Element tool is also a good time to talk about using the Reset (right mouse button) to help you select the right element. There are often situations where elements coincide or are in close enough proximity so that your data point may select the wrong element. Instead of moving the cursor and trying another data point, you can just click Reset. It will toggle you through all of the elements that are close enough to the data point that you might have meant instead. When the element you wanted highlights, then you can accept with a data point. This is illustrated below.

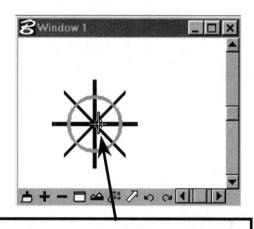

Assume you want to delete the horizontal line, but the data point is close enough to the center of the circle that it highlights instead.

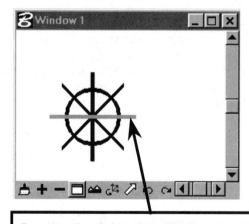

By using the **right mouse button** to perform a Reset, the horizontal line is selected for deletion instead. If you keep using Reset it will toggle through all the lines.

If elements have already been selected using the Element Selection tool, they're deleted as soon as you click on Delete Element. This can be used to your advantage, but it also can be disastrous if you're not careful. Remember, the undo buffer limitations may not let you undo all the deletions.

CURSOR

The type of cursor displayed is a visual aid to what input the program is expecting.

 This plus symbol is the cursor displayed at the start of a command when a data point is needed to specify a beginning coordinate.

 A circle is displayed (often superimposed on the plus symbol) when a data point is needed to identify an element.

 An X symbol appears while executing a command when a data point is needed to specify coordinates that will result in the completion of an element or operation.

 A larger single line plus symbol indicates that a tentative point has been specified. Here the endpoint of the line has been snapped to using a tentative point.

QUESTIONS

① Explain why the Color Attribute set to 0 (zero) is special. *Zero is set to white but is actually black on paper.*

② When doing a Window Area for view control, what actually determines the window area shape? *the first and second pts ie at opposite corner*

③ What is the mouse action for a tentative:
  a) For a 2-button mouse? *left right together*
  b) For a 3-button mouse? *centre*

④ What is the **Keypoint** snap setting and why is it so useful? *snaps to end point, centre, quarter, Text BRIA, & insertion pts* *use with tentative*

⑤ The following questions deal with the View Levels Settings shown:
  a) What level(s) are ON? *5-63*
  b) What level(s) are OFF? *1,2,3*
  c) What is the current **Active Level**? *④*
  d) What **View** would these View Levels settings be applied to? *3 ▼*
  e) Which View Levels pull-down menu gives you access to the Level Names dialog box? *EDIT DEFINE or SETTING/LEVEL/NAME*

⑥ Name two different methods that allow the Element Selection tool to select more than one element at a time. *Use element select then hold down CTRL key & select elements* *ie to delete more than Select all in edit commands*

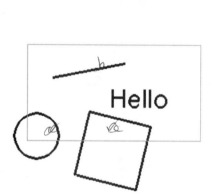

⑦ Name this PowerSelector Method (including whether it is set to Inside or Overlap) and this PowerSelector Mode.

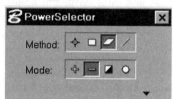

⑧ The following questions refer to the light gray rectangular fence and the four elements (Line, Circle, Square, Text) shown at the right.
  a) Which element(s) would be included in an **Overlap** fence mode?
  b) Which element(s) would be included in an **Inside** fence mode?

⑨ What should the settings be if you want the reference grid to show up at every 5 master units and the master grid points to show up at every 0.25 master units?
  a) Value for Grid Master *0.25*
  b) Value for Grid Reference *20*

⑩ Name the tool shown on the right. (It is found in the Fence tool box.) Explain what the tool is used for. *to delete an element*

delete

Chapter 5:
Printing a Hard Copy

SUBJECTS COVERED:

- Plot Dialog Box
- Entity
- Plot
- Preview Refresh
- Page/Print Setup
- Plot Layout
- Plot Options
- Plotter Driver
- Printing to Scale
- Summary of Printing Sequence

Your design file contains a great deal of information. In order to distribute that information to other people, you will often want to produce a hard copy of your design file. A hard copy refers to transferring your drawing onto paper versus the electronic version, which is the .dgn file. Producing this hard copy can be done with a plotter or a printer. A plotter uses pens to draw the graphics and a printer is usually an inkjet or a LaserJet. The software uses the terms plot and print interchangeably regardless of which device is actually used. So the Plot dialog box has a Print Preview tool and a printer can be chosen for the plotter—confusing, huh? Just remember that both terms refer to the same task—getting a hard copy. This book will match the software's terms and use print when it isn't specific.

PLOT DIALOG BOX

This is command central of the printing process. It controls the what, where, and how of getting a hard copy. Entity and options specify what you want printed. Where you want the print to be on the paper is found under Plot Layout. The Print/Page setup takes care of how the hard copy is going to be made.

Accessing the Plot Dialog Box

In the Standard tool box, there is a printer icon. In most Windows programs, using this tool button will immediately start printing. You don't get a chance to change the settings; it goes ahead and uses the last known settings. However, MicroStation is different. This Print tool will bring up the Plot dialog box. You can also get to the Plot dialog box from the main pull-down menu of FILE>PRINT/PLOT.

Tools and Menus of the Plot Dialog Box

Most of the items in the Plot dialog box pull-down menu are duplicated as tools. When this is the case, the tools will be discussed. The Entity menu and its options can be found only in the Plot dialog box pull-down menu. Entity is very important since it controls what portion of the design file is going to be printed. Just remember that you can get to most controls one way (tool buttons) or another (pull-down).

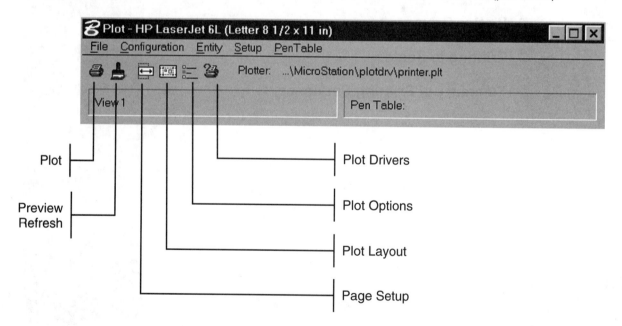

Printing a Hard Copy   95

Status Fields of the Plot Dialog Box

The Plot dialog box has its own status area that you can read to quickly view some of its current settings. As always, it is very important to pay attention to the status fields.

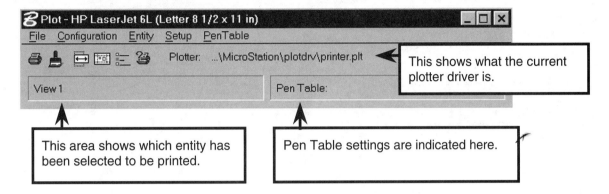

This shows what the current plotter driver is.

This area shows which entity has been selected to be printed.

Pen Table settings are indicated here.

Defaults of the Plot Dialog Box

When you make changes to the various settings in the Plot dialog box, they will remain changed even if the Plot dialog box is closed and reopened during a drawing session. However, the Plot dialog box settings will return to the defaults when the design file is closed. These are not among the "Save Settings" items. So every time you open a design file, you'll need to make the wanted changes to the Plot dialog box.

ENTITY

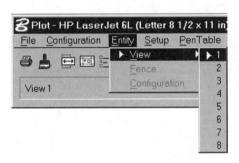

What portion of the design file is to be plotted is determined by the Entity setting. This is available only by the pull-down menu of the Plot dialog box. You have 3 choices: View, Fence, and Configuration.

View

This refers to the window displaying a view of the design file. There can be as many as 8 different windows open to display different portions of the design file. When using View, the entire view is printed, including any surrounding blank space.

Fence

This option will be available only when there is an existing fence. This option allows you greater control over the portion printed. The Fence option is recommended for beginners.

Configuration

This will be available only if a configuration file has been set up. A configuration file standardizes printing. The file specifies which portion of the design file to print based on set coordinates. All other settings are automatically determined by the configuration file. In large production work, a configuration file is worth the time it takes to set it up because it can save workers' time in the long run. However, for beginners it is not recommended. Often the coordinates that you want to print vary from design file to design file and therefore the View or Fence method is much more convenient.

Chapter 5

For Example: Here are two windows displaying different portions of the design file. Window 1 has the entire truss displayed. There is also a fence placed (block) that is visible in Window 1. Window 2 is zoomed in to show a close-up of the joint.

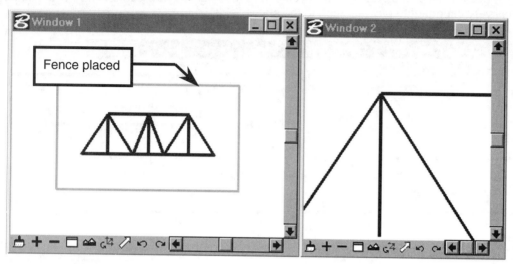

Printing each of these different entities on the same size sheet, without setting the scale of the print, would result in the following three prints. Each of the three entities, View 1, Fence, and View 2, are printed to fill the page. They are each shown as depicted when previewing the plot.

View 1

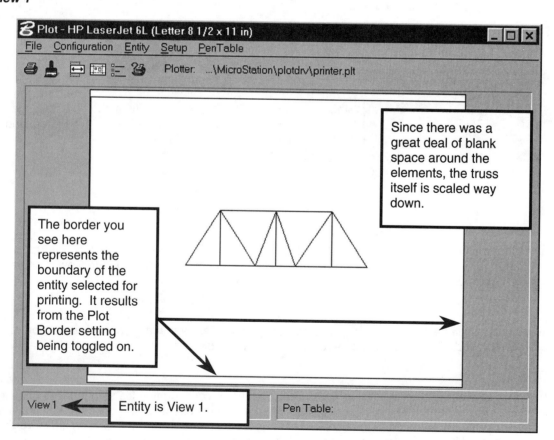

Printing a Hard Copy 97

Fence Placed in View 1

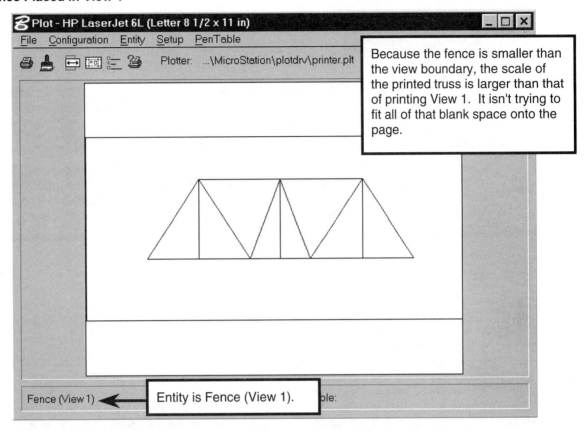

View 2

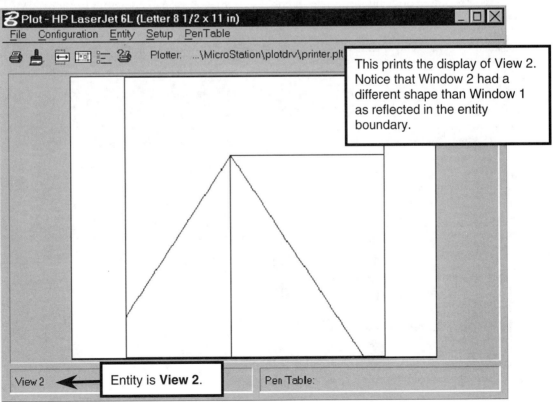

PLOT

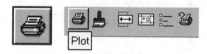

Now this Plot tool with the printer icon in the Plot dialog box is the one that will immediately start printing. Usually you go through the other tools first to get the settings adjusted, so do a preview to see what you'll get, and finally after everything's just right, you use this tool to actually print.

 Don't be too quick on the click or you'll have numerous prints! Click on the button once (remember it will change color when clicked on but won't stay that way). The main status bar of the MicroStation window will indicate "Finished Creating Plot" in its lower right corner. Wait for that and then check the printer for your hard copy.

PREVIEW REFRESH

The Preview Refresh tool will give you a preview of what is to be printed. It shows the actual sheet of paper and its orientation. If you make changes to the settings, they will be shown here. The Plot dialog box will enlarge to include this preview and then it is titled as Plot Preview. You can resize the dialog box down to the point where the preview doesn't appear and it will revert to the Plot dialog box. Clicking on the tool will again enlarge the box to include the preview or just resize the dialog box. There are limitations to the preview. You cannot zoom in to check on closer details such as hidden lines, etc. It is still recommended that you do a preview before printing in order to catch the obvious mistakes.

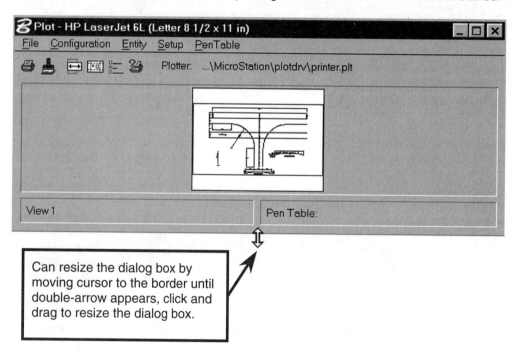

Can resize the dialog box by moving cursor to the border until double-arrow appears, click and drag to resize the dialog box.

PAGE/PRINT SETUP

The tool tip comes up as Page Setup, but the dialog box can be called either Page Setup or Print Setup. It depends what plotter driver is being used. Plotter drivers are covered later in this chapter. Assuming that the Windows default printer is specified, the Print Setup dialog box will appear and is illustrated below. The dialog box is where you set up the actual printer you want to send the information to, and the orientation of the page is also controlled in this dialog box. Two important players in getting a hard copy.

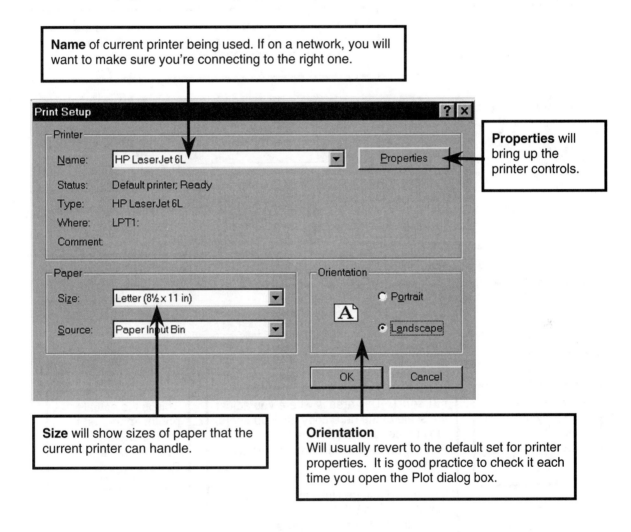

Name of current printer being used. If on a network, you will want to make sure you're connecting to the right one.

Properties will bring up the printer controls.

Size will show sizes of paper that the current printer can handle.

Orientation
Will usually revert to the default set for printer properties. It is good practice to check it each time you open the Plot dialog box.

A common setup is to use the Windows printer as the default. The examples in this chapter use this setting. See the discussion on plotter drivers later on in this chapter or refer to the software documentation.

PLOT LAYOUT

This dialog box is important because it is where you can adjust the scale of the print. In engineering, hard copies are usually not "printed to fit"; instead they are printed at a **specific** scale. Printing to **scale** allows measurements to be obtained directly from the hard copy.

Margins will adjust automatically in response to changes in the **Scale to** fields. Set them if necessary.

% of normal is the percentage of the printable area that is being used. <u>Let it set automatically.</u> 100% means Maximize, but does NOT mean 1:1 scale.

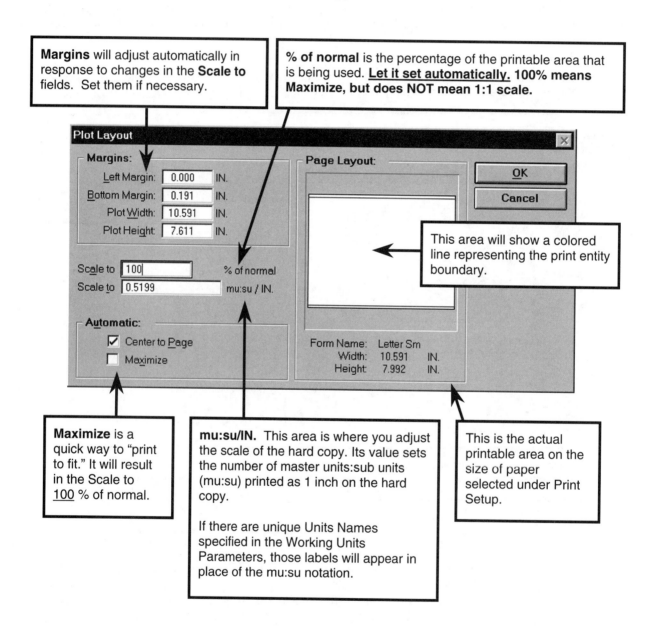

This area will show a colored line representing the print entity boundary.

Maximize is a quick way to "print to fit." It will result in the Scale to 100 % of normal.

mu:su/IN. This area is where you adjust the scale of the hard copy. Its value sets the number of master units:sub units (mu:su) printed as 1 inch on the hard copy.

If there are unique Units Names specified in the Working Units Parameters, those labels will appear in place of the mu:su notation.

This is the actual printable area on the size of paper selected under Print Setup.

PLOT OPTIONS

These settings are quite similar to those available in the View Attributes dialog box. When printing a fence or view, the Plot Options will reflect the same setting as in the View Attributes. If it is being displayed in the view, it will be toggled on to be printed. The View Attributes settings take precedence; therefore, those options are grayed out in the Plot Options dialog box. You would have to make changes under the View Attributes dialog box. There are some exceptions, with the Plot Border and Fence Boundary being the most obvious. These don't apply to a view but only to printing.

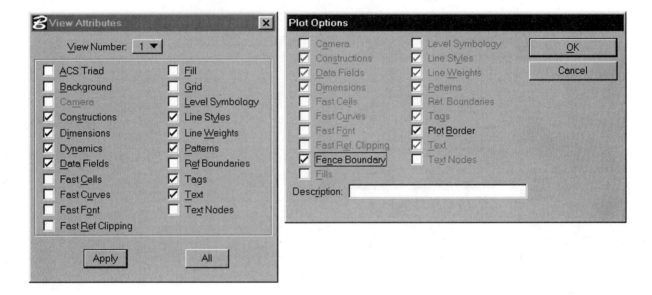

Plot Border
If the Plot Border setting is checked ON then a rectangular border will be drawn around the entity printed. Another result of plotting the border is that the text in the Description field will be printed below the border.

Fence Boundary
The Fence Boundary setting is important when there is an existing fence in the drawing. With this checked ON, when you print a view that has a fence displayed, its boundary will also be printed. If the entity you are printing is a fence and both the Fence Boundary and the Plot Border are checked ON, then you can end up with two borders—which can look funny.

For Example: Shown below are the Plot Options settings for a landscape print of View 1. Plot Border is checked ON but Fence Boundary is unchecked (OFF).

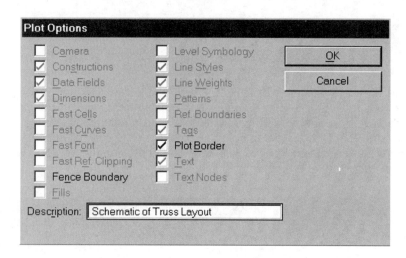

Shown below is the resulting hard copy.

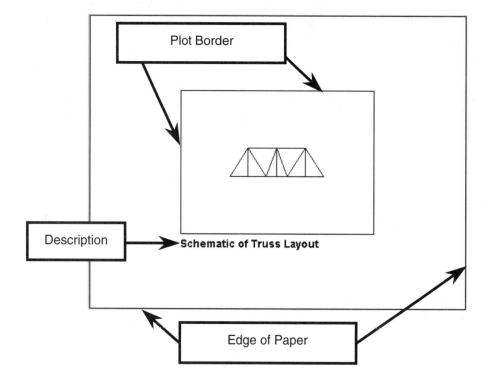

PLOTTER DRIVER

A plotter driver converts the graphics to printer language. When set to printer.plt it will let Windows take control—which is most common. **If you have trouble with it, such as line styles not printing with the correct line weights, open up the printer.plt file as a text file and the read additional instructions there.** If you don't want to use the Windows printer, please refer to the software documentation because the driver you will use will depend on the language of your printer. Shown below is an example of using a plotter driver other than printer.plt.

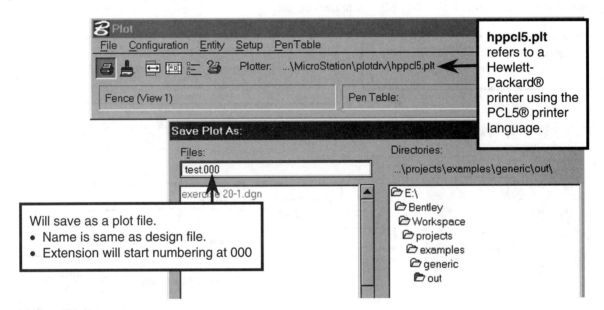

hppcl5.plt refers to a Hewlett-Packard® printer using the PCL5® printer language.

Will save as a plot file.
- Name is same as design file.
- Extension will start numbering at 000

 If you don't want to save as a plot file (which then needs to be submitted to a printer/plotter), enter **LPT1** in the field for the plot file name. This will automatically send the information to the printer attached to LPT1. Of course, if your printer isn't hooked up to LPT1 then this won't work! This may not work with network connections either—consult your network administrator.

Page Setup Dialog Box for Non-Windows Printer

When you choose a printer other than the one defaulted to in Windows, the Page Setup tool will bring up the Page Setup dialog box shown here.

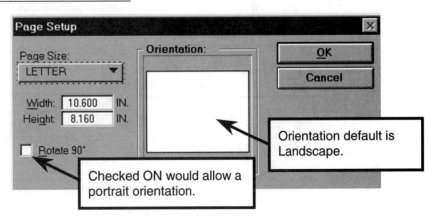

Orientation default is Landscape.

Checked ON would allow a portrait orientation.

PRINTING TO SCALE

As stated previously, one of the fundamental concepts of CAD is that elements are drawn at the real size of the object they represent. Therefore, you do not change the size of elements just to get them to fit on a piece of paper—that would defeat one of the purposes of CAD! "Printing **to scale**" means that the hard copy will maintain a specific ratio between the element size printed on the paper and the actual size of the object. **This is accomplished without destroying the precision of the elements in the design file.** The same design file can be used to produce hard copies of varying scales on different sizes of paper. Since the elements are "to scale," measurements can be made on the hard copy by using the appropriate engineer's/architect's scale.

Illustrated below is a typical road cross-section drawing. In order to fit on an 8½ x 11 inch paper, it has been printed so that 10 feet in the design file will be equivalent to 1 inch on the paper. The scale of printed hard copy would be 1" = 10 FEET as shown below.

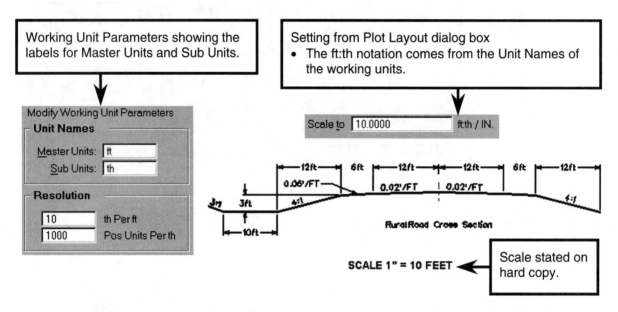

The same design file could also be used to print the typical road cross-section on B size paper, which is 11" x 17". Since the paper size is larger—the same elements can be printed so that 5 feet in the design file could be equivalent to 1 inch on the paper. The scale stated on the hard copy would now be 1" = 5 FEET.

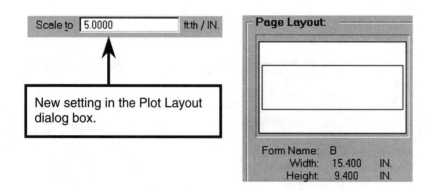

Printing a Hard Copy   105

Specific Ratio for Scale to

As seen in the previous illustrations, when printing to scale you need to pay special attention to the ratio found in the Scale to [   ] **mu:su/IN.** setting found in the **Plot Layout** dialog box. The mu:su just indicates the generic names for the working units. This ratio controls the scale of the hard copy. The value it has stands for master units:sub units per INch on paper. In engineering graphics a hard copy print can have a scale assigned to it. The scale can be stated as a ratio of unit length (on paper) to actual length such as Half Size (1:2), Full Size (1:1), or Double Size (2:1). The relationship between the unit length (on paper) and the actual length of the object can also be stated as an equality such as 1" = 5 feet or 1" = 2". Keep in mind that elements in a design file are placed to reflect the actual size of the object they are to represent. That means that the master units actually represent the **right** side of the relationship (actual length). So when you set the value of **mu:su/IN.**, it is actually the ratio of actual length to 1 unit length on paper, which is the INVERSE of what the scale printed on the hard copy will state.

On hard copy: Scale to [   ] mu:su/IN.
1 inch = 3 master units   3 master units/inch

Shown below is an example of getting a **Half Size** hard copy of your design file.

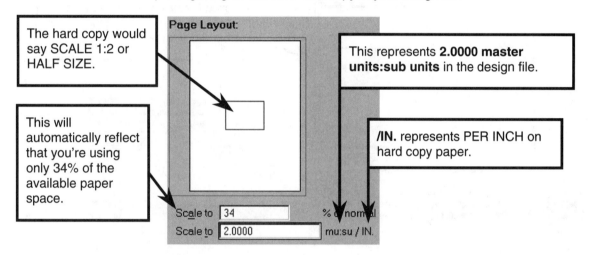

The hard copy would say SCALE 1:2 or HALF SIZE.

This will automatically reflect that you're using only 34% of the available paper space.

This represents **2.0000 master units:sub units** in the design file.

/IN. represents PER INCH on hard copy paper.

So some of the more common settings would be
Full Size hard copy   Scale to **1.0000** mu:su/IN.
Double Size hard copy Scale to **0.5000** mu:su/IN.

Now all of this is assuming that the entity you've chosen will actually fit on the paper size that was specified **and** at the scale specified! How can you make sure it will all work? Pay attention to details and remember these concepts:
- The entity you've chosen to plot will not be adjusted. The scale will adjust to get the entity to print.
- There is an inverse relationship between the Scale to [   ] **% of normal** and the Scale to [   ] **mu:su/IN.** If one value gets smaller the other will automatically get larger. When maximized to Scale to 100% of normal the entity will print as large as it can and still fit on the paper! So when Scale to is 100% of normal, the value shown in the Scale to [   ] mu:su/IN. **CANNOT** be made **smaller**. If you try to make it smaller, it will reset back to whatever it was and leave Scale to 100% of normal. This is because you can't get any more than 100% of normal.
- If you need to make the Scale to [   ] **mu:su/IN.** smaller, then you'll need to change the entity that is being printed. Quite often it is a case of mistakenly printing a view instead of an existing fence. Including the surrounding blank space of the view means having a larger value of master units printed per inch on the paper than you really need.

When making changes to the Scale to values—be careful of using ENTER when finished typing. It will act like an OK and actually close the dialog box. Instead, type in one field, then use the mouse to click in the other Scale to field. The first field will change and the second field will be adjusted automatically. Since the box doesn't close, you can see if the change in
Scale to [   ] mu:su/IN. was actually made! You can also use the TAB key to move from one field to another.

For Example:
Here is a standard border with a titleblock that will fit onto an 8½ x 11 inch paper oriented as landscape.

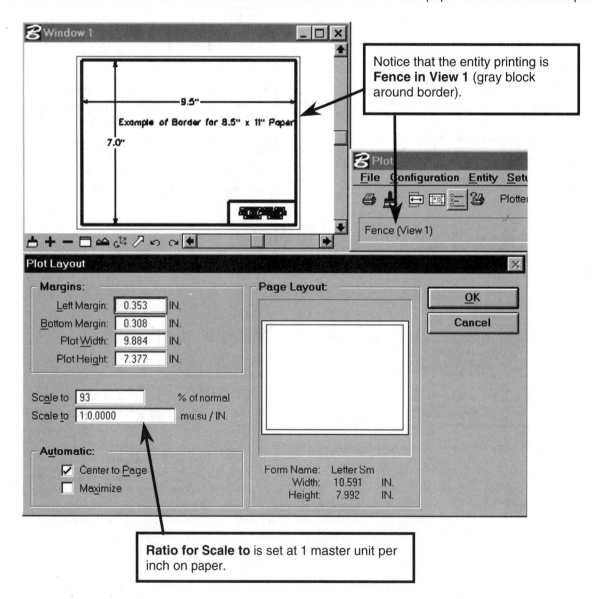

Notice that the entity printing is **Fence in View 1** (gray block around border).

Ratio for Scale to is set at 1 master unit per inch on paper.

But what would happen if the entity was accidentally set as View 1 instead of Fence? In that case, you would need to adjust the entity by either printing the fence or changing the view size. No matter how you try to change the Scale to 1.0000 mu:su/IN., it will always reset it back to the value calculated from having Scale to 100% of normal.

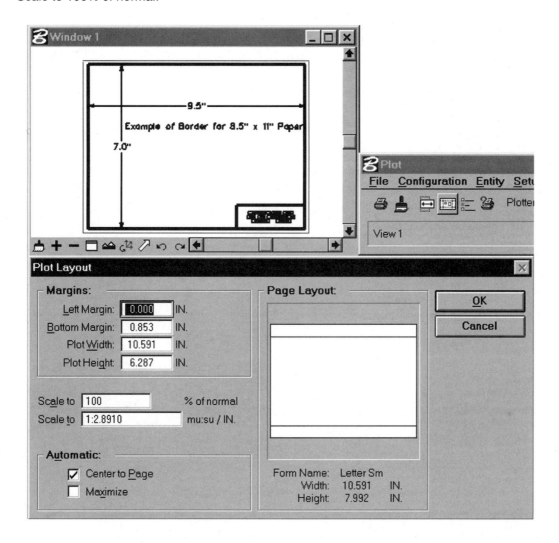

In this example, when the entity was changed from Fence (where the 1.0000 mu:su/IN. value worked) to View 1 (where 1.00000 mu:su/IN. was invalid), there was only a beep informing you that there was an error. The percentage went to 100% of normal and 1:2.8910 was automatically entered for mu:su/IN. without letting you know otherwise.

SUMMARY OF PRINTING SEQUENCE

Here is a brief summary of a typical sequence to follow when making a hard copy:

- Draw elements to the exact size of the objects they represent.
- Select an entity to print. If using a fence as the entity to print, make sure that one exists in your design file. If using a view as the entity to print, make sure that what you see in the view is what you want.
- Open up the Plot dialog box (use the printer icon in the Standard tool box or the main pull-down menu FILE>PRINT/PLOT).
- Use the Page Setup tool to open the Print Setup/Page Setup dialog box. Check the Printer, Paper, and Orientation settings.
- Check the settings in the Plot Layout dialog box. If printing to scale, pay close attention to which field you are entering a value in—you want the one using a ratio, **not** the % one.
- Modify the Plot Options to suit your needs. You may need to go the View Attributes dialog box and make changes there.
- Use the Preview Refresh to see a preview of the print you will get.
- Make any changes needed if the preview was not to your liking. Use Preview Refresh again if necessary.
- Once the preview is acceptable, use the Plot tool in the Plot dialog box—being careful to click **only once**.

QUESTIONS

① What is the difference between the Print button in the Standard tool box and the Plot button in the Plot dialog box? They both have the printer icon.

② You want a printed hard copy to have the different scales listed below. For each different scale, give the correct value to be specified in Field A or Field B in the Plot Layout. State that the other field would be automatically calculated.

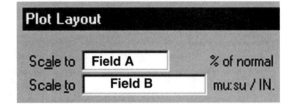

a) SCALE 1" = 100 MILES considering the design file working units are 1:5280:100 (miles:feet:position).
b) SCALE DOUBLE SIZE considering the design file working units are 1:10:1000 (inches:tenths:position).
c) SCALE 1" = 50 FEET considering the design file working units are 1:10:100 (feet:tenths:position).
d) As specified by your instructor.

③ Use the View Windows 1 and 2 shown, with a circle, triangle, text elements, and the light gray rectangular fence (block, inside). Sketch the print that would result from the following three different entity plot settings:

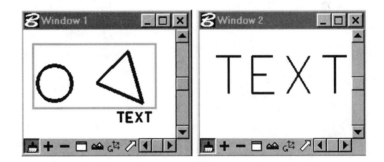

↓ a)   ↓ b)   ↓ c)

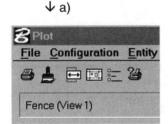 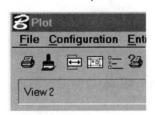 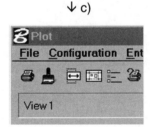

④ You are going to print a fence or a view but decide that you don't want the dimensions to be printed. Looking at the Plot Options shown, note that Dimensions is checked ON so that they will print. However, since the option is grayed out, no matter how many times you click on the ✓, you can't uncheck the Dimensions check box. What should you do so that the dimensions will **not** be printed (without deleting the dimensions!)?

Chapter 6: Lines, Circles, Arcs, and Polygons

SUBJECTS COVERED:

- Coordinates
- Linear Elements Tool Box
- Place Line
- Ellipses Tool Box
- Place Circle
- Place Ellipse
- Arcs Tool Box
- Place Arc
- Polygons Tool Box
- Place Block
- Place Regular Polygon

COORDINATES

Coordinates were discussed in detail in Chapter 2 following the discussion of working units. Let's review some of the information again since we're ready to start placing elements at a precise location and an exact size. Precise coordinates can be specified using key-ins, or graphically with the aid of snaps.

Absolute and Relative

Coordinates can also be specified as relative or absolute. Relative coordinates means they are calculated from a coordinate that has already been specified (or specified with a data point). Absolute coordinates means they are calculated from the drawing's origin. Relative coordinates are more useful since you usually know sizes in relation to other elements rather than to the origin.

Key-Ins

This table reviews the key-ins generally used for specifying coordinates or distances. Key-ins are entered into the Key-in field. Remember that they aren't case-sensitive, but the syntax is important. There are no spaces separating any of the characters.

ABSOLUTE	
Rectangular system	⌨XY=*xcoordinate,ycoordinate*

RELATIVE	
Rectangular system	⌨DL=△*xcoordinate,*△*ycoordinate*
Polar system	⌨DI=△*distance,angle*

Coordinate Entry Example

Here are 3 different lines on a grid whose spacing is 1 unit. Shown are the key-ins that were used to enter line A-B, and the key-ins to enter lines C-D and D-E, which share a common endpoint (D).

Line A-B
A entered as ⌨XY=1,1 (absolute rectangular)
B entered as ⌨XY=5,6 (absolute rectangular)

Line C-D
C entered as ⌨XY=7,1 (absolute rectangular)
D entered as ⌨DL=4,2 (relative rectangular)

Continuing on from D:
Line D-E
E entered as ⌨DI=2,90 (relative polar)

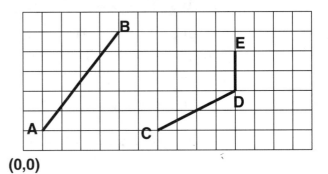

(0,0)

Using the tentative point in conjunction with a relative key-in is a handy way to offset a coordinate. In the example shown above, C could have been placed by first using a tentative point snapped to A and then a key-in of ⌨DL=6,0. The starting point of line CD would have been offset relative to the tentative point.

LINEAR ELEMENTS TOOL BOX

Linear elements are just that—lines. Placing a linear element requires at least two points: a first point and an endpoint (but an endpoint of one line can be the first point of the next line). However, there are more complex elements, such as multi-lines, which use lines as just one of their components. These more complex tools will not be covered here, but see the software documentation for more information.

The Linear Elements tool box is found in the Main tool box. The default tool on top, which looks like a line and a light bulb, is Place SmartLine. The light bulb means that it is a "Smart" tool. Smart tools are supposed to be more intuitive and therefore handier to use. For beginners, learning the regular tools first will enable a better understanding of the software. In this case, Place SmartLine is a tool that works well when using AccuDraw, so it will be covered in the chapter on AccuDraw. Right now, let's work with the basic regular tools such as Place Line.

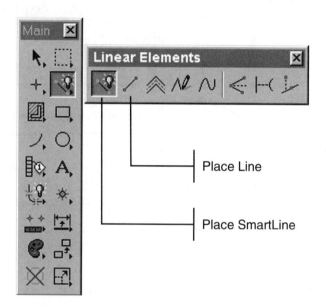

PLACE LINE

The basic element in any CAD drawing is a line. Either two points or one point and a length and an angle specify a line. These specifications can be done graphically using a data point, or values can be entered using key-ins. This tool stays active until you reset it (using the right mouse button) or pick another tool. Each line that you draw will be its own individual element, even if it shares an endpoint.

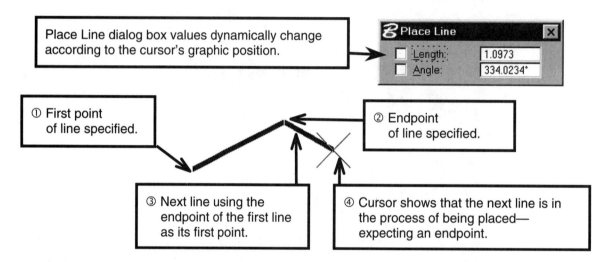

Constrained Values

The values in the Place Line dialog box can be constrained so that they don't change dynamically according to the position of the cursor. A checkmark indicates that the constraint is ON. It can be toggled off and on by clicking on the check box. The constraint is also checked **ON** if you type a number in the field of the dialog box and then hit the ENTER key. The value itself can be edited while leaving the constraint checked ON. When values are constrained, the cursor no longer appears at the actual endpoint of the line.

Length Constrained

The line shown below has had the first point specified (left one) and is being prompted for the endpoint. This line has been constrained to having a length of 2.0000 master units. The angle is free to vary according to the location of the cursor on the design plane.

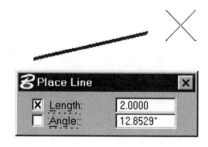

Angle Constrained

Instead of its length being constrained, now the Angle of the line is constrained to 22°. Again the first point has been specified at the left and the endpoint needs to be specified.

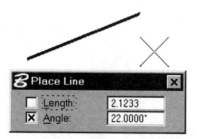

Setting the Angle constraint also changes the Active Angle setting. Since Active Angle controls the orientation of elements such as text, this constraint can have unwanted results. To set the Active Angle back to zero use the key-in ⌨AA=0.

Both Length and Angle Constrained

If both Length and Angle are constrained, then all that needs to be specified is where to start the line. The cursor will appear at the start point of the line that needs to be entered. Here the length of the line has been constrained to 1.0000 master unit and its angle to 325° with the cursor showing the location of the first point.

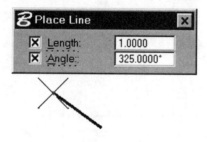

Often the use of constrained values becomes more frustrating and time-consuming than just using the regular key-ins such as ⌨XY= or ⌨DL=.

Example of Drawing Three Lines

This example illustrates the steps taken to draw three lines that share endpoints. The first point is entered graphically, the next two endpoints are specified using relative coordinate key-ins, and the last point uses the tentative snap to go to the first point of the first line.

Step 1: **Graphically place the first point.**
Cursor indicates graphic position.
Data point specifies first point graphically.

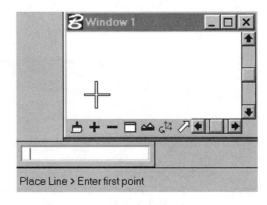

Looking for endpoint:
Moving cursor graphically will cause values to change in dialog box. They are relative to first point.

Step 2: **Use relative rectangular key-in for endpoint of the first line.**
⌨DL=.5,.2 will specify endpoint of first line. Endpoint will be 0.5 master units in X-direction and 0.2 master units in Y-direction from the first point.

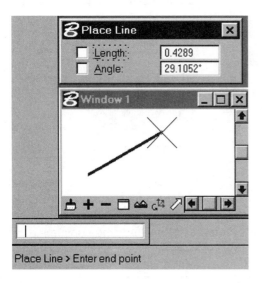
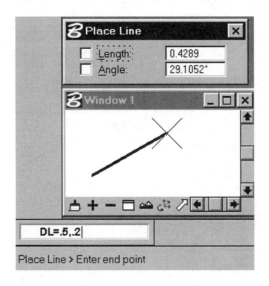

Lines, Circles, Arcs, and Polygons

Example Continued

Continuing Place Line:
Using end point of first line as first point of second line—so now looking for endpoint of second line.

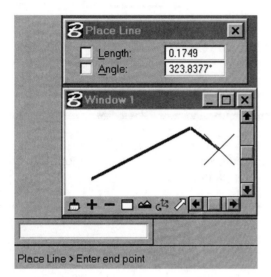

Step 3: **Use <u>relative polar</u> key-in for endpoint of second line.**
⌨DI=.25,–90 will specify endpoint of second line. Endpoint will be a distance of 0.25 at an angle of –90°.

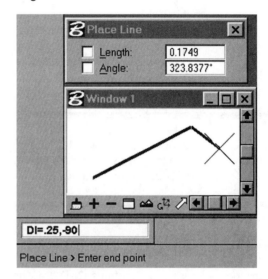

....

Continuing Place Line:
With the second line placed, will now continue using its endpoint as the first point for the third line.

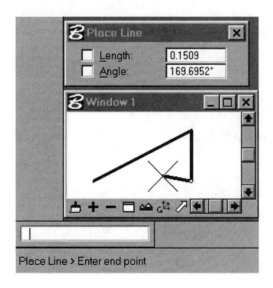

Step 4: **Use tentative snap.**
Snap to keypoint of existing first line to specify endpoint of third line.

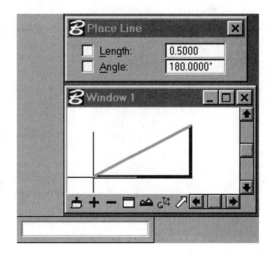

Example Concluded

Step 5: Accept tentative and use Reset.
Need to accept the tentative snap with a data point (left button) to actually finish the third line. Reset (right button) will then start the Place Line tool again. A first point for the new line must be entered.

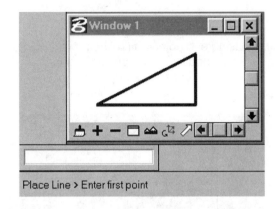

 A common problem when starting out is remembering that **tentative points must be accepted with a data point.** If you tentatively snap to another element, you get to "see it before you buy it" ... just don't forget to buy it!

ELLIPSES TOOL BOX

Circles are the next major element in a CAD drawing. The Place Circle tool is actually found in the Ellipses tool box and is the default tool on top. Mathematically a circle is just a specialized ellipse, so that is why it is in the Ellipses tool box along with the Place Ellipse tool. There are various ways to place a circle whether you want to work with a diameter or a radius, or specify its center or points on its edge. Ellipses are not as common as circles, but are still used and will also be covered in more detail.

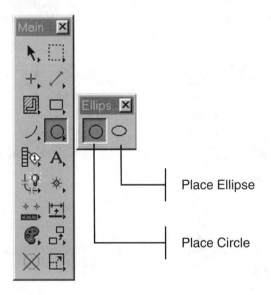

PLACE CIRCLE

Here is the Place Circle dialog box that comes up when you click on the Place Circle tool.

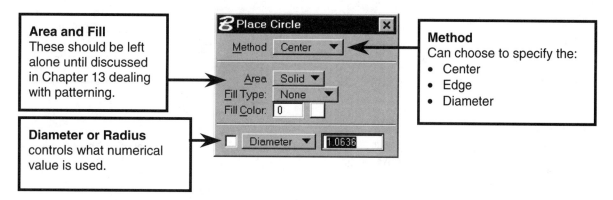

Area and Fill — These should be left alone until discussed in Chapter 13 dealing with patterning.

Diameter or Radius controls what numerical value is used.

Method — Can choose to specify the:
- Center
- Edge
- Diameter

Diameter/Radius Size Constraint

Each Method works a bit differently depending on whether or not the Diameter/Radius size has been constrained. If it is checked ON then the circle's diameter or radius will already have been specified. If it is unchecked (OFF) then the diameter or radius can be specified graphically or with a key-in. Without the constraint, the value displayed will dynamically change to reflect the size resulting from the cursor position.

Center Method

The Center Method of placing a circle will require that you identify a coordinate for the center of the circle. Additional information may be required depending on if there are any size constraints.

Center Method *without* Size Constraint

This is the simplest method to use in placing a circle. The center (radius point) of the circle is specified and then you'll be prompted to Identify Point on Circle, as shown in the illustration below.

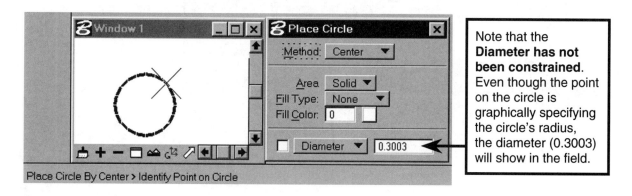

Note that the **Diameter has not been constrained**. Even though the point on the circle is graphically specifying the circle's radius, the diameter (0.3003) will show in the field.

Center Method with Size Constraint

With the Radius or Diameter option checked ON for the Center Method, the circle will actually appear on the drawing. Only the center remains to be specified in order to place the circle, as indicated by the Identify Center Point prompt.

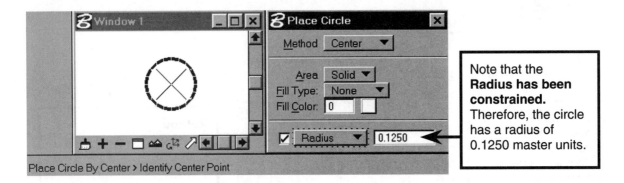

Edge Method

Placing a circle using the Edge Method will require at least two points to define the circle. The coordinate points specified lie on the perimeter of the circle.

Edge Method without Size Constraint

This method is similar to the three-point method of drawing circles by conventional hand-drafting techniques. There will be 3 points to be specified and the circle will pass through all three.

In the illustration below, 2 points have already been identified and are marked with a ■. The cursor is shown waiting for the third point on the circle to be identified. The circle will dynamically be displayed and will change sizes according to the cursor's position.

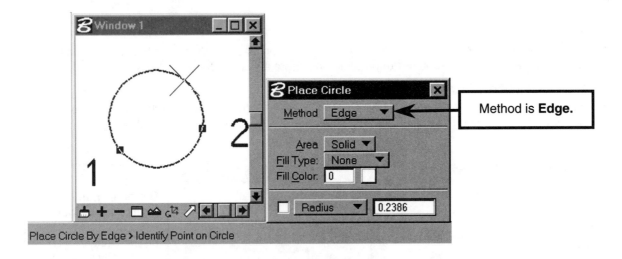

Edge Method with Size Constraint

Since the size has been predetermined, only 2 points on the edge of the circle need to be identified.

In the illustration below, the first point has been identified and the circle will then appear. Its position will dynamically change according to the cursor's position.

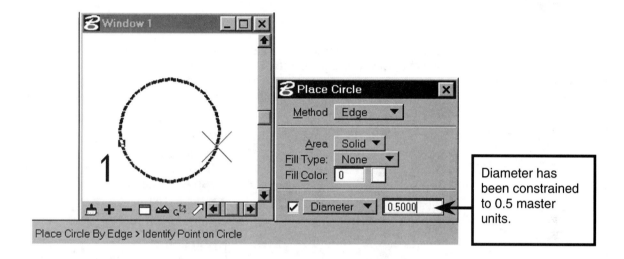

Diameter has been constrained to 0.5 master units.

Diameter Method

In this method, the option to constrain the size is not available. Instead, 2 points will be identified to define the diameter of the circle. The center of the circle will be at the midpoint of an imaginary line passing through the 2 points. The use of the Diameter Method is limited to very specific situations.

In the illustration below, Points 1 and 2 have been identified and the circle has been placed.

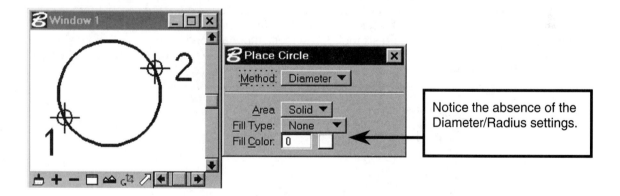

Notice the absence of the Diameter/Radius settings.

PLACE ELLIPSE

The Place Ellipse tool has a large number of variations, and not all of the different combinations will be covered here. Ellipses are not used extensively, but when they are, it is usually straightforward. You need to be familiar with the terminology of an ellipse. Then you can experiment with all the different variations on your own.

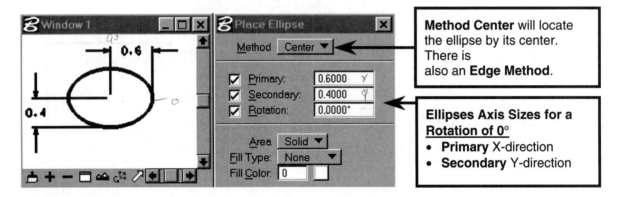

Method Center will locate the ellipse by its center. There is also an **Edge Method**.

Ellipses Axis Sizes for a Rotation of 0°
- **Primary** X-direction
- **Secondary** Y-direction

ARCS TOOL BOX

If you need only part of a circle, then you may place an arc. For some reason, most users will place a circle and then modify it rather than use an arc in the first place. Either that or they'll use the Fillet modification tool, which will be covered later. Sometimes those techniques are handier, but there is still a time and a place for arcs. The Arcs tool box is a prime example of one with lots and lots of tools, but with only one being used frequently. There are six tools. The most important one is on top and quite simply places an arc. The others deal with parts of ellipses and modifying arcs—try them on your own when you get some experience under your belt.

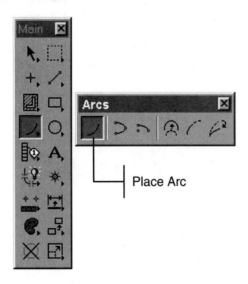

Place Arc

PLACE ARC

 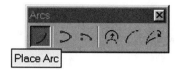

Since an arc is part of a circle, it will have some of the same characteristics as a circle. An arc must also

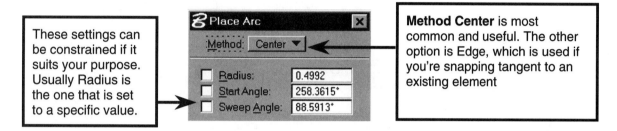

These settings can be constrained if it suits your purpose. Usually Radius is the one that is set to a specific value.

Method Center is most common and useful. The other option is Edge, which is used if you're snapping tangent to an existing element

have a start point (actually called First Arc Endpoint) and an endpoint (called Second Arc Endpoint).

Center Method

The arc will be placed by specifying its first endpoint, its center, and its second endpoint <u>in that order</u>. If Radius is not constrained (unchecked) it will be determined by the first point and center defined. It is very important to remember that the **arc will be swept in a POSITIVE angle direction** from the first to the second endpoint. This is (by default) **counterclockwise**.

Example of Placing an Arc Using the Center Method

This example will not use any constraints; nor will key-ins be done. Instead all points will be picked graphically.

Step 1: **Identify First Arc Endpoint.** The cursor indicates the graphic position, and a data point will make it so.

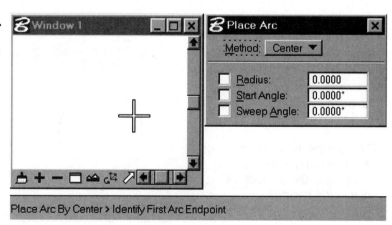

Example Continued

Step 2: **Identify Arc Center.**
The ■ will appear to visually indicate the first endpoint that was identified. The cursor indicates the location for the Arc Center.

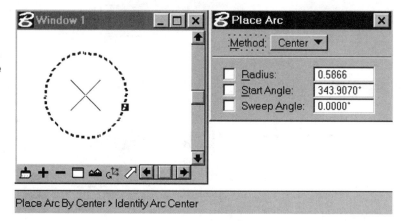

Step 3: **Identify Second Arc Endpoint.**

Once the center has been identified then the Radius and Start Angle have been constrained—a checkmark magically appears to say so!

The Second Arc Endpoint will be prompted for, but the cursor will not necessarily appear at the end of the arc that it defines.

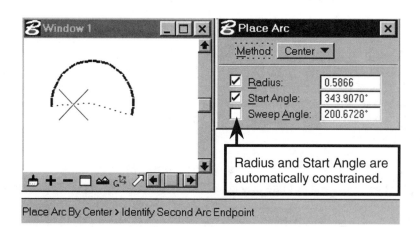

Finished:

When the arc's second endpoint has been identified, then the arc will be placed.

The constraints will be removed (checkmarks will magically disappear), and the tool will reset to the beginning.

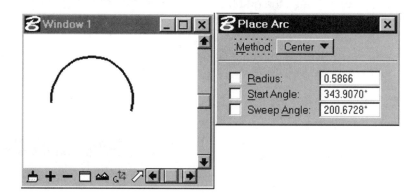

Edge Method

For the Edge Method of Place Arc, you'll be prompted for its first endpoint and then for a point on the edge of the arc instead of its center. The arc is finished when the second endpoint is then identified. Unlike the Center Method, the arc is not necessarily swept counterclockwise; instead **its direction will be determined by the points selected.**

Shown below is an arc being placed using the Edge Method. Point 1 is the First Arc Endpoint, and Point 2 is the Point on Arc Radius that was then identified. The prompt indicates that the Second Arc Endpoint still needs to be identified. <u>In this case</u>, the angle will be swept **counterclockwise**.

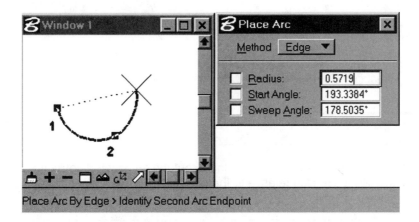

The illustration below shows a clockwise orientation. The first two points are the same as shown previously. The Second Arc Endpoint was specified clockwise from Point 2. The angle is swept **clockwise** so that the arc can go from Point 1 to Point 2 and then on to the endpoint.

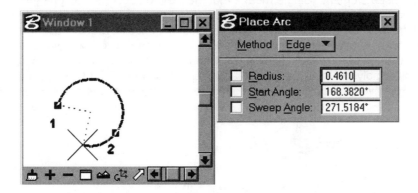

 If you plan on snapping to a **tangent** then you'll want to place the arc using the Edge Method since the Center Method doesn't allow a tangent snap.

Example of Arc with Its Radius Constrained

In this case, the arc is being placed with a constrained (checked) Radius of 2. Using the Edge Method, the first endpoint has been identified and the ■ marks its location. The cursor indicates that the second endpoint needs to be identified. Only 2 points are needed since the radius was already established.

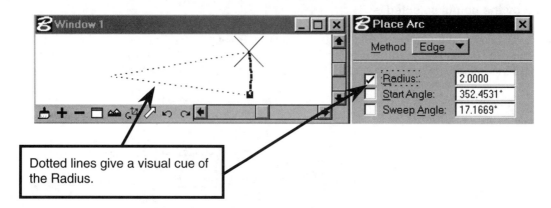

Dotted lines give a visual cue of the Radius.

 You may need to move the cursor around to get the arc located correctly. The movement and its direction relative to the first point will determine which "side" of the arc the radius point will be at.

POLYGONS TOOL BOX

A polygon is a closed shape that is made up of straight line segments. Rather than drawing each line individually, the polygon tools will place them all at once. There are different variations of polygons, but each will place a single closed element. Single means that the polygon acts as only one element even though it is made up of line segments. Closed means that the first endpoint of the first line segment will also be the last endpoint of the last segment—the shape encloses a specific area. If all the sides are of equal lengths then you'll want to use Place Regular Polygon. The Place Block tool creates a closed four-sided shape with unequal lengths—a rectangle! This by far the most frequently used polygon tool, so it shows up on top. The other polygon tools available place irregular shapes that use more than four unequal line segments. However, this type of shape configuration is better accomplished using AccuDraw and SmartLine.

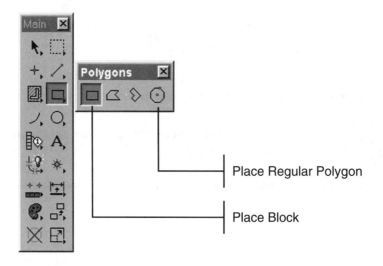

PLACE BLOCK

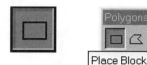

The Place Block tool is used for placing a rectangular (or even square) closed shape. The actions required are similar to that of placing a simple fence. In fact, make sure that you pick the correct tool because it even looks like the Place Fence tool! One corner of the block is placed and then the opposite corner is specified. It's a very simple tool and often used with relative coordinate key-ins to get a specific rectangle size.

The block shown below is in the process of being placed. The first point (lower left corner) has been identified with a data point. The cursor indicates where the opposite corner is being placed.

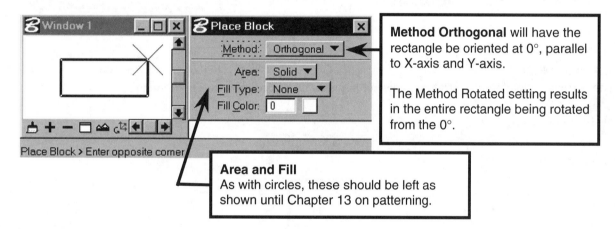

Method Orthogonal will have the rectangle be oriented at 0°, parallel to X-axis and Y-axis.

The Method Rotated setting results in the entire rectangle being rotated from the 0°.

Area and Fill
As with circles, these should be left as shown until Chapter 13 on patterning.

In engineering graphics there is often a standard border that goes around the outside edge of the drawing paper. In the lower right corner of that standard border is a title block that includes information about the drawing. These two items are a perfect application for Place Block.

For Example: A standard border and title block will be drawn using Place Block.

Step 1:
Graphically Enter first point.
Cursor indicates location and data point will specify it.

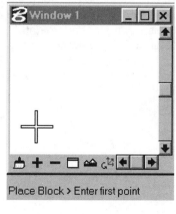

Step 2:
Use relative rectangular key-in to Enter opposite corner.
DL=9.5,7 will specify a rectangle that fits nicely on an 8½ x 11 inch landscape paper.

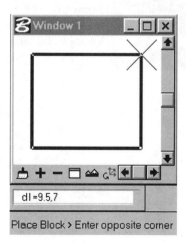

Now a title block will be placed in the lower right corner of the standard block.

Step 3:
Graphically Enter first point.
Tentative point to snap to the first block's exact lower right corner. **Don't forget to accept with a data point!**

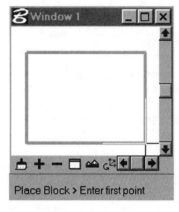

Step 4:
Use relative rectangular key-in to Enter opposite corner.
DL=–3,1 will go a minus 3 master units in the X-direction and a positive 1 master unit in the Y-direction.

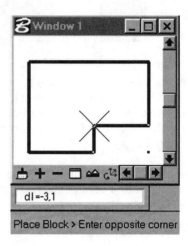

Results:
The resulting two blocks illustrated below serve nicely as a bordered area for printing.

 Sometimes it's important that the lower left corner not be placed arbitrarily in the drawing. It may be entered at XY=0,0 but when this is keyed in, you may not see the block because the window may not be displaying that coordinate!! Finish placing the block with key-ins and then fit your design with the view controls.

PLACE REGULAR POLYGON

Shown below is the Place Regular Polygon dialog box that comes up when you click on the tool.

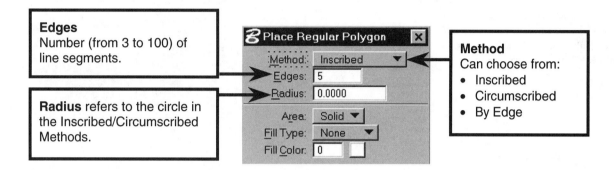

Edges
Number (from 3 to 100) of line segments.

Radius refers to the circle in the Inscribed/Circumscribed Methods.

Method
Can choose from:
- Inscribed
- Circumscribed
- By Edge

Inscribed/Circumscribed Methods

Regular polygons can be **in**side a circle, and then the endpoints of the edges will touch the circle. This is called an **inscribed** regular polygon. If the midpoints of the edges touch the circle, then the polygon is outside of the circle. The polygon goes around the **circum**ference of the circle, so it is called a **circumscribed** regular polygon.

Shown here are two circles with the same radius (light line weight). Two regular 5-sided polygons (heavier line weight) were placed using the same radius, but the two different methods. The one on the left was Inscribed and the one on the right was Circumscribed.

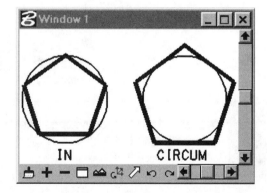

When placing a regular polygon with either of these methods, you'll be prompted to Enter point on axis. This is really just asking for the center of the circle that is related to the polygon. Then you'll be prompted to either Enter first edge point (Inscribed Method) or Enter radius or pnt on circle (Circumscribed Method). This point specified will control two things: 1) the **size** of the polygon (unless Radius was already specified in the dialog box) and 2) the **rotation** of the polygon.

The easiest technique is to enter a value for Radius in the dialog box and toggle Axis Lock ON. With this technique, the second step of specifying the point (edge point or point (pnt) on circle) will just involve moving the cursor and watching the polygon "jump" to a certain angular orientation. Both of the pentagons in the example above had the point in the second step specified vertically using Axis Lock ON.

By Edge Method

The By Edge Method places the polygon and sizes it according to one of its edges (line segment). This makes it a bit more difficult to use. Since the polygon's edge also controls its orientation, the use of a relative coordinate key-in (polar or rectangular) is handy for specifying both the edge length and the orientation. Using the By Edge Method is especially good for equilateral triangles.

The first step is to Enter first edge point. A data point will graphically select the point at the cursor (no ■ will show up—sorry!).
Notice that the Radius setting has been grayed out because there is not a circle related to this method.

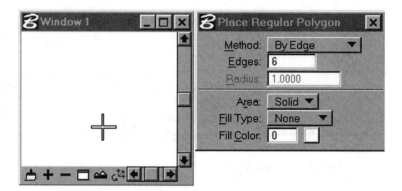

Then a prompt to Enter next (CCW) edge point appears. Without Axis Lock ON (or relative key-in), the point specified will control both edge length and orientation.

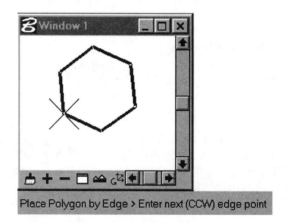

 With any of the Method settings, the polygon will **not** have a center keypoint. The line segments of its edges do retain their keypoints to snap to even though they are not single elements.

Lines, Circles, Arcs, and Polygons   131

QUESTIONS

① For **each** of the three coordinate systems stated below, answer parts a, b, and c.
  1.1) Absolute Rectangular system
  1.2) Relative Rectangular system
  1.3) Relative Polar system

  a) Name the key-in that is used to input the coordinates stated.
  b) Briefly explain the coordinate system (include a sketch if you like).
  c) Assume that you have a line starting at the coordinates (25,–20) and the key-in is used with a value of (12,90). State the coordinates of the endpoint that results.

② Assume the grid shown to the right has a grid spacing of 1 unit. For each line state the specific values to use for the key-in given (in the order stated) that will place the line as shown.

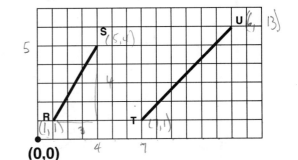

  2.1) Placing R as endpoint 1 and then S as endpoint 2
  a) XY= for endpoint 1 then DL= for endpoint 2.
  b) XY= for endpoint 1 then DI= for endpoint 2.

  2.2) Placing T as endpoint 1 and then U as endpoint2.
  a) XY= for endpoint 1 then DL= for endpoint 2.
  b) XY= for endpoint 1 then DI= for endpoint 2.

③ Name the two settings in the Place Line dialog box. Also explain what it means to constrain their values.

④ What two sizes can you constrain when using Place Circle?

⑤ Explain the Edge Method of Place Circle without a size constraint.

⑥ Sketch the default 0° angle and the corresponding **positive** angle direction.

⑦ What is one of the main differences between drawing a rectangle using Place Line four times and drawing a rectangle using Place Block?

⑧ Sketch the difference between placing two regular 4-sided polygons (with equivalent circle radii) by these two Methods:
  a) Inscribed
  b) Circumscribed

⑨ Using the Analyze Element tool, you've selected a **circle** and Element Information for ELLIPSE as shown at the right appears—Why?

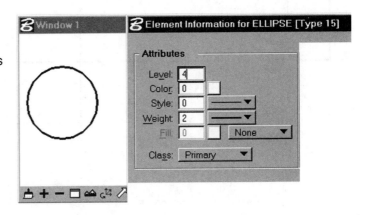

EXERCISE 6-1   Computer Screen

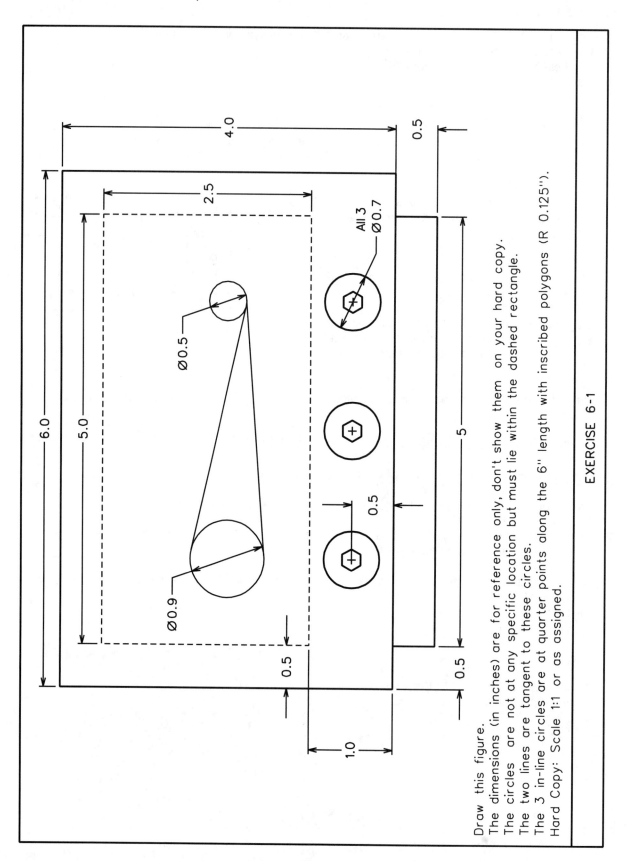

EXERCISE 6-1

EXERCISE 6-2   Hinged Elliptical Plate

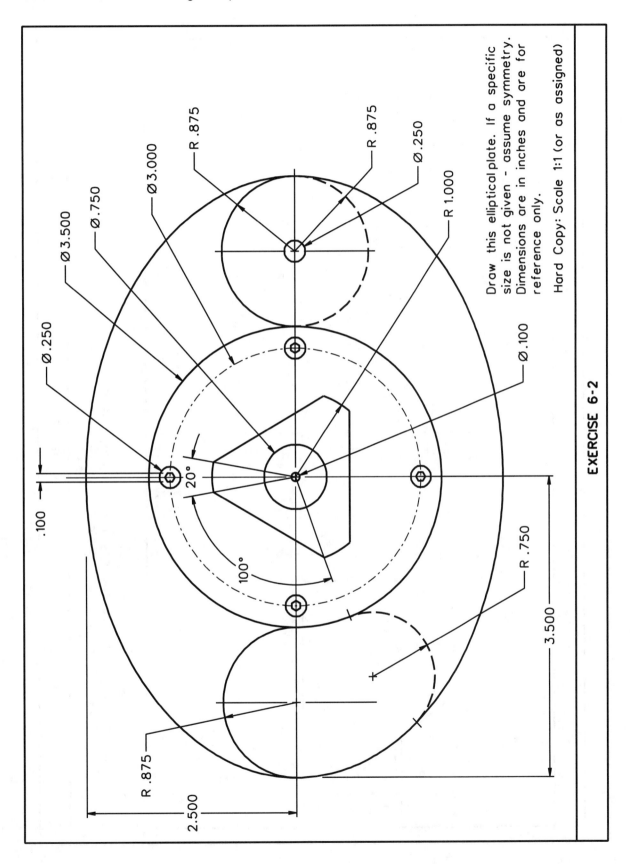

EXERCISE 6-2

EXERCISE 6–3 Basketball Court Layout

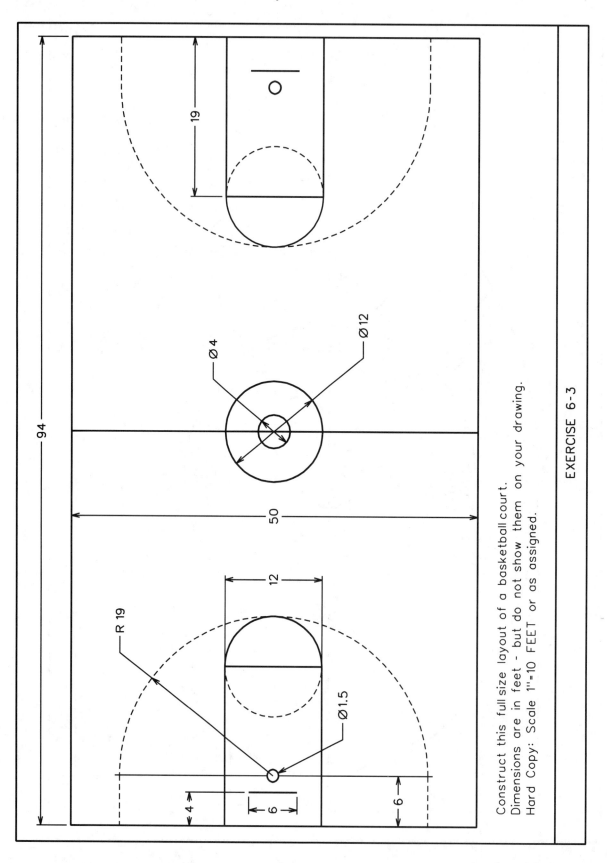

EXERCISE 6-3

Construct this full size layout of a basketball court. Dimensions are in feet - but do not show them on your drawing. Hard Copy: Scale 1"=10 FEET or as assigned.

EXERCISE 6-4   Adjustable Angle Bracket

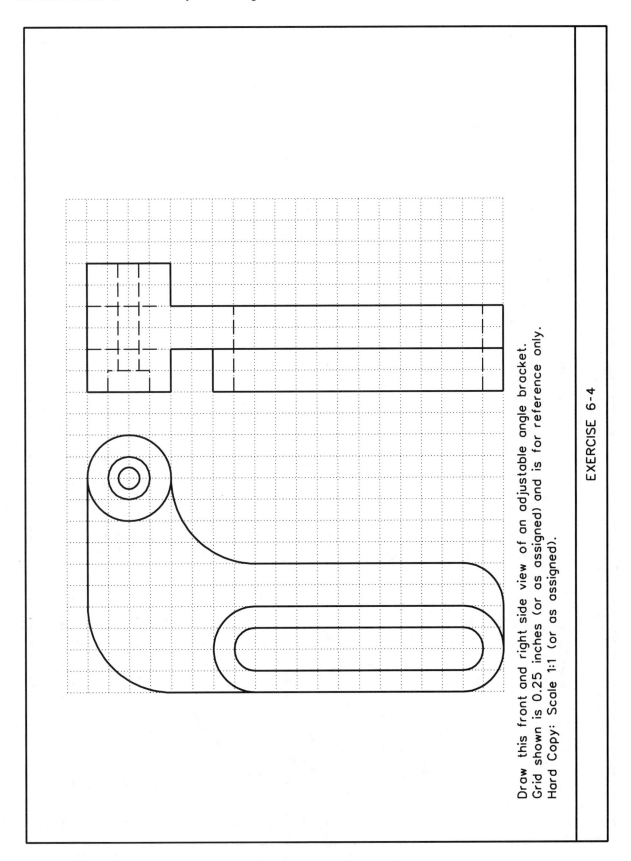

Draw this front and right side view of an adjustable angle bracket. Grid shown is 0.25 inches (or as assigned) and is for reference only. Hard Copy: Scale 1:1 (or as assigned).

EXERCISE 6-4

EXERCISE 6-5 Golf Course Layout

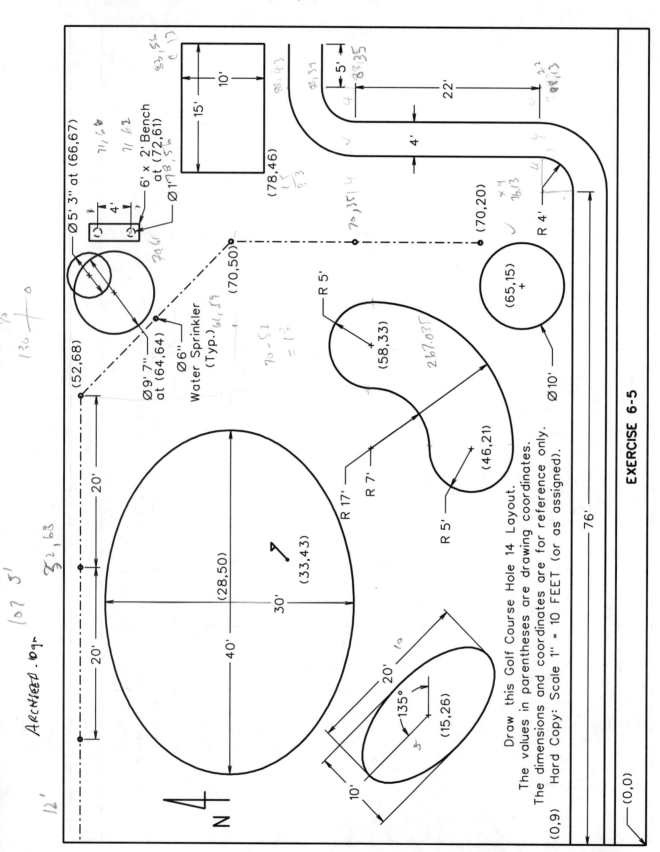

EXERCISE 6-5

Chapter 7: Text

SUBJECTS COVERED:

- Text Attributes
- Text Tool Box
- Place Text
- Edit Text
- Match Text Attributes
- Change Text Attributes
- Enter Data Field
- Fill In Single Enter-Data Field
- Determining Text Size for Printing to Scale

Even though it's been said that a picture is worth a thousand words, you still need words—text. From title block information in a standard plan to identifying landowners on a map, the use of text is an integral part of a drawing. There are settings that control the look and size of the text that is placed. Actually placing text in the design file can be done in numerous ways, from specifying an origin point to locating it in relation to an existing element. Once text has been placed, there is always a need to edit it (unless you type better than I do—but that wouldn't be saying much!). For us hunt-and-peck typists, excellent text editing tools are available. The height and width of existing text are easily changed, which is an important feature when you're printing at various scales for different paper sizes. All in all, text is a very important component of a design file.

TEXT ATTRIBUTES

There are a lot of text attributes that help determine the appearance and size of the text. These settings are called text attributes and apply only to text and are not to be confused with the regular element attributes. The typical element attributes such as Color, Level, and Line Weight still apply since text acts as an individual element (even if it is made up of more than one word). The exception to this is Line Style; text will normally come in as a solid line style. **The current text attribute settings will be applied to the text when it is placed. If the attributes are changed, existing text will not be affected—only the text that is subsequently placed.**

Accessing the Text Attributes Settings

Since each text (single word or string of words) is an individual element, you can access the settings under the main pull-down menu ELEMENT>TEXT. This will bring up the very large Text dialog box shown below.

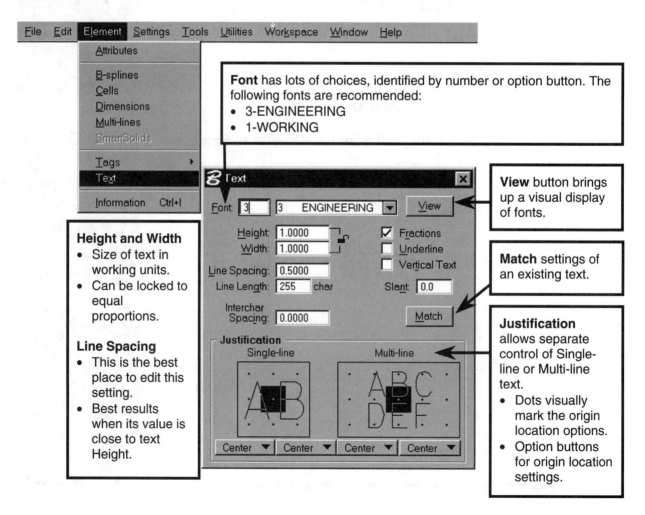

Fonts

There are numerous fonts available with the software. Each Font has a number assigned to it. The simple ones of 3 ENGINEERING and 1 WORKING are the best choices for beginners. Fonts 100 and 101 are often used when doing dimensioning since they have engineering-related symbols too. For you font collectors out there, TrueType® fonts can also be loaded for use (consult the software documentation to see how it's done).

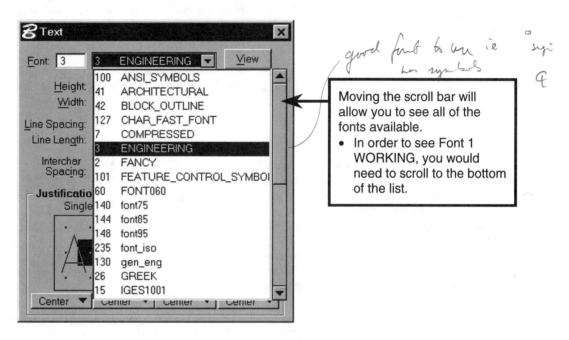

Moving the scroll bar will allow you to see all of the fonts available.
- In order to see Font 1 WORKING, you would need to scroll to the bottom of the list.

View Button

If you click on the View button of the Text dialog box then you can see what the various fonts look like in the Fonts dialog box. By double-clicking on a Font Name in the list box, you'll set it as the Active Font.

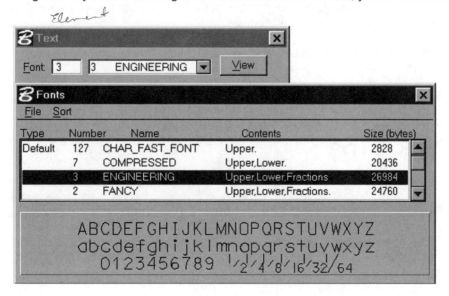

Height and Width

The Height and Width of the text can be set independently or locked to the same proportion. The Height value is in working units and is the actual height. The Width value includes the letter width <u>and</u> the space around it (allowing for white space between letters) and is in working units. Neither has to do with point sizes such as those used in word processing.

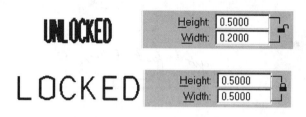

Here the Height is larger than the Width. The padlock is open, indicating that is unlocked to allow unequal values.

If the padlock is closed, then the Height and Width values are locked to the same value. If you change one, the other automatically changes.

Line Spacing

The Line Spacing setting will affect the spacing (in working units) between Multi-line text. Multi-line text refers to placing more than one line of text with a **single** Place Text operation. When typing and you reach the end of the first line of text, hit the ENTER key. This will start the next line of text. Once characters are typed in that second line, it becomes Multi-line text. Single-line text occurs when only one line of text is typed. Even if you hit the ENTER key after typing the characters of the first line, it will remain Single-line text until characters are typed in the second line of text.

The Line Spacing setting also affects a single line of text that is placed using one of the different Method options such as Above Element, Below Element, or Along Element. Please refer to the Methods Options section to see an illustration of this.

Justification

When text is placed, it has an origin point. An origin point is the coordinate point that corresponds with the text (where to grab onto the line of words). The position of the text in relation to this origin point is called Justification. The quickest way to change the Justification is to click on a dot that you want to be the origin location. The option button settings will change accordingly.

Shown below is a Single-line text that has a Left Bottom Justification. The cursor is at the origin point.

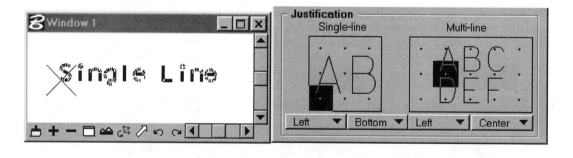

An example of Multi-line text being placed with Left Center Justification is shown below. Notice that the cursor marks the origin, which is different from that of the Single-line.

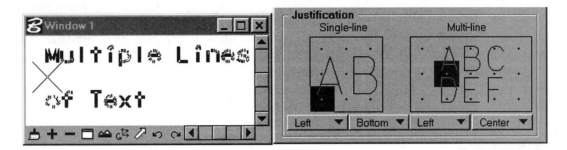

Underlining Text

In order to underline text, the Underline setting in the Text dialog box must be checked ON. However, the **entire** string of words must be underlined. There are no control characters that you can type in the text string that allow for turning underlining on and off. The same goes for diameter symbols and degree symbols. There aren't corresponding control characters. However, some fonts do have these symbols available.

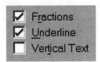

 Each text element has an associated **Text Node**. This will actually appear (and be printed) at the origin of a text— **if the Text Node setting has been checked ON under View Attributes**. Here it looks like a + and a 0 in between the lines of the Multi-line text. If you don't want text nodes displayed or printed, go to View Attributes (CTRL-B is the shortcut) and toggle Text Nodes OFF.

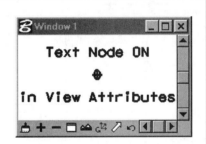

TEXT TOOL BOX

Once the settings have been adjusted to your specifications, placing text is a breeze. The Place Text tool is the default tool on top in the Text tool box. Other tools available that are related to text include Edit Text and Change or Match Text Attributes, all of which are used extensively. The creation of enter data fields will be discussed and the use of the Fill In Single Enter-Data Field tool will be covered. However, the other tools involving data fields will be left to advanced users.

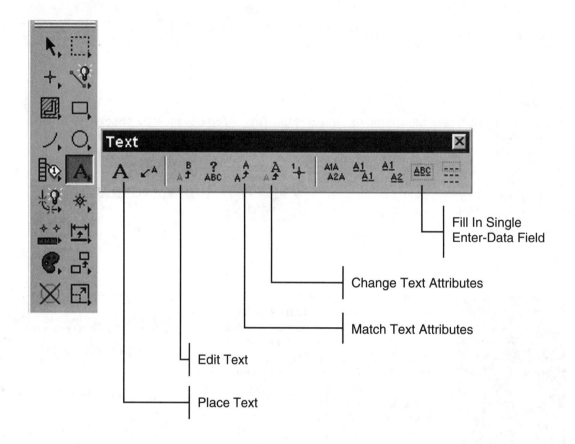

PLACE TEXT

When the Place Text tool is clicked, two dialog boxes appear. One is the actual Place Text dialog box. It contains new stuff, but some of the most-used text attribute settings are available too. This is very handy. The other is the Text Editor dialog box where you actually type the text wanted. **Don't type text into the main Key-in field!**

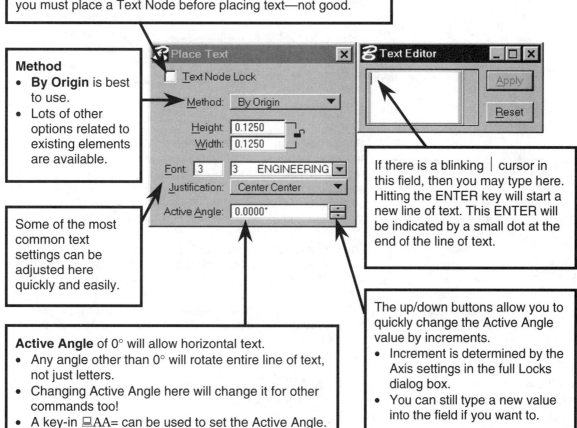

Text Node Lock should be left unchecked (OFF). Otherwise, you must place a Text Node before placing text—not good.

Method
- **By Origin** is best to use.
- Lots of other options related to existing elements are available.

Some of the most common text settings can be adjusted here quickly and easily.

If there is a blinking | cursor in this field, then you may type here. Hitting the ENTER key will start a new line of text. This ENTER will be indicated by a small dot at the end of the line of text.

The up/down buttons allow you to quickly change the Active Angle value by increments.
- Increment is determined by the Axis settings in the full Locks dialog box.
- You can still type a new value into the field if you want to.

Active Angle of 0° will allow horizontal text.
- Any angle other than 0° will rotate entire line of text, not just letters.
- Changing Active Angle here will change it for other commands too!
- A key-in ⌨AA= can be used to set the Active Angle.

To actually Place Text (By Origin Method) the steps are as follows: Click on the Place Text tool. Make sure that the blinking cursor appears in the field of the Text Editor dialog box. If it isn't there then click in the field. Once you have the blinking cursor, proceed to type the text wanted in the Text Editor dialog box's field. If you want Multi-line text you must use the ENTER key to end one line of text and start the next. **Once you are done typing, move the mouse so that the cursor is in the drawing area (using the ENTER key will just start another line of text).** The text will appear with the cursor at the origin (if Dynamics is checked ON in View Attributes). The text is not actually placed in the design file until the origin point is specified graphically or with a key-in.

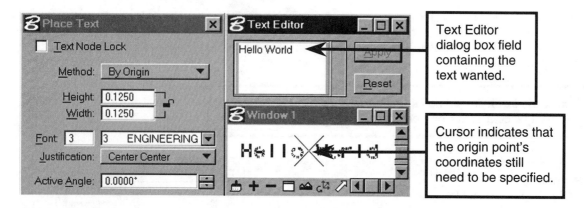

Method Options

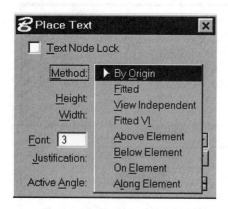

Shown at the left is a list of the other Method options. Fitted, View Independent, and Fitted VI (View Independent) require two coordinates to fit the text between. In these methods, the size of the text will be adjusted in order for the text to fill out the distance between the points. That's why they aren't preferred methods—usually you really do care what size the text is!

The bottom four methods involve placing the text in relation to an existing element. The element will need to be selected; **where** the element is selected is used to indicate the origin (and its related justification), except that it is offset to lie Above, Below, On, or Along the element. **Read the prompts**. You get to see it before you buy it, so don't forget the data point to accept it!

These different examples of text illustrate some of the different methods. Each was placed using a Center Center Justification and snapping to the midpoint of the line. The Line Space setting for text determines how far the text is from the element.

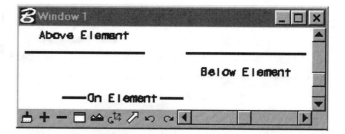

 Be careful with the On Element Method. It cuts the element! In this case, the one long line is now two shorter segments. However, this method is very useful for labeling contour lines.

With the Along Element Method, the Place Text dialog box undergoes some minor changes. As shown below, it no longer has the Active Angle Setting but instead the Interchar. Spacing and Line Spacing settings can be adjusted.

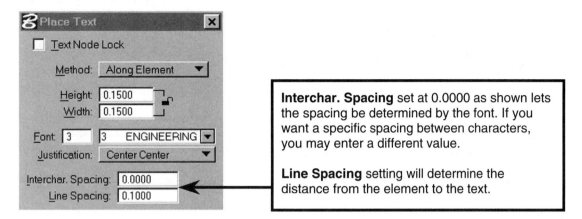

Interchar. Spacing set at 0.0000 as shown lets the spacing be determined by the font. If you want a specific spacing between characters, you may enter a different value.

Line Spacing setting will determine the distance from the element to the text.

When placing text using the Along Element Method, the data point to accept the text placement also determines whether to put the text above or below the element. This allows you to see which side looks better. Shown on the right is the middle step of placing text with the Along Element Method. The status bar informs you that a data point is needed to Accept, select text above/below. A data point where the cursor is shown here would result in the text being above the element.

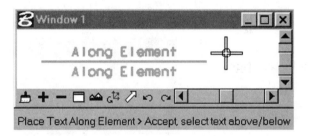

 When using Place Text with the Along Element Method the text is no longer a string of words—**each letter is an element**. Just thought you should know because it is really tough to edit or delete this type of text.

EDIT TEXT

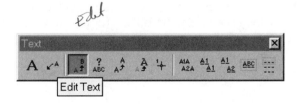

The Edit Text tool is one everyone needs sooner or later. It is quite simple to use—data point on the text you want to edit (accept the selection with a data point) and it will be brought up in the Text Editor box. Type your corrections or changes and then use the Apply button to guess what—yep, apply it! The Reset button is used to cancel out of the editing session. Most of the typical word processing editing tools can be used in the text editor. For instance, you can highlight more than one word and use the keyboard's DELETE key to get rid of the highlighted text. Other keyboard shortcuts that are handy are CTRL-X to Cut and CTRL-V for Paste.

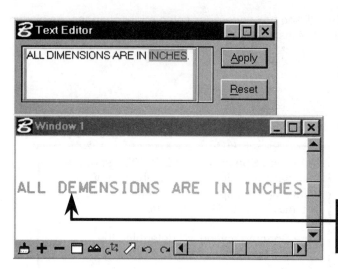

First of all, the text to edit needs to be selected with a data point. Shown at the right, the text to edit has been selected and accepted. It highlights in the view and the Text Editor dialog box opens. The E in DEMENSIONS has been corrected to an I and the INCHES is highlighted ready to be changed to FEET.

The E in DEMENSIONS has not been corrected yet.

Once you've made the changes you want, the Apply button **must be used** to apply the correction to the text in the design file. After you have clicked on the Apply button, the Text Editor field will empty and wait for you to select another text element. This is illustrated below.

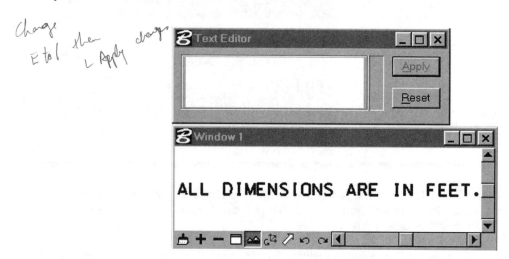

Often, in addition to makiing corrections to the text itself, you may want to move single lines of text so that they line up on one edge. There is a handy-dandy tool in the Manipulate tool box called Align Elements by Edge that works great for this. See Chapter 10 for a detailed explanation of this tool.

MATCH TEXT ATTRIBUTES

 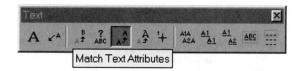

This Match Text Attributes allows you to automatically change the Text Attribute settings to those of an existing text. It does not change the selected text element. You may want to use this tool to match existing text when working in a group or using someone else's design file.

Illustrated below are two separate text elements. The current text attributes settings are shown. The Match Text Attributes tool has been started, and the status bar indicates that you need to identify the text element to match.

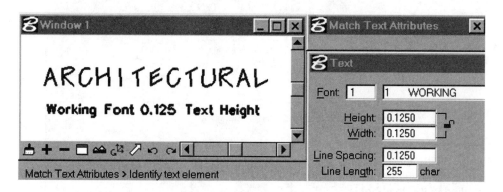

Now the "architectural" text has been selected to match and the Text Attributes settings have changed accordingly. Not only the Font but the sizes are matched also. Any new text placed will have the same text attributes as "architectural."

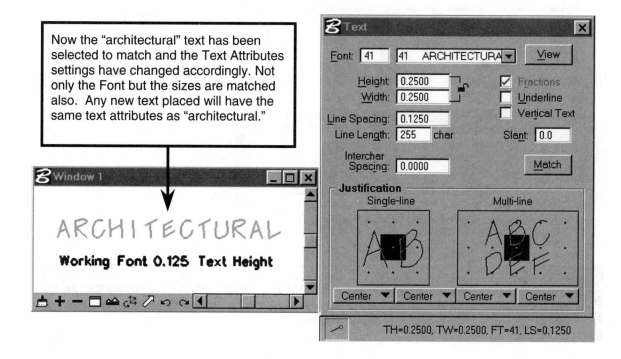

CHANGE TEXT ATTRIBUTES

 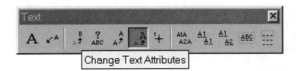

This tool brings up a dialog box that looks a lot like the Text Attributes settings. If you need to change the typical element attributes of the text (Color, Line Weight, etc.) then use the Change **Element** Attributes tool—**not** this one. Only the settings that are checked ON will be applied—the others will not be affected. Not only will the text be changed but the actual Text Attributes settings are also changed. Therefore, any subsequently placed text will have the new text attributes.

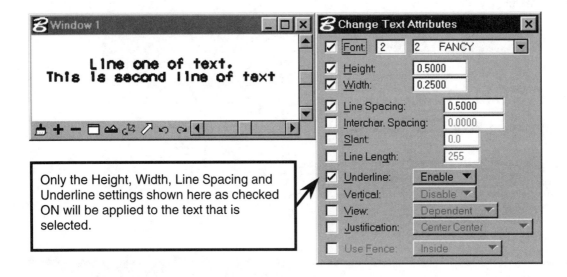

Only the Height, Width, Line Spacing and Underline settings shown here as checked ON will be applied to the text that is selected.

The result of this Change Text Attributes will look like this:

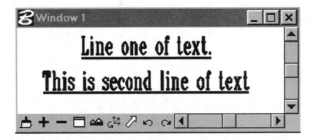

 One challenging text attribute to change is Justification. That's because the text will actually "move" to justify itself correctly around the same origin—the origin doesn't move. This can give unexpected results.

ENTER DATA FIELD

An enter data field can be used to help standardize drawings. It is common terminology to drop the "enter" and just call it a data field, which we'll do from now on. However, if you are using MicroStation's Help make sure to include the word enter. A data field contains a specified text character that acts like a placeholder. This placeholder indicates that information needs to be filled in. For instance, a title block with data fields for the title, drawn by, checked by, or scale information can be included in the seed file so that it shows up on each of your drawings. This information could then be filled in when the drawing is completed. Other uses of data fields are bill of materials, special plan notes, etc. **There is NOT a special tool such as Place Enter Data Field. You have to use the Place Text tool to place the special data field character that creates the empty data field.** The Fill In Single Enter-Data Field tool in the Text tool box is then used to select the empty data field and fill in its specific information.

Data Field Character

The default data field character is an underscore (_). This data field character is determined by the ED Character setting, which can be found under the main pull-down menu WORKSPACE>PREFERENCES and the Text category. We'll just stick with the default. To create the data field, use the Place Text tool and just type the data field character for each character of the text that will be filled in later. The number of characters doesn't need to be exact because you can edit the text later if necessary, but it's a good idea to have enough characters in the data field to take care of the typical entries.

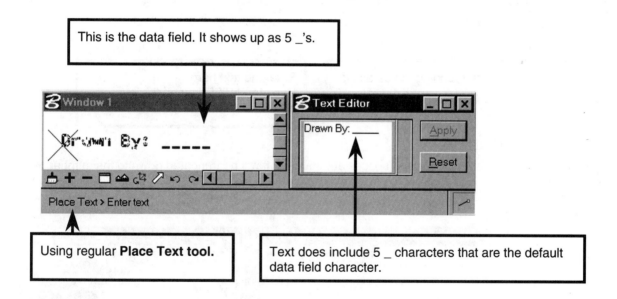

 The Data Fields setting can be toggled on or off in the View Attributes dialog box. If Data Fields is unchecked (OFF), filled-in data fields will appear without the _ character, but empty data fields won't be visible at all.

FILL IN SINGLE ENTER-DATA FIELD

 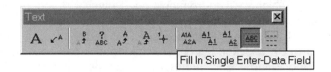

Now that there is a data field, we can use the Fill In Single Enter-Data Field tool. You will be prompted to select the data field. Be sure to data point near the _ _ _ so that it recognizes the data field. There will be a rectangle indicating the data field. The Text Editor dialog box will open up, and that is where you enter the text for filling in the data field.

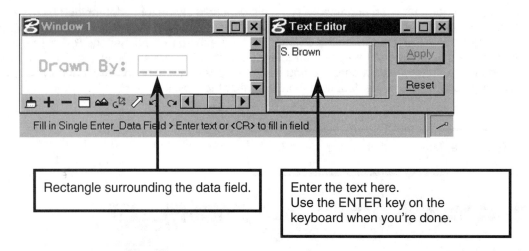

Rectangle surrounding the data field.

Enter the text here.
Use the ENTER key on the keyboard when you're done.

Now since S. Brown has 8 characters (spaces are counted) it will be truncated down to the number of characters in the data field. In order to correct this, you need to use the Edit Text tool. Shown below is an illustration of this process.

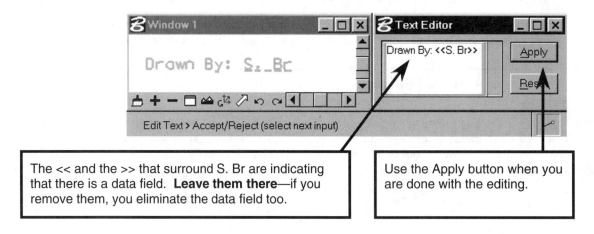

The << and the >> that surround S. Br are indicating that there is a data field. **Leave them there**—if you remove them, you eliminate the data field too.

Use the Apply button when you are done with the editing.

Shown here is the result of editing the text. The number of special characters in the data field has increased accordingly.

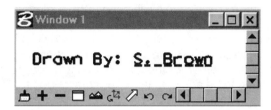

It is important to note that if you see the _ character, then it will show up in printing too. Be careful not to toggle OFF the display of data fields until just prior to printing. That way the empty data fields are visible, so they are easier to select.

 If you want to, you can skip using the Fill in Single Enter-Data Field tool and go directly to the Edit Text tool. It can be a bit tricky editing the data field information this way.

The use of data fields in a cell can really streamline the process of using standard notes on various plans. As a cell, a note can be included in numerous drawings and you don't have to worry about having inconsistent wording. The same note cell can be used with the specific information filled out on an individual basis. Cells are covered in detail in Chapter 12. Shown below is an example of the use of data fields in a standard note. Notice that there are 4 data fields interspersed thoughout the note. There can be more than one data field in a string of text.

```
Sta. _____
Build ___" x ___' Corr. Metal Pipe
with Concrete Flume Type ___.
```

Here the note cell has been placed two times in a drawing. The data fields have been filled in to reflect the specific information needed in the note. The display of the data fields has been toggled OFF, so you don't see the underscore.

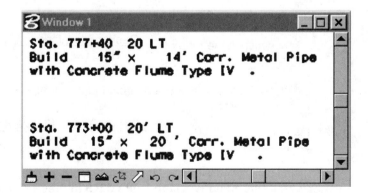

DETERMINING TEXT SIZE FOR PRINTING TO SCALE

When a design file is printed to a hard copy, often it will need to be scaled up or down in order to fit on the paper. A hard copy marked HALF SIZE or 1:2 will have its design file elements printed at half the size that they are in the design file. A hard copy marked DOUBLE SIZE or 2:1 will print its design file elements twice the size that they are in the design file. The text height on the hard copy (paper) is usually a standard size such as 1/8 of an inch. In order for the text on the hard copy to be a specific size then the <u>text height in the design file</u> will need to be adjusted larger or smaller depending on the scale of the hard copy.

Calculating the text height size to use in the design file requires knowing what the scale the final hard copy will be. This means we need to know the Plot Layout dialog box Scale to settings, especially the value of the Scale to [   ] **mu:su/IN.** setting. It may help to go back and review the material in Chapter 5 dealing with printing to scale. Just remember that the IN notation in that setting indicated working with a paper size in inches (so hard copy unit length was an inch). The ratio of the design file working units to hard copy unit length must equal the ratio of the text size in design file working units to the text size on the hard copy. With this in mind, by knowing the Scale to [   ] **mu:su/IN.** setting and the text size on the hard copy you can then determine the text height needed in the design file.

Trying to calculate this setting.

$$\frac{\text{Design File Working Units}}{\text{Hard Copy Unit Length}} = \frac{\text{Text Size in \textbf{Design File} Working Units}}{\text{Text Size on Hard Copy}}$$

This **entire** left side of the equation is the value of the Scale to [   ] **mu:su/IN.** setting in the Plot Layout dialog box. It was determined when printing to scale.

The Text Size on Hard Copy is specified.

For Example: The text size on the Full Size hard copy is specified to be 0.5 inches high. The design file working units are (inches:tenths:positional).

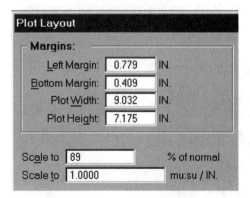

For a Full Size hard copy the Plot Layout Scale to [   ] **mu:su/IN.** is as illustrated at the left.

Scale to **1.00** mu:su **/** IN.

which is the ratio of: $\frac{\text{Design File Working Units (mu:su)}}{\text{Hard Copy Unit Length (IN.)}}$

so the left side of the equation equals **1.00**
therefore:

$$1.00 = \frac{\text{Text Size in Design File Working Units}}{0.5 \text{ inches}}$$

Doing the math means that the

Text Size in Design File Working Units = 1.00 x 0.5 inches
Text Size in Design File Working Units = 0.5 inches = 0.5 master units

Yes, that was very simple—but you get the idea.

Text   153

Another Example: Now let's look at a **Half Size** hard copy. **This time the text size on the hard copy is specified to be 0.125 inches.** The design file working units are (inches:tenths:positional).

For a Half Size hard copy the Plot Layout Scale to [ ] **mu:su/IN.** is as illustrated at the left.

Scale to **2.00** mu:su **/** IN.

which is the ratio of: $\dfrac{\text{Design File Working Units (mu:su)}}{\text{Hard Copy Unit Length (IN.)}}$

so the left side of the equation equals **2.00** therefore:

$$2.00 = \dfrac{\text{Text Size in Design File Working Units}}{0.125 \text{ inches}}$$

Doing the math means that the

Text Size in Design File Working Units = 2.00 x 0.125 inches
Text Size in Design File Working Units = 0.250 inches = 0.250 master units

Again that seems pretty simple. Remember that a Half Size hard copy means that the elements will be printed at half their actual size–then logically the text elements need to be **larger** in the design file than their size on the hard copy.

Another Example: In this last example, the design file working units are (**feet**:tenths:positional). The Unit Names settings have been changed from mu:su to FT:th so they will be labeled in the Scale to setting. **The scale stated on the hard copy is 1 INCH = 500 FEET.** The text height on the hard copy is specified to be **0.125** <u>inches</u>.

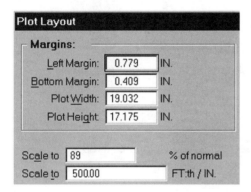

For a hard copy with SCALE 1 INCH = 500 FEET, the Plot Layout Scale to [ ] **FT:th/IN.** is as illustrated at the left. Remember that its value is the <u>inverse</u> of what the scale printed on the hard copy will state.

Scale to **500.00** FT:th **/** IN.

which is the ratio of: $\dfrac{\text{Design File Working Units (FT:th)}}{\text{Hard Copy Unit Length (IN.)}}$

We need to keep track of units since they don't cancel each other out as before. We will work only in master units.

So the left side of the equation equals $\dfrac{500 \text{ FEET}}{1 \text{ INCH}}$

and the entire equation is $\dfrac{500 \text{ FEET}}{1 \text{ INCH}} = \dfrac{\text{Text Size in Design File Working Units}}{0.125 \text{ inches}}$

Doing the math:

Text Size in Design File Working Units = $\dfrac{500 \text{ FEET}}{1 \text{ INCH}}$ X 0.125 inches

Text Size in Design File Working Units = **62.5 FEET** = 62.5 master units

 In summary, to determine the text height in **master units** of the design file, you can take the text height desired on the hard copy and multiply it by the value used in the Scale to [ ] **mu:su/IN.** field of the Plot Layout dialog box.

Chapter 7

QUESTIONS

① What is a Text Node and where do you find the setting to toggle on/off the display of Text Nodes?

② Name the two Fonts often used in dimensioning since they include engineering-related symbols.

③ Describe what locking the Height and Width of text means.

④ Name the Place Text Method that will result in each letter of the text string becoming its own element. For instance, DOG would actually be 3 separate elements.

⑤ Name the Place Text Method that would be best suited to label the contour line shown on the left so that it looks like the illustration on the right. What is a drawback to this method?

⑥ Determine the text size in the design file (in master units of inches) if the following is required:
- Text size on the hard copy is to be $3/8$ inch.
- Design file working units are 1:10:1000 (inches:tenths:positional).
- Scale to _____ mu:su/IN. setting value is 4.00

⑦ Determine the text size in the design file (in master units of inches) if the following is required:
- Text size on the hard copy is to be ¼ inch.
- Design file working units are 1:10:1000 (inches:tenths:positional).
- Hard copy scale is ¼" = 1'

⑧ Determine the text size in the design file (in master units of **feet**) if the following is required:
- Printed text size is to be 0.375 inches.
- Design file working units are 1:12:5000 (ft:inches:position).
- Hard copy scale is 1 INCH = 200 FEET

⑨ State what these two tools (A and B) are and what they do. Explain a typical situation that would require using both tools together and in what order.

A   B

⑩ Explain what an **enter data field** is and how you create one.

Text   155

EXERCISE 7-1   Store Layout Schematic

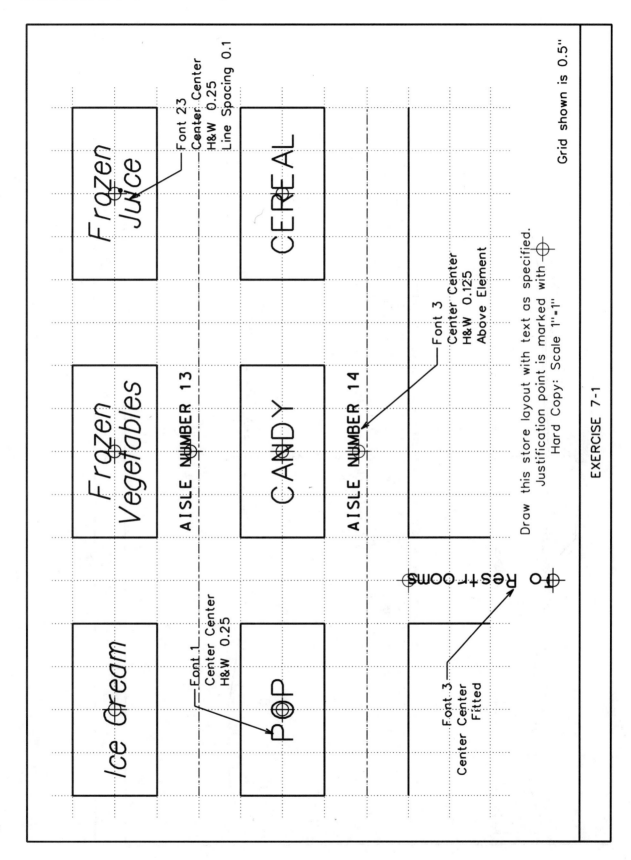

EXERCISE 7-2 Cross-Section

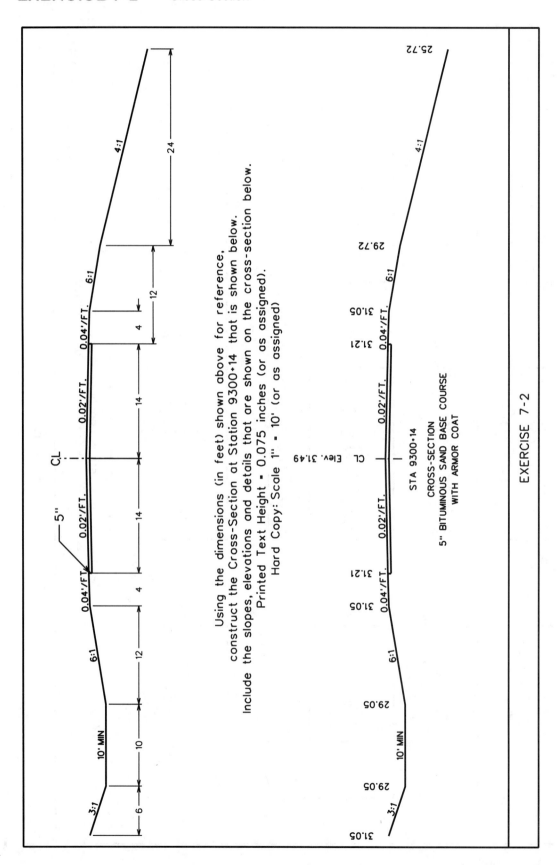

Text   157

EXERCISE 7-3   Golf Course Layout with Text

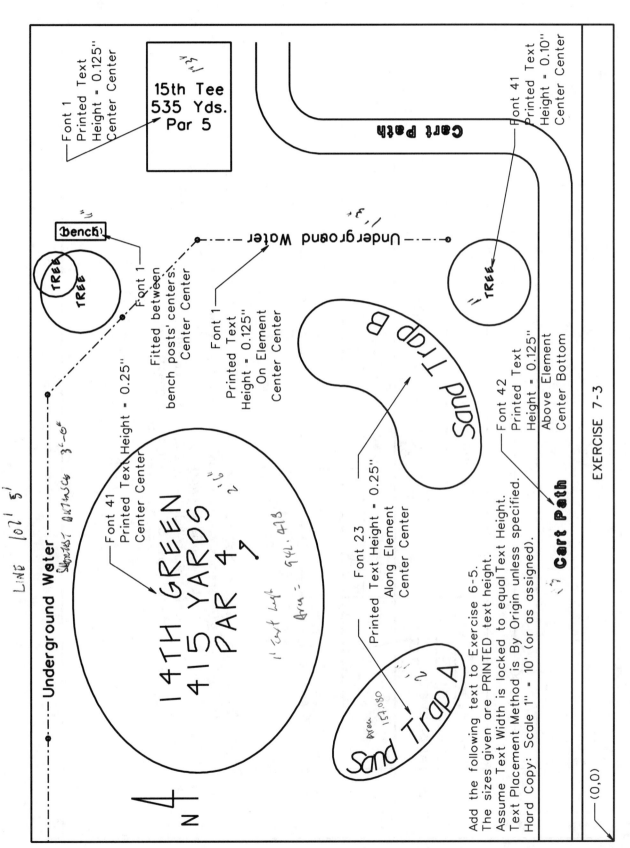

EXERCISE 7-3

EXERCISE 7-4   Title Block

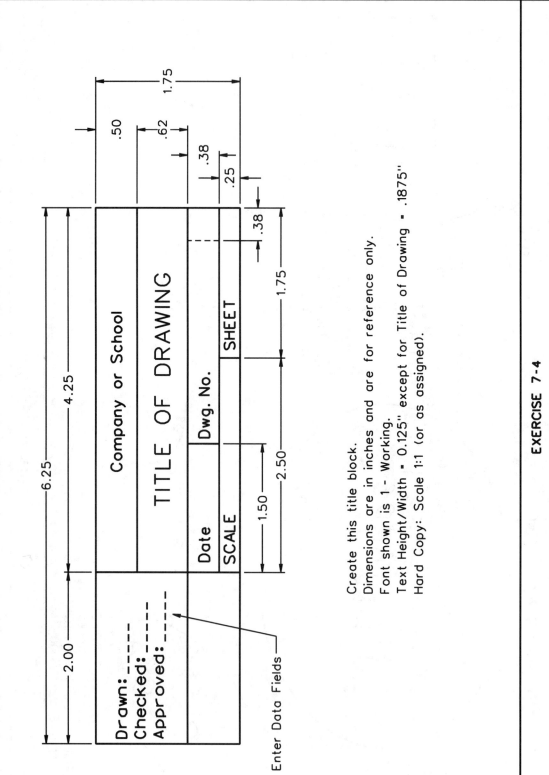

EXERCISE 7-4

Chapter 8: Measurements and Information

SUBJECTS COVERED:

- **Measure Tool Box**
- **Measure Distance**
- **Measure Radius**
- **Measure Angle**
- **Measure Length**
- **Measure Area**
- **Analyze Element**

When you're looking at an existing design file, it is sometimes necessary to get information about its elements. There are measurement tools that will determine the distance between two coordinate points, the length of a line, or the area inside a closed shape. There is also an Analyze Element tool that gives detailed information about graphical elements. The measurements and information that can be obtained are useful in design because they relate back to the real-life object that the elements represent. Once again, the importance of creating elements at the actual size and precise coordinates of the objects they represent is reinforced.

MEASURE TOOL BOX

The Measure tool box contains the tools necessary to make measurements based on coordinates that are selected (usually snapped to) as well as existing elements. The results of these measurements are displayed in the right side of the status bar. The default is to display the results as master units (mu) so that is what the examples in this chapter will use.

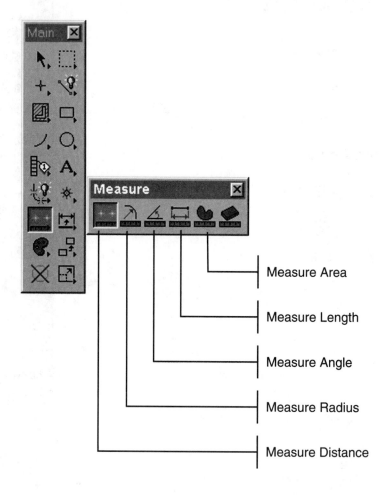

Measure Area

Measure Length

Measure Angle

Measure Radius

Measure Distance

Most of these tools use the same premise: you either 1) identify graphically with a data point the points to measure between or 2) identify the elements involved in the measurement. All of the tools involve **reading the prompts—to see what is required.**

MEASURE DISTANCE

The Measure Distance tool is the most basic one. It has different options available with the option button as shown. Between Points and Perpendicular are most commonly used. The others, Along element and Minimum Between, are for specialized cases.

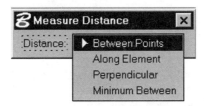

Between Points

This will measure the linear distance between any two coordinate points. The points can be on an element or anywhere in the graphic area. The start point will be entered, and then the distance to be determined needs to be defined. If you use the Reset button (right mouse) then the process starts over with needing a first point. If you don't use Reset, then you can continue on and define another distance that will be added to the first—you get a cumulative distance.

 Don't forget that if you use a tentative point to snap to an existing element, **it must be accepted with a data point.** Otherwise the status bar will continue to show the coordinates of the point—<u>not</u> the distance measured.

Here is a case of measuring between the center of the circle and the endpoint of the line:

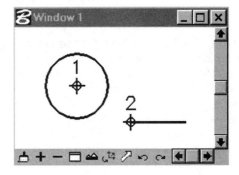

1) Specifies the first point by snapping to the center of the circle.
2) Defines the distance by snapping to the endpoint of the line.

The result in the status bar will read:

| Measure Distance Between Points > Define distance to measure | Dist = 0.6217mu |

 You can control the units displayed by adjusting the settings in the Coordinate Readout category of the Design File Setting dialog box, which is accessed by the main pull-down menu SETTINGS>DESIGN FILE.

The Measure Distance tool can also be used to measure distances along an existing element.

If points along Line A are measured:
1) Specifies the first point by snapping to the **left endpoint** of Line A.
2) Defines the distance by snapping to the **intersection** of Line A and Line B.

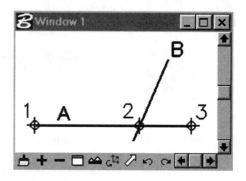

The result in the status bar will read:

| Measure Distance Between Points > Define distance to measure | Dist = 1.0000mu |

If you want only the distance between the intersection of Line A and Line B and the right endpoint of Line A, then you would need to use the Reset button to restart the distance measurement. But if you want to keep measuring to the right endpoint of Line A, then don't use the Reset button but keep going instead.

3) Defines the next distance by snapping to the right endpoint of Line A.

The status bar reading will show the **cumulative** distance.

| Measure Distance Between Points > Define distance to measure | Dist = 1.5000mu |

It is interesting to note that if you continue and snap back to Point 1, the distance will be 3.000 mu, even though the distance between Point 1 and itself is 0.

Perpendicular

This option will measure the Perpendicular Distance from an element to another graphical point. Using a circle and a line, the shortest distance (which would be perpendicular) between the line's endpoint and circle could be determined using this option of the Measure Distance tool.

In the illustration at the right, the first prompt was for a start point, which actually isn't a specific coordinate point but instead is a data point to select the circle element. The circle will highlight as shown and a dynamic line between the circle and the cursor will give you a visual indication of the Perpendicular Distance. It is waiting for an endpoint to be entered as indicated in the status bar.

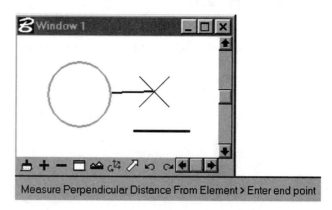

If the endpoint of the line is snapped to as shown below then the distance will be measured and displayed. Again, the option to continue specifying more points is available, or you can do a Reset to start over. The perpendicular line is just a visual aid; it is not an element and will not stay on the drawing.

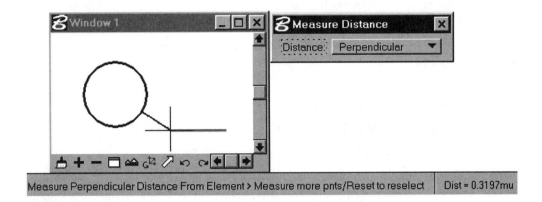

MEASURE RADIUS

 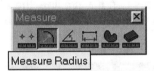

This is a fairly easy to use tool that can be used to measure the radius of a circle or the length of the major and minor ellipse axis. You just data point on the element (it will highlight) and then data point again to Accept, Initiate Measurement. If the element identified was a circle, the Radius measurement is then displayed in the status bar. Nope, there isn't a switch for diameter. If the element identified was an ellipse then the Major and Minor axis lengths are displayed in the status bar. The tool is that simple—no need to illustrate it.

MEASURE ANGLE

This tool doesn't have any settings. It measures the **smallest** angle between two linear elements. Even though the usual angle convention is to go in the counterclockwise direction, that doesn't matter here because it will measure only the smallest angle.

The sequence is as follows:
1) Identify first element.
2) Accept, Identify 2nd Element/Reject.
3) Accept, Initiate Measurement.

The smallest angle is measured in degrees and is shown in the status bar.

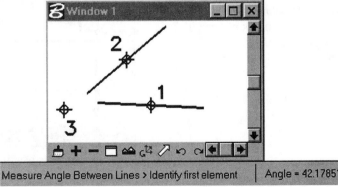

This is a good place to remind you to notice the prompts and the difference between a comma (,) and a slash (/). A slash divides the difference between what a left mouse button action will do and what a right mouse button action will do. The comma indicates that two things will happen with that single mouse action. For example, in the second step in using the Measure Angle tool, the prompt will read **Accept, Identify 2nd Element/Reject**. A left mouse button will do two things: both Accept the first identification and Identify 2nd Element. If you use the right mouse button, it will Reject the first identification.

MEASURE LENGTH

With the Measure Length tool, you only have to data point on the element and then Accept, Initiate Measurement with another data point. Specifying coordinates is not necessary—it uses the actual element or shape.

Single Element

The length of an arc or a line will be displayed in the status bar. If it is a line, the angle from a 0° orientation will also be displayed. Shown below is a line whose length was measured and the result that the status bar displays.

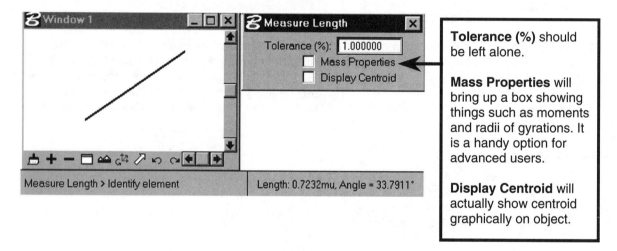

Tolerance (%) should be left alone.

Mass Properties will bring up a box showing things such as moments and radii of gyrations. It is a handy option for advanced users.

Display Centroid will actually show centroid graphically on object.

Shape

The Measure Length tool can also be used to measure the perimeter of a closed shape such as the block shown. It also will give the circumference of a circle.

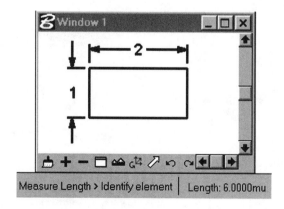

MEASURE AREA

 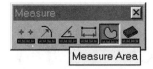

The Measure Area tool uses a shape or a closed element. It measures the area within the shape. With the Element Method, you identify the element with a data point and then Accept/Initiate Measurement with another data point. However, there are methods other than Element. These other methods can be used when two or more shapes or closed elements are involved. The Boolean operations of Union, Differencel, and Intersection allow you to specify areas of overlapping shapes. The Fence Method automatically measures the area contained within an existing fence. The Flood and Points methods can be used in special cases.

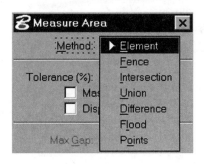

Element Method

Here is a simple circle whose area was measured with the Element Method. The status bar displays the Area (A) as square master units (SQ mu). The perimeter of the circle (circumference) is also stated as P=.

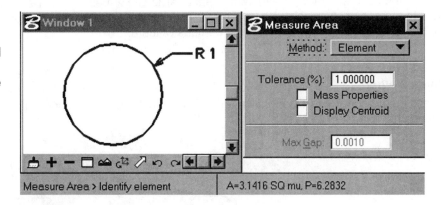

Methods Involving Boolean Operations

When using a method that involves a Boolean operation, you will need to select at least two elements that are closed shapes. As each closed element is identified, it will highlight. When done identifying elements, data point in a blank part of the view window and a display of the area that will be measured will appear highlighted. The prompt of Reset to complete appears; this means that the Reset button will complete the area measurement and display the results in the status bar.

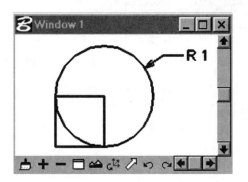

For illustrating the three Boolean operation methods, let's use this simple circle of radius 1 and a 1 x 1 square block with its corner at the circle's center. That way you can check the numerical area and perimeter values if you want to!

Intersection Method

Here is the middle of measuring the intersection of the two elements. The area that will be measured is shown highlighted. A left mouse button will identify additional elements and a right mouse button will complete the measurement.

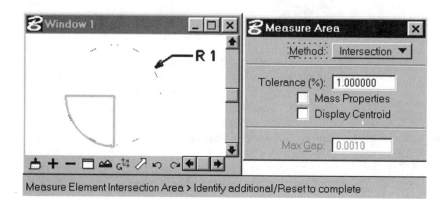

Using Reset, the status bar will display: A=0.7854 SQ mu, P=3.5708

Once the measurement has been made, that quarter circle area will not be displayed. Instead it just goes back to the original two elements unharmed. This applies for all of the Boolean methods. They don't modify the elements, they just measure portions of them.

Union Method

This is the area that would be highlighted when using the Union Method on the same circle and square closed shape block.

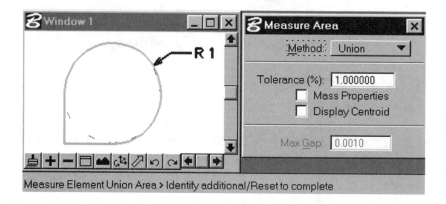

The measurements that result would appear in the status bar as: A=3.3562 SQ mu, P=6.7124

Difference Method

With this method, the order in which you select elements is very important. Mathematically it could be stated as Element 1—Element 2. Shown below are the two different results you get, depending on which element is selected first (considered Element 1).

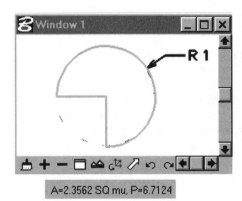
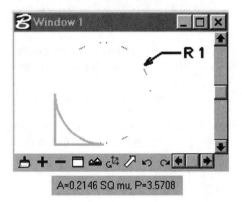

Circle first (Element 1), then Square (Element 2)   Square first (Element 1), then Circle (Element 2)

Flood Method

The Flood Method requires a data point inside the area you want to measure. It doesn't have to involve closed elements, **but there has to be some sort of defining boundary**. Illustrating the Flood Method of measuring an area is a roadway with an intersecting driveway shown below. The data point is inside the area that we want to measure—the defining boundary highlights as shown.

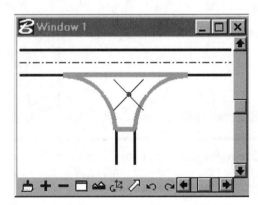

The measurements that result appear in the status bar as: A=1973.0092 SQ ft, P=293.0796

ANALYZE ELEMENT

Detailed information about a graphic element can be obtained by using the Analyze Element tool. It is found on the Primary Tools tool box (which is usually docked). The tool looks like an i for information (not an A for Analyze—go figure). Click on the tool and then you'll be asked to identify what element you want information about. Data point to select it, and do another data point to accept the highlighted element. A large dialog box will appear showing all sorts of information about the element—from standard attributes to a lot of numbers dealing with coordinates. Shown below is the information about a circle (it says ellipse, but a circle is a simple ellipse—don't get confused).

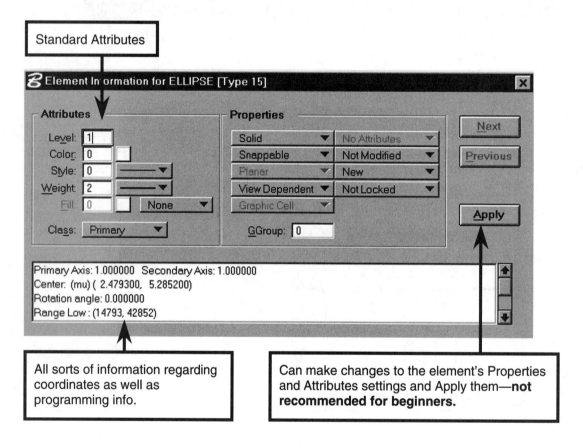

Standard Attributes

All sorts of information regarding coordinates as well as programming info.

Can make changes to the element's Properties and Attributes settings and Apply them—**not recommended for beginners.**

All this detailed information is impressive, but normally you will use other tools to modify the element. However, it is good to know that this information is available.

QUESTIONS

① Where do you find the <u>results</u> of most measuring tools? *Measure Tool Box.*

② Name the 4 different Distance options of the Measure Distance tool. *Between points / Along Element / minimum between*

③ When using the Measure Angle tool and the two lines shown here, what angle would actually be measured, A or B? *measures smallest angle.*

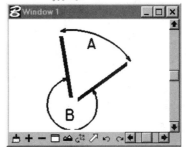

④ Name a Measure tool that will give you the circumference of a circle. *measure Area*

⑤ For the two closed shape elements shown, sketch the area that would be measured with the Measure Area tool and the different Boolean operations stated:

a) Intersection
b) Union *both*
c) Difference (rectangle selected first and then the triangle is selected)
d) Difference (triangle selected first and then the rectangle is selected)

⑥ What Method of the Measure Area tool would give you the area **between** the arc and the line shown below? *intersection*

⑦ When angles are measured, are they stated in degrees or radians?

⑧ For this tool shown, answer the following questions:
a) What tool box is it found in? *Primary tools*
b) What is it called by its tool tip? *Analyze*
c) What does it do?

let give information ie tools on element.

EXERCISE 8-1   Circular Plate with Hexagons

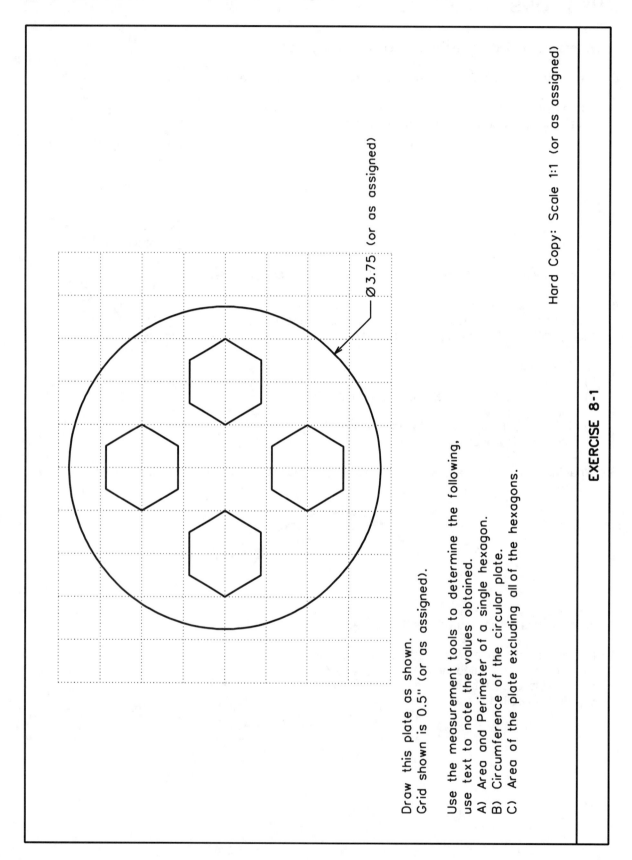

EXERCISE 8-1

EXERCISE 8-2   Golf Course Measurements

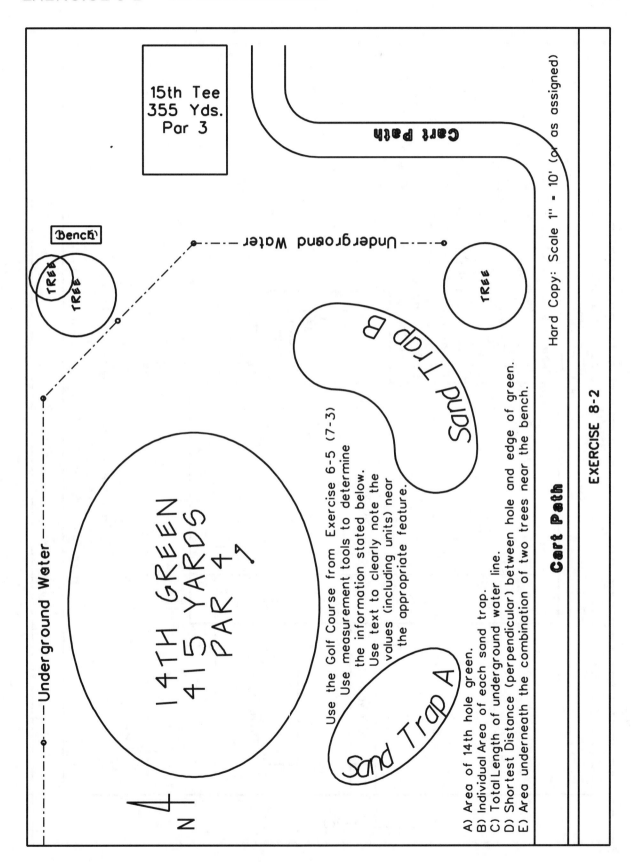

EXERCISE 8-3   Residential Lot Layout

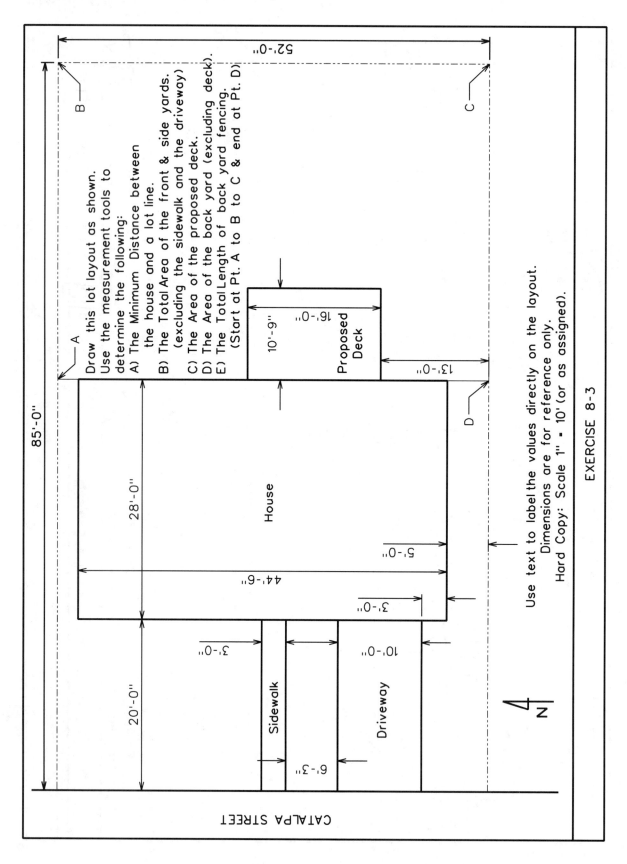

EXERCISE 8-3

Chapter 9: Changes and Modifications

SUBJECTS COVERED:

- Change Attributes Tool Box
- Change Element Attributes
- Match Element Attributes
- SmartMatch
- Modify Tool Box
- Modify Element
- Partial Delete
- Extend Line
- Extend Element<u>s</u> to Intersection
- Extend Element to Intersection
- Trim Elements
- Intellitrim
- Construct Circular Fillet
- Construct Chamfer

Changing an element refers to changing the general attributes and properties of the element, but not its graphical characteristics such as coordinates and size. Modification of an element generally refers to leaving the attributes but adjusting the size and coordinates of the element itself. Knowing how to make changes or modifications to existing elements is necessary for different reasons. The most obvious one is that everyone makes mistakes. Instead of deleting the element and starting over (which is good practice for beginners but tedious), it is often easier to just edit it. Another less obvious reason for mastering modification tools is that they may actually speed up the drawing process. Often you can place elements with the preplanned intention of modifying them to suit your needs. For example, it is easier to draw a rounded corner by first drawing two sharp intersecting lines and then using the modification tool that constructs circular fillets, rather than jumping to the final product and drawing two shorter lines with an arc placed between them. This ability to conceive of less time-consuming methods comes only with experience. There may be more than one sequence or technique to accomplish the same final product—and after you've done it the hard way a few times, you'll start thinking about shortcuts! The more tools that you know how to use—the better you get at preplanning.

CHANGE ATTRIBUTES TOOL BOX

Using the Element Selection tool for changing an existing element's attributes was discussed way back in Chapter 4 dealing with drawing aids. That method works, but there is a better way and that is with the Change Elements Attributes tool. It is found in the Change Attributes tool box. Another tool available in the Change Attributes tool box is Match Elements Attributes, which is often used in conjunction with changing an element's attributes. There is also a very simple SmartMatch tool. Other tools are found there also—the ones dealing with Area and Fill attributes will be covered in Chapter 13, which discusses patterning concepts.

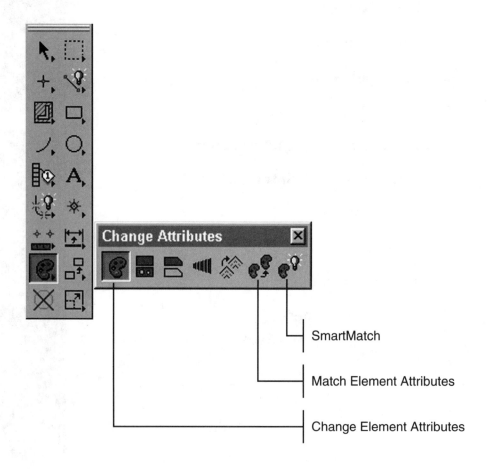

CHANGE ELEMENT ATTRIBUTES

When the Change Element Attributes tool is selected, its dialog box appears as illustrated here on the left. The Change Element Attributes dialog box has check boxes for the standard attributes that can be changed. If the setting is checked ON then an option button appears. The option selected **will become the Active setting** for that attribute and will also be applied to the elements, which are then identified. The elements to be changed can be identified by a data point, with PowerSelector, or by an existing fence when the Use Fence setting is checked ON. If there isn't an existing fence, the Use Fence setting is not available (grayed out). The identified element(s) will highlight, and a data point is needed to accept the change.

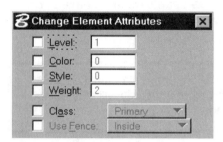

When identifying elements using a data point, remember that after the first one, you can keep selecting so that the next data point does two things: 1) accepts the highlighted selection and 2) identifies the next element. When you've identified everything, data point in an empty area of the view to accept the last highlighted element.

For Example: Let's look at using Change Element Attributes on a circle, block, and line that all started with a Line Style of 0. Illustrated below is the middle of a Change Element Attributes sequence. The Style setting was checked ON and changed to 3. The first element (the circle) was identified with a data point. The next data point was on the block, it accepted the change to the circle and highlighted the block as shown. The cursor indicates that a selection can be made.

Here are 3 **different** actions possible for the next step:
    1) A data point on the line will change the block's Line Style and select the line.
or
    2) A data point in a blank area will change the block's line style and reset the command.
or
    3) A right mouse button will reject the block selection and reset the command.

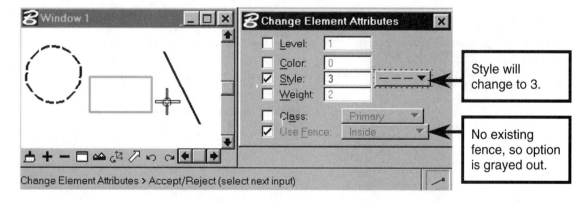

The Active Attribute settings in the Primary Tools tool box will reflect the settings in this dialog box. Any new elements placed will then have these attributes.

MATCH ELEMENT ATTRIBUTES

This tool allows you to automatically change the Active Attribute **settings** (checked ON) to match those of an existing element. **It does not automatically change the attributes of an existing element.** You would have to go and use the Change Element Attributes tool to do that.

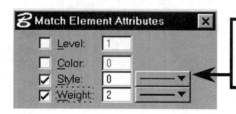

Only the Style and Weight of the element identified will be matched. The other attributes will stay at their current active settings.

SMARTMATCH

The SmartMatch tool will actually bring up the Match All Element Settings dialog box illustrated below. This tool quite completely matches all the attributes of the elements and automatically changes the settings associated with that element. This works on Level, Color, Style, and Weight but also deals with things such as Fill Type and Area settings. There isn't a Use Fence option because it can only match one element's settings, otherwise there could be a conflict.

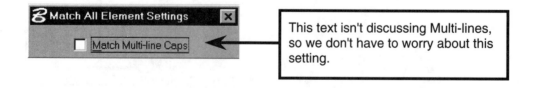

This text isn't discussing Multi-lines, so we don't have to worry about this setting.

MODIFY TOOL BOX

This tool box contains some really good stuff—from a general Modify Element tool to some more specific tools such as Extend Elements to Intersection. These modification tools deal with changing the actual size and shape (and therefore the coordinates) of an element. Some, like Construct Circular Fillet and Construct Chamfer, also construct a new element in addition to modifying the existing elements.

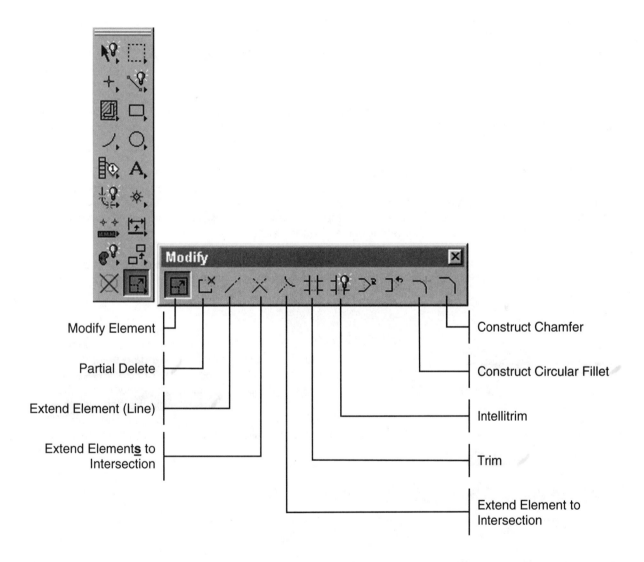

MODIFY ELEMENT

The Modify Element tool lets you graphically select an element with a data point and then modify it into a different size and/or shape graphically (by moving the cursor) or by key-in. This is often referred to as "dragging" the element into its new geometry. It isn't a true Windows drag, which (if you recall the discussion of the Element Selection tool and handles) involves keeping the left mouse button depressed. In this situation (sometimes referred to as a dynamic drag in MicroStation), when you data point you don't have to keep the left mouse button depressed while moving the cursor. You just data point (making sure to release the left mouse button), move the cursor, and then data point at the new location (or type a key-in if you prefer).

By using the Modify Element tool, you can move an endpoint of a line, modify the radius of a circle, move the vertex of a shape, or move a line segment of a polygon. **In all of these cases, when identifying an element for modification, it matters <u>where</u> you data point on the element to select it.** With the Modify Element tool, if the data point is near an endpoint or a vertex, the cursor will jump to this point on the element (without using a tentative point to snap there). If there isn't an endpoint or a vertex nearby then the element will be attached to the cursor at the point of selection. Then when you move the cursor, the element is modified accordingly. We'll refer to this concept of the element being attached to the cursor at the point where you did a data point to select it as "grabbing" on to the element. It is a simple but effective description that will also be used when we discuss manipulation of elements.

The Modify Element tool has different settings depending on what type of element you are modifying. Key-ins can also be used to make the desired modifications. Even though this is the most generic modification tool, it is not necessarily the best choice. More likely one of the other Modify tools will be easier to use to get consistent results. Later on in the text there are certain situations, like closed shapes and dimensions, where the Modify Element tool may be the best choice.

Let's look as some examples of modifying different elements. In these examples, the element before modification will be a lighter gray, and the black one with the cursor on it will represent the modification in progress.

Modifying a Block

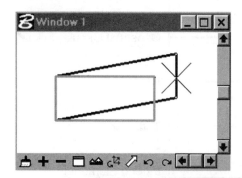

In the illustration on the left, a block has been selected (lighter gray) to be modified by a data point on its right side (away from a vertex). The block's line segment has been grabbed (the element is attached to the cursor at the point where you did the data point), and it is being dynamically dragged to a new position. When the cursor moves, the left side of the block remains fixed, but the rectangular block modifies its shape into a rhombus.

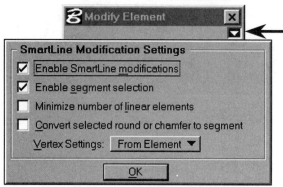

The Modify Element dialog box is skimpy, but if you click on the ▼ button, the SmartLine Modification Settings will show up.
- These are the default settings.
- The examples and discussion in this text will assume they are left set as shown.

Now if the data point to select the block had been closer to a vertex, the Modify Element tool would have behaved differently. The cursor would have jumped to the vertex and the Modify Element dialog box would appear as shown below.

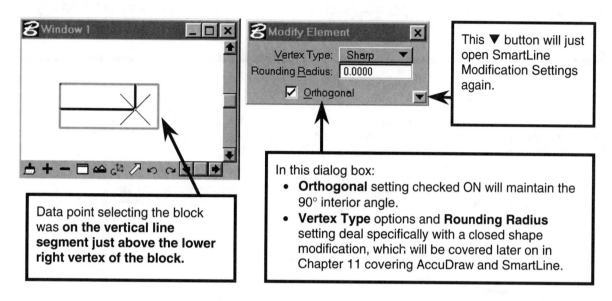

Data point selecting the block was **on the vertical line segment just above the lower right vertex of the block.**

This ▼ button will just open SmartLine Modification Settings again.

In this dialog box:
- **Orthogonal** setting checked ON will maintain the 90° interior angle.
- **Vertex Type** options and **Rounding Radius** setting deal specifically with a closed shape modification, which will be covered later on in Chapter 11 covering AccuDraw and SmartLine.

The illustration at the right shows what happens if the check box for Orthogonal is unchecked (OFF). The block's shape can be modified drastically by changing the coordinate of the vertex.

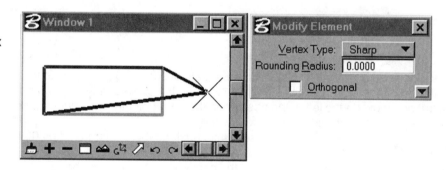

It is extremely important to keep in mind that with the Modify Element tool, there isn't necessarily a tentative point that actually snaps to the vertex. If the data point is close to the vertex, the cursor just jumps to the vertex and allows its modification. This causes problems if you try to use relative key-ins to specify the new coordinates of the vertex. The key-in coordinates would be relative to the **original data point** that selected the block element—**not** to the original coordinates of the vertex. To illustrate this, look at the following result of a relative rectangular key-in of ⌨DL=1,0 to complete this modification of the block.

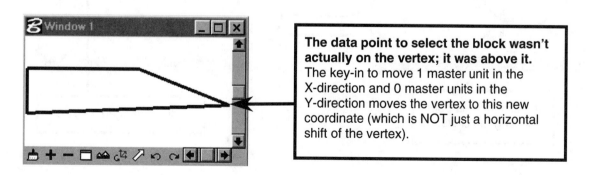

The data point to select the block wasn't actually on the vertex; it was above it. The key-in to move 1 master unit in the X-direction and 0 master units in the Y-direction moves the vertex to this new coordinate (which is NOT just a horizontal shift of the vertex).

When making precise modifications, play it safe and use a tentative point to actually snap to precise coordinates of the element.

Modifying a Line or a Circle

Lines and circles are modified fairly easily. A line's endpoint can be moved anywhere. A circle's size is moved, but its center remains anchored. In both cases, there are no additional settings in the Modify Element dialog box.

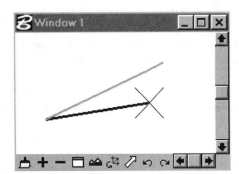

Let's look at an illustration that shows the benefit of snapping to a specific location on the element when identifying it for modification. We want to increase the circle's radius by 0.5 master units. Shown in the illustration below on the left is the position of the data point that is selecting the circle for modification. The modification was specified by using a key-in of ⌨DL=.5,0. The illustration on the right shows that the radius was **not** increased by 0.5 master units (instead it was increased by 0.281). This is because the relative key-in was relative to the data point location when selecting the circle. Obviously, grabbing the circle at a "random" spot makes it hard to specify a precise radius change.

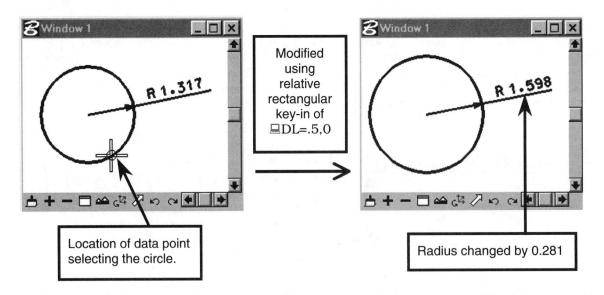

For a precise increase in the size of the radius, you should snap to a keypoint on the circle. Shown below on the left is the original circle, but this time the keypoint at the right quadrant of the circle was snapped to using a tentative point. The tentative point was accepted with a data point, and then the relative rectangular key-in of ⌨DL=.5,0 was used to specify the modification. As you can see at the right, the radius was increased from 1.317 to 1.817, a precise increase of 0.5 master units.

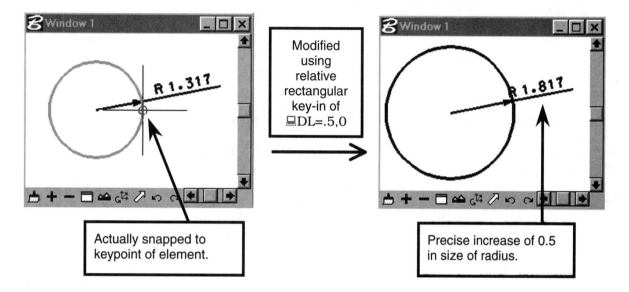

Modifying Arcs

When you're using the Modify Element tool on an arc there are three different Method options. These options show up in the Modify Element dialog box only after it knows you're modifying an arc. They are Angle, Radius about Center, and Radius preserve Ends. The different methods of modifying an arc are illustrated below.

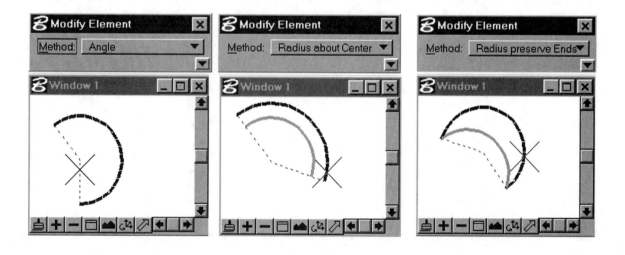

PARTIAL DELETE

This tool does just what it says—it deletes a portion of the element. The sequence of steps for selecting the modifications will depend (again) on the element: Read the prompts!

- For open shapes (like lines), first point and then endpoint.
- For closed shapes (like circles), first point, **then direction to go in**, and then endpoint.

Shown below are the steps taken to delete part of a circle. Since it is a closed shape, there are three selections done.

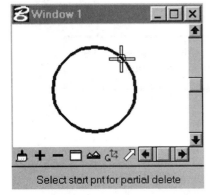 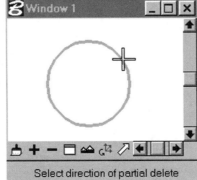 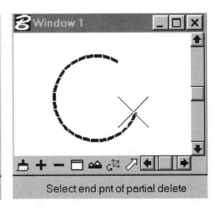

You can use other elements to snap to if necessary. However, if you are trying to delete a portion of an element that lies between two other elements (like shown below), it may be handier to use the Trim modification tool. As mentioned before, figuring out the best tool to use comes with experience. The end result can be accomplished in numerous ways—some techniques are better than others. Draw and learn.

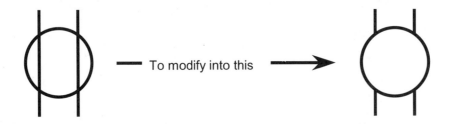

 Be careful of "leftovers" when using Partial Delete. If you try to use Partial Delete to remove the end portion of an element, often a very tiny piece will remain. To avoid this, snap to the end of the element or use a different modification tool.

EXTEND LINE

 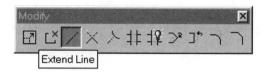

This tool is used to lengthen **or shorten** a line. When you use the Extend Line tool, the orientation of the line will stay the same. A data point can be used to graphically specify the distance, or a value (+ or −) can be entered in the dialog box. The extension (or retraction) will occur at the end of the line that is closest to the selection data point; the coordinate of the other endpoint of the line remains the same.

Shown below is one situation that lends itself to using Extend Line. The vertical hidden line needs to be retracted at the top in order to leave the gap necessary to distinguish itself from the visible line. By using Extend Line you can modify the line shorter so there is a gap.

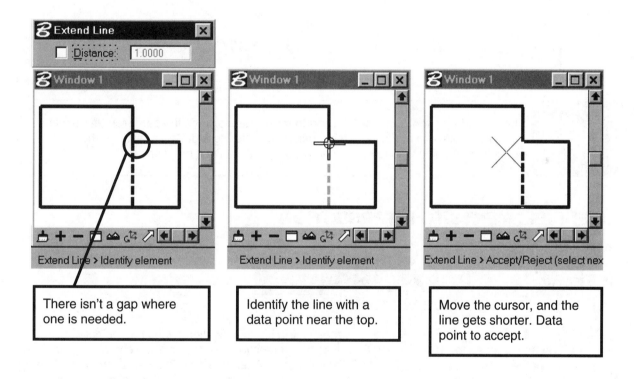

There isn't a gap where one is needed.

Identify the line with a data point near the top.

Move the cursor, and the line gets shorter. Data point to accept.

The Extend Line tool is also useful if the line needs to be adjusted by a known distance. In this case Distance is checked ON and a negative value is entered. The line is identified near its upper right endpoint (snapping is not necessary—just a data point will do) and then it highlights showing the shorter line. The modification is accepted with a data point (off the element or it will take off another 0.25). The result is a line cut off by a distance of 0.25 master units from the selected endpoint.

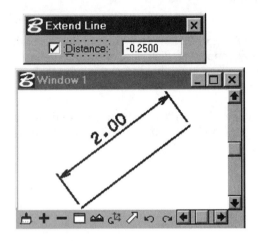
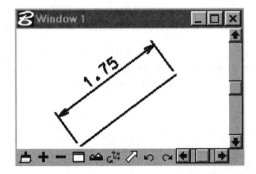

> The Extend Line tool can also be used to make parallel lines the same length. Identify the line you want to modify and then instead of entering a distance, snap to the endpoint of the line whose length you want to match. Try it on your own.

EXTEND ELEMENTS TO INTERSECTION

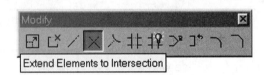

Now with this tool, the tool tip says **Elements** and means it. This modification tool is not restricted to lines; arcs can be extended too. This Extend Element**s** to Intersection tool requires 2 elements and they will **both** be lengthened or shortened to their intersection point. Again the orientation of each of the lines will not change. You identify the first element and then the second element. Then it will show a preview of the result and you can accept the modification with a data point. This tool can be used to actually extend (lengthen) the elements or the elements can be shortened so that they stop at the point of intersection. If you are using the Extend Elements to Intersection tool in a situation that causes an element to shorten, then **where** you data point to select the element makes a big difference. The side of the element that you data point on will remain, and the other portion of the element will be removed.

Example of Extending Linear Elements

Let's use the two lines shown at the right to illustrate the Extend Elements to Intersection tool.

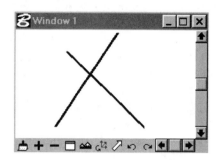

The illustration below on the left shows the order and position of the data points that select the two lines. The result of the modification is shown on the right.

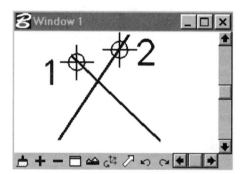 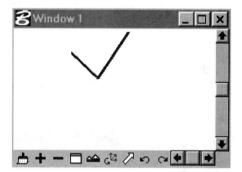

Use the same two original lines, but this time let's data point in a different position on the lines. The two data points shown on the left will result in the modification illustrated at the right.

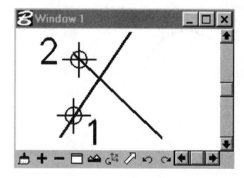 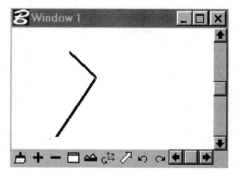

Example of Extend Elements to Intersection on Arc and Line Elements

Here are the before and after views of using Extend Elements to Intersection on an arc and a line. The selection was done at 1 and then at 2. If you select two elements and they don't modify, there probably isn't an intersection point. For instance, the arc could be so small that it would never intersect the line.

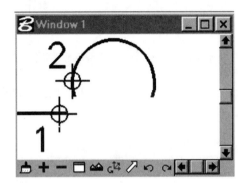 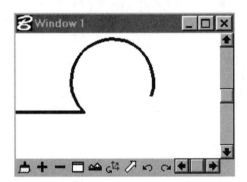

EXTEND ELEMENT TO INTERSECTION

 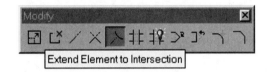

The Extend Element to Intersection tool differs from the previous tool because only **one** element is actually modified. Don't get it confused with the previous tool in which "Elements" is plural. With Extend Element to Intersection, the first element is selected for extension (the one to modify). The other element is selected to be the intersection (it won't be modified). As usual, read the prompts.

Shown below are the same line and arc as in the previous example, but this time only the first element (the line) will be extended to where it would intersect with the arc. The arc is not modified.

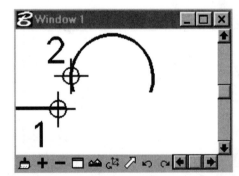 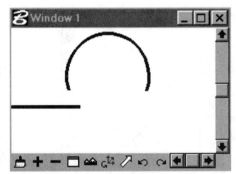

As with the other Extend tools, it will work to retract (or shorten) an element too. The side of the element to extend that you data point on will remain, and the other portion of the element will be removed. You can try this out on your own.

TRIM ELEMENTS

 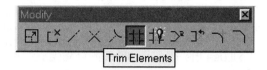

The Trim tool allows you to remove part of an element, by using another element to "cut" it. First, you select the element that will serve as the cutting edge, then you data point on the element that is to be trimmed. With this Trim tool, the portion that you data point on will be the part that is removed. This differs from the previous modification tools.

First you will be prompted to Select Cutting Element.

The element selected will be highlighted (the circle). You will then be prompted for the element to trim
Notice that you can pick only one cutting element.

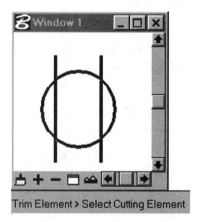 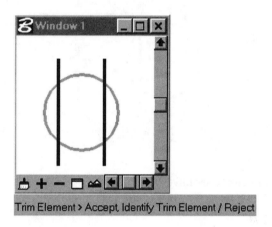

As shown below, the left line inside the circle was identified with a data point. It highlights and you can identify another element to trim.

To complete the task, the right line has also been identified for trimming with a data point on the portion of the line that is inside the circle. Another data point to accept, and you're finished.

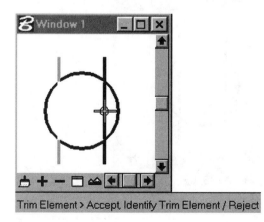 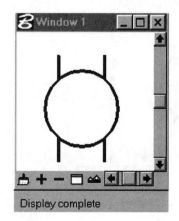

Noun/Verb or Verb/Noun Order

This is a good time to discuss the order of things. So far in these examples, the action is done first (modification tool activated), then the elements are selected. This can be thought of as verb/noun order. However, you can also do things in a noun/verb order, in which case the elements can be selected first and then the action. When using the noun/verb method, a tool may behave differently. This doesn't just apply to modification tools, it also works with other tools such as the manipulation tools of copy, array, and mirror (discussed in Chapter 10.)   We'll use the Trim modification tool as a prime example.

The effect of order is easy to see when using the Trim modification tool in a noun/verb order as shown below. First the elements will be selected with the Element Selection tool (remember, it's the arrow in the uppermost corner of the Main tool box). You can also use PowerSelector, but since we're only looking at two elements, the Element Selection tool is just fine. Those elements will then be used as the cutting edges because that's the first thing that's required for the Trim tool! When the Trim tool is then activated, it will jump right into prompting you for the element to trim (it already knows what the cutting edges are).

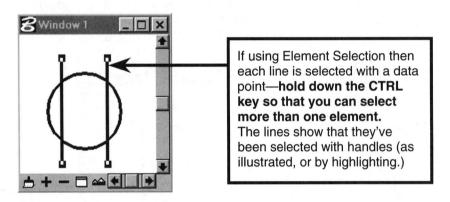

If using Element Selection then each line is selected with a data point—**hold down the CTRL key so that you can select more than one element.** The lines show that they've been selected with handles (as illustrated, or by highlighting.)

Now that the elements have been selected—it's time for the action.

The Trim modification tool is clicked and now the prompt for the Trim tool **automatically goes to Select Element to Trim** as shown below.
(Don't worry about the handles on the lines disappearing.)

Shown here is the result of identifying the circle as the Element to Trim. The portion between the two lines is trimmed away.

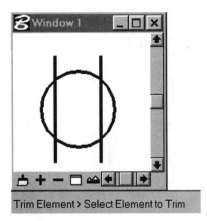

INTELLITRIM

The IntelliTrim® tool makes the trimming <u>and</u> extension of multiple elements easier **once you get the hang of it.** It looks like the regular Trim tool but has a light bulb to show that it's "smart"! The idea is to use both the PowerSelector and IntelliTrim tools together to control the Trim, Cut, and Extend Operations. If there is a selection set of elements active when the IntelliTrim tool is used, then you will get more options in the **Advanced** Mode of the IntelliTrim tool, such as illustrated below on the left. If there aren't any elements selected, then it defaults to the **Quick** Mode setting, as shown below on the right.

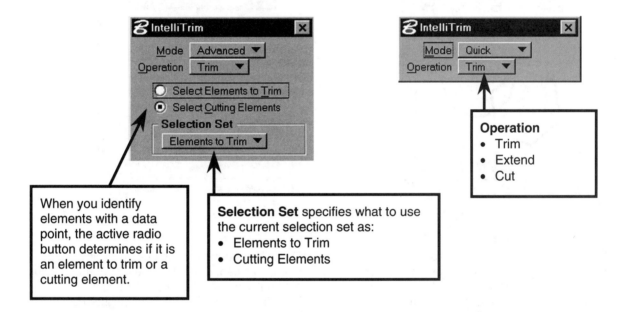

When you identify elements with a data point, the active radio button determines if it is an element to trim or a cutting element.

Selection Set specifies what to use the current selection set as:
- Elements to Trim
- Cutting Elements

Operation
- Trim
- Extend
- Cut

Quick Mode

The Quick Mode has three different Operation options: Trim, Extend, or Cut. The use of IntelliTrim in the Quick Mode to trim **or extend** an element gives it some advantage over the regular Trim tool. The options of Trim and Extend are carried out in the same manner; you select an element to be your "boundary" and then you specify an imaginary line to identify what to trim off up to the boundary (or extend up to the boundary.) The Cut option is carried out a bit differently and the end result is not very obvious at first. If you're like me, you may wonder what the difference is between trimming and cutting—they seem like the same thing! Trimming actually deletes part of the element while cutting just cuts the element up into smaller elements and leaves the pieces there. Let's look at the Trim Operation in Quick Mode and then consider the Cut Operation in Quick Mode.

Trim Operation

When you start the IntelliTrim tool and there aren't any elements selected, you will see that it defaults to the Quick Mode and the Trim Operation as illustrated at the right.

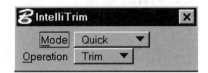

Let's look at using the circle element to trim off the polygon and the line. The status bar prompts to Identify element. This would be the "cutting" boundary.

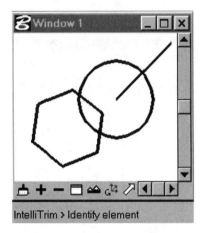

The circle was identified with a data point and it highlights as dashed because it is the cutting element. The status bar prompts to Enter start point of the line. This line will identify which elements will be trimmed.

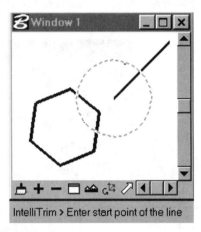

The first point of the line was specified just inside the polygon. You can see the darker dashed line—this is indicating the imaginary line's location. The status bar prompts to Enter end point of the line. This will be done with a data point located such that the imaginary line will pass over the existing line element as illustrated.

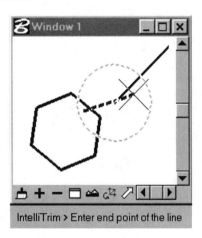

Once the endpoint of the imaginary line has been specified, the polygon element and the line element are trimmed off up to the circle boundary. The circle stays highlighted because you could identify additional elements to trim by specifying another imaginary line. See how the status prompts for Enter start point of the line again.

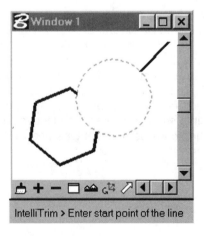

Cut Operation

The Cut Operation in Quick Mode doesn't use a cutting element; instead you specify an imaginary line at which any intersecting elements will be cut. To illustrate this let's look at 4 block elements that need to be cut.

The status bar prompts to Enter start point of the line. This will be done by snapping to the upper left corner of the largest block as illustrated below.

Once the tentative snap has been accepted, the imaginary line will show up as dashed. The status bar still prompts to Enter start point of the line—but think of it as the endpoint of the line! It will be specified with a data point at the location shown.

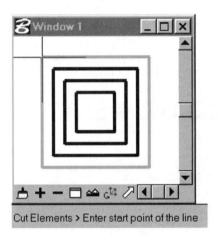
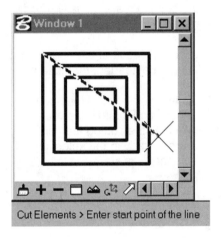

Once the data point is done, the imaginary line is no longer displayed. It also looks as if nothing changed! Don't be fooled. It really did cut all of the blocks (at two points since the line intersected them at two different spots).

If you use the Delete Element tool and data point on one of the corners of the former block, you will see that it was actually cut!

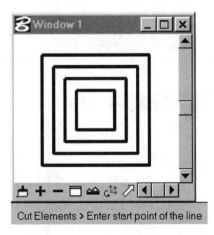
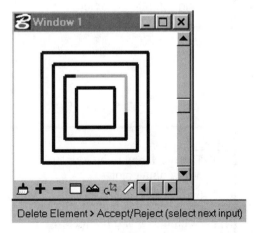

Advanced Mode

This tool has so many different combinations of settings that it is tough to illustrate all of them. The Advanced Mode seems more useful than the Quick Mode. We'll look at the different scenarios:
1) Trim Operation when there is an existing selection set and 2) Extend Operation when there isn't an existing selection set.

Here is a brief rundown of concepts involved in using IntelliTrim. With the Advanced Mode, you can only trim or extend, you aren't allowed to cut.

- If the Selection Set is specifying Cutting Elements, the elements will be highlighted **and dashed**. These elements will not be modified. You will be required to identify the elements to be trimmed.

- If Selection Set is specifying Elements to Trim, the elements will be highlighted. These elements will be the ones that are actually modified (made longer if extending and shorter if trimming). You will be required to identify the elements to act as cutting elements.

- A preview of the result of the operation will be displayed. A Reset will accept the operation. If modifications are needed, then use a data point to switch the element's behavior.

To illustrate the Trim Operation in Advanced Mode, let's trim off five horizontal lines in one shot by using a block as the cutting element. The horizontal lines were already in the selection set when the IntelliTrim tool was started.

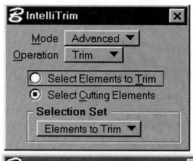

Step 1:
- The **Selection Set** option is set to Elements to Trim.
- The horizontal lines are highlighted (not dashed because they aren't cutting elements).
- You are being prompted to Identify cutting elements.

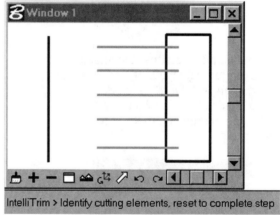

Step 2:
- The block has been identified with a data point.
- It highlights and is dashed because it is a cutting element.
- Potential cutting points are indicated with small dots.

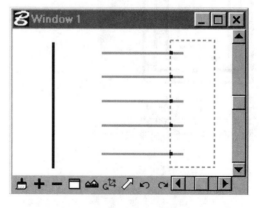

We don't have any other cutting elements to identify, so a Reset will be used to complete the step. This will result in a preview of the trimming operation as shown below. If you don't like it, you can modify it.

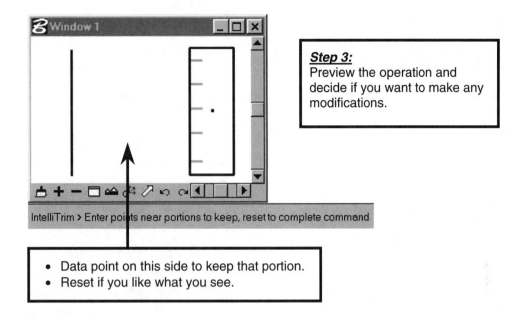

Step 3:
Preview the operation and decide if you want to make any modifications.

- Data point on this side to keep that portion.
- Reset if you like what you see.

Assume we didn't like what we see. Instead we wanted to keep the other portion, so a data point was done at the left side of the block. The preview will now show that portion of the lines as illustrated below.

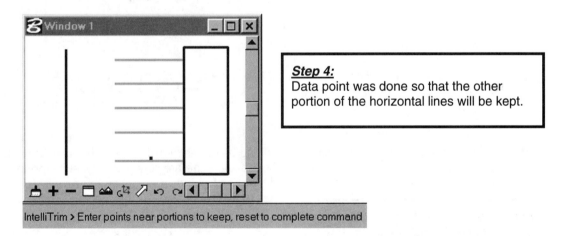

Step 4:
Data point was done so that the other portion of the horizontal lines will be kept.

Step 5:
The Preview was acceptable, so a Reset is used to complete the command. The trimming of the horizontal lines is completed.

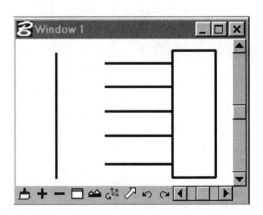

Now let's see how to use the Extend Operation by extending those same horizontal lines to the left using the vertical line as the element to extend to. The tool is started without any elements selected.

Step 1:
- Operation set to **Extend**.
- Identified the horizontal lines as the Elements to Trim (higlighted but no dash) by a data point on each. **This actually makes them the elements that will extend.**
- Reset needed to complete step.

Step 2:
- After using Reset, it automatically switched to the Select Cutting Element option.
- Identified the vertical line (highlights and is dashed) with a data point. **This actually makes it the element to extend to.**
- Potential extension points are indicated by small dots.

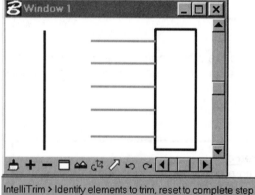

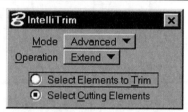

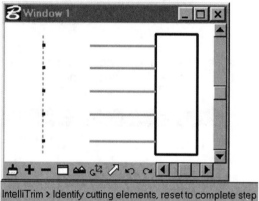

Once again, upon completing the step of identifying cutting elements, you'll get to see a preview of the Extend Operation to see if you like it. Shown below on the left is the preview of the Extend Operation. It looks good, so a Reset completes the command, and the result is shown below on the right.

Previewing the command, the lines are highlighted to show that it needs to be accepted or modified.

The command is completed. All the horizontal lines have been extended to the vertical line.

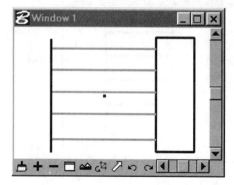

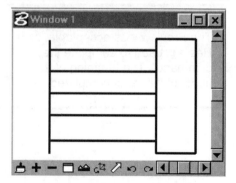

CONSTRUCT CIRCULAR FILLET

 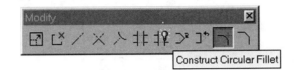

The name of the tool includes "construct" for a reason—it actually places a new element (at the active attribute settings) in addition to being able to modify an existing element. The Truncate options deal with how the existing elements are modified. In mechanical applications, fillet refers to eliminating sharp interior corners by adding material. A round is similar except that it removes material to round exterior corners. The Construct Circular Fillet tool can be used to draw either one of these representations. However, it has many uses other than what its name implies. It can be used to draw tangent circular arcs quickly, such as in a case of a turning radius for a roadway edge. In any of these cases, a value is entered for the Radius and then the two elements that will have the arc between them are selected. You get to see the results highlighted and another data point is needed to accept.

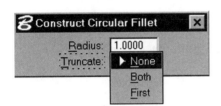

 WARNING: The value entered for the radius must be valid for the elements selected. If the size of the radius is too large and one of the elements would end up being eliminated by the construction of the fillet, then it will prompt that it is an "Illegal definition."

Shown below are the result of Construct Circular Fillet on the upper interior corner and lower exterior corner of an object using the different Truncate options. The order of selection of the two elements matters in the Truncate First Method because the first element selected will be truncated. In each of the examples, the vertical line was selected first and then the horizontal line was selected.

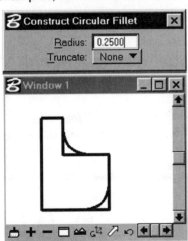 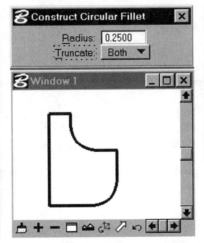 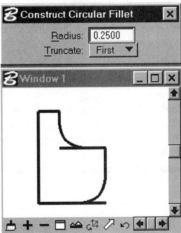

 The actual circular arc that is placed when using the Construct Circular Fillet is at the **Active** attribute settings—it does NOT use the attributes of the existing elements. Pay attention or you'll be using the Change Element Attributes tool!

CONSTRUCT CHAMFER

 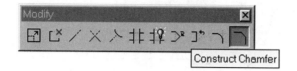

A chamfer is a straight line segment that is used to eliminate sharp corners. The Construct Chamfer tool requires two elements to be selected. They will be truncated according to the distance specified for each of them (so the order of selection matters unless, of course, you have them set to the same distance). A line, with the active attributes, will be drawn.

Shown below is the construction of a chamfer on the right corner of an object. The lines were selected in the order shown. Since the distances are different, you can see that the element selected first will be truncated by Distance 1 and the element that is selected second is cut shorter by Distance 2.

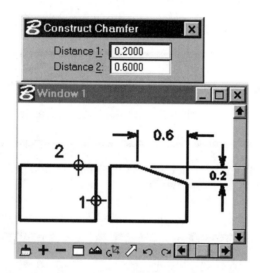

 If you see "Illegal definition" in the status bar after selecting the elements, that means at least one of the distances is too large. It won't allow you to totally eliminate one of the elements. Check your distances and try again.

Both of the last tools for constructing fillets and chamfers may also be used on shapes and other closed elements. When a fillet is constructed, the arc becomes its own individual element—it doesn't remain part of the shape (no matter what the Truncate option is set to). However, when a chamfer is applied, the integrity of the closed element is maintained, since the chamfer line segment becomes part of the whole closed shape element.

Changes and Modifications

QUESTIONS

① Will the Match Element Attributes tool have any effect on existing elements' attributes? If not, what tool can you use to modify an element's attributes?

② What are the **three** different Method options of modifying an arc using the Modify Element tool?

③ What is the sequence of steps to use Partial Delete to remove part of a closed shape like a block? How do the steps differ from those used to partially delete an open element such as a line?

④ Assume that you are using the **Extending Elements to Intersection** tool on these two elements (arc and line). Sketch the result of the data points in the order and position shown.

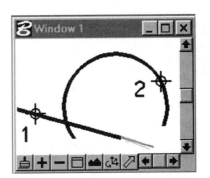

⑤ Assume that you are using the **Trim** tool on these two elements (arc and line). Sketch the result of the data points in the order and position shown.

⑥ What is the difference between a Trim Operation and a Cut Operation for the IntelliTrim tool in Quick Mode?

⑦ Shown is the status bar that results from using the Construct Circular Fillet tool on the two highlighted lines. Why does it read "Illegal definition"?

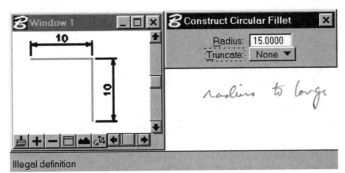

⑧ Name the **three** different Truncate options of the Construct Circular Fillet tool. For <u>each</u> option, sketch the result of using Construct Circular Fillet on these two lines with the data points in the order and position shown.

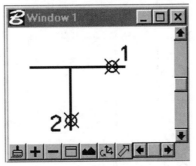
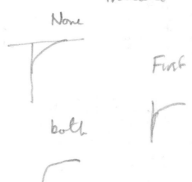

EXERCISE 9-1   Rod Brace

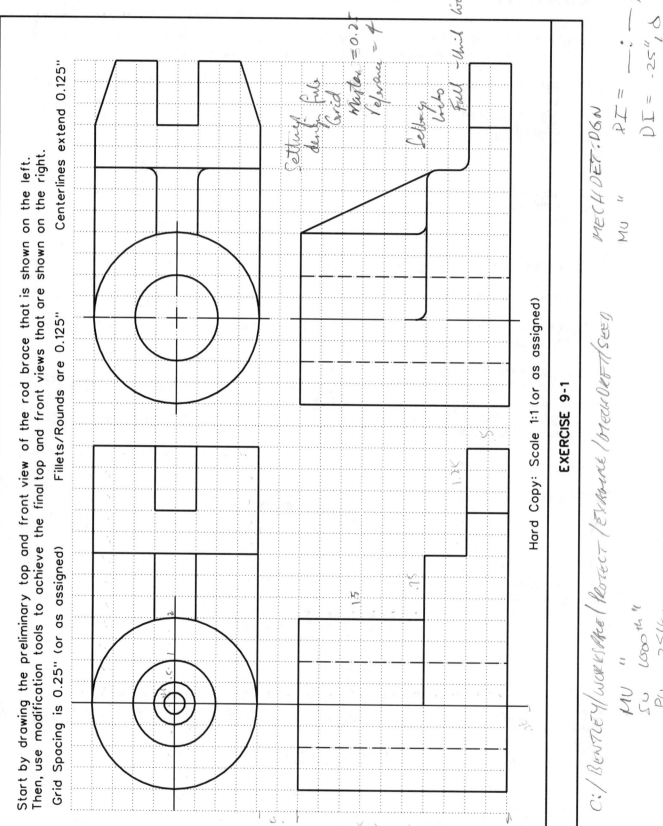

Exercise 9-1

EXERCISE 9-2   Pole Guide

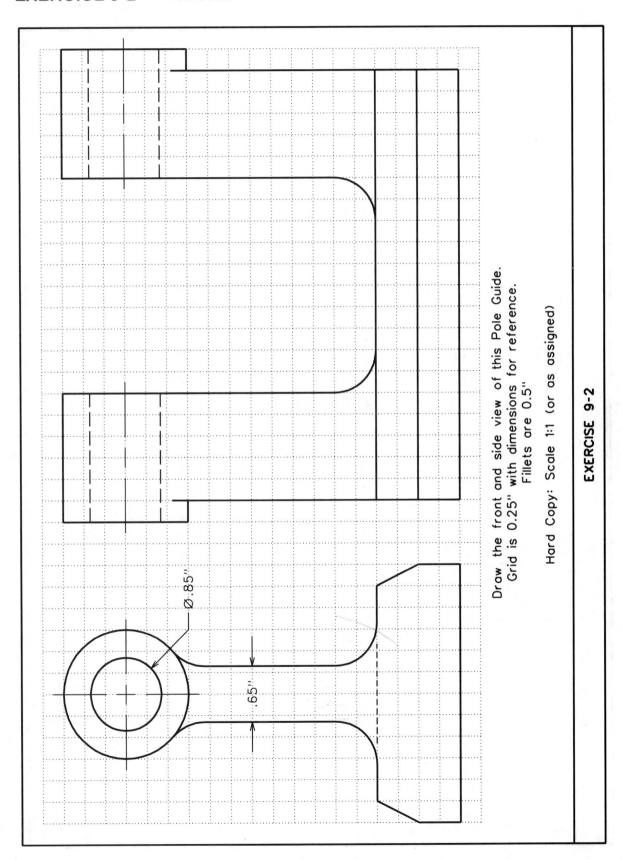

EXERCISE 9-2

EXERCISE 9-3 — Typical Rural Intersection and Drive

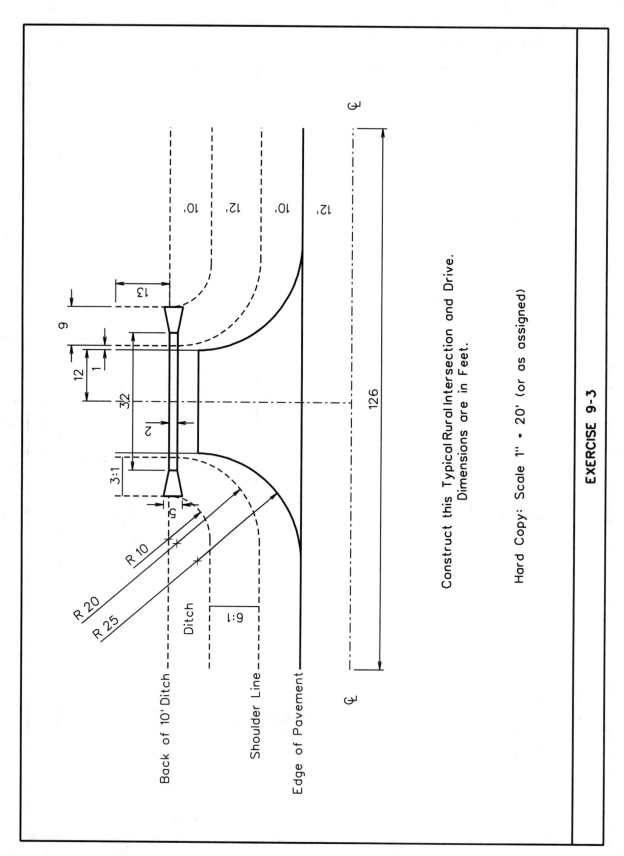

EXERCISE 9-4   T-Beam Bridge and Rail

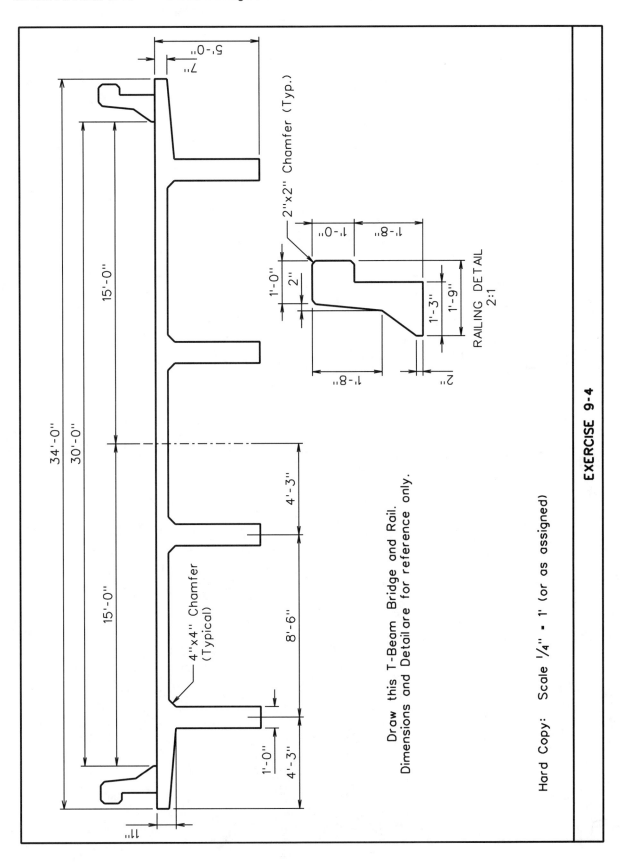

Chapter 10: Manipulation

SUBJECTS COVERED:

- Manipulate Tool Box
- Copy
- Move
- Move Parallel
- Scale
- Rotate
- Mirror
- Align Edges
- Construct Array

In this chapter, the elements will be manipulated in order to create new elements. This lets you create one element and then manipulate it to quickly create others. With experience, the use of these tools will help speed up the process of placing elements. Instead of drawing ten bolt holes individually, you may place one and copy it or array it—this definitely saves time.

MANIPULATE TOOL BOX

Each of the tools will be covered since each is extremely useful. Also, most of these tools have a Use Fence setting. This allows you to manipulate the entities selected by an existing fence. The setting is grayed out if there is not an existing fence. If there is an existing fence, the setting is available, but that doesn't mean you have to use it. You may also select elements by using PowerSelector **before** using a manipulation tool. If you already have elements selected, then you're working in the **noun/verb** order. If you've forgotten what this means, go back to the section on the Trim tool in Chapter 9. In any case, remember to read the prompts—they'll help you keep track of what's going on!

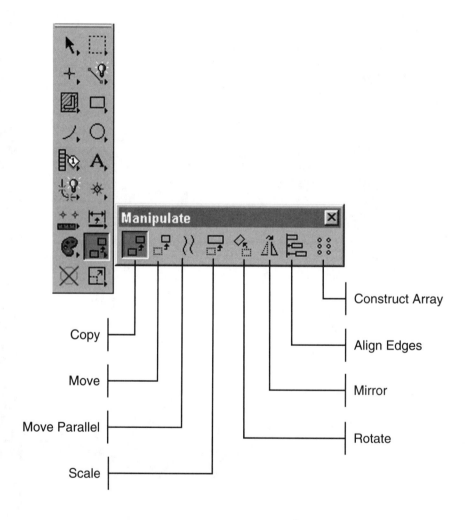

COPY

This tool makes a copy of the existing element and then allows you to move the placement of the copy. The original element stays put. The sequence of events is to select the element to copy with a data point. Once again, the location of the data point will indicate where to "grab" onto the element. Then the copy is placed with another data point (or quite often a key-in) and where you grabbed onto the element will be placed at that location. It will continue placing new copies until you use the Reject to start over. In order to do more than one element at a time (and stick with the verb/noun technique), the Use Fence setting is checked ON. This allows you to copy many elements at one time. When using a fence to select the elements to be manipulated, there are certain things to look out for, so pay close attention to that discussion.

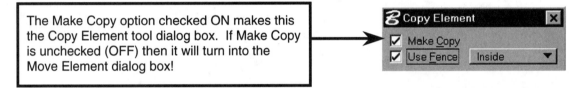

The Make Copy option checked ON makes this the Copy Element tool dialog box. If Make Copy is unchecked (OFF) then it will turn into the Move Element dialog box!

The following example will show how a simple copy manipulation can do a lot of work. It goes through a lot of steps, but pay attention to how the fence behaves and how its mode affects the copy.

Step 1:

A fence has been placed (Type-Block, Mode-Inside) around the three elements that need to be copied—the circle, line, and text.

The prompt reads Enter first point. Since the elements to copy have already been identified with the fence, you still have to specify where to grab onto the fence contents. The fence contents will be attached to the cursor at this first point (sometimes referred to as the origin). It can be defined graphically or with a key-in. Illustrated here is snapping to the center of the circle to specify the first point.

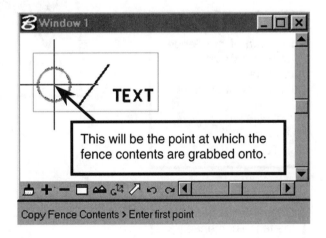

 If you want to copy to a precise location, it is good practice to use a tentative point and snaps when selecting the element(s) to copy or when defining the origin to use for the fence contents.

Accept the tentative and now we can continue on with the copying of the fence contents.

Step 2:
Once the first point is defined, then moving the cursor around will let you see the fence with an X for the origin point. This will move dynamically around the screen. (No, you don't get to see the elements being copied that are inside the fence.)

The prompt reads Enter point to define distance and direction. This will specify where you want to place the copy. The fence contents will be copied the distance and direction defined graphically or with a key-in.

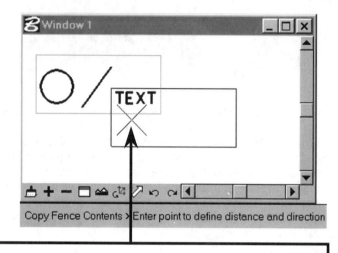

This indicates the new location for the origin defined for the fence contents. The copied elements will be located based on their positions relative to this origin.

Once the point has been entered, then the new elements show up in the display, but the copying can continue on. It is important to note **the fence relocates with the copied elements.**

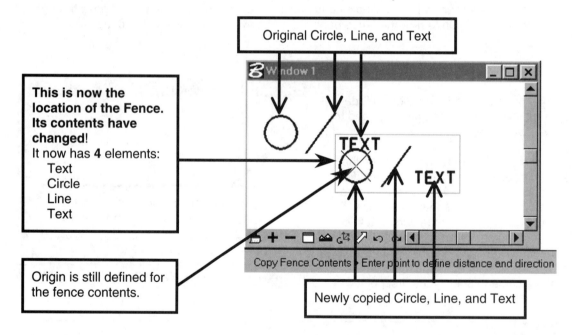

Original Circle, Line, and Text

This is now the location of the Fence. Its contents have changed!
It now has **4** elements:
  Text
  Circle
  Line
  Text

Origin is still defined for the fence contents.

Newly copied Circle, Line, and Text

Let's place another data point to make another copy.

Step 3:
By moving the cursor, you can dynamically see where the next copy of the fence contents will be placed.

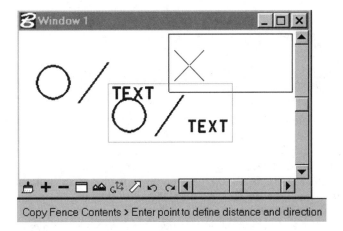

 Since you can't see the fence contents during this time, it is easy to mistakenly believe that the contents are the same as when you originally started out—**not so! You can easily have added elements. Be careful!**

Finally done! Here is the result of the last copy of the fence contents. We now have the next 4 elements placed and the new location of the fence (shown in light gray). All in all, 7 new elements have been added very quickly with this copy sequence. That's not too shabby!

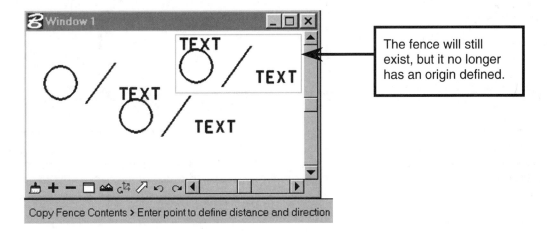

The fence will still exist, but it no longer has an origin defined.

 Oblique pictorials can easily be created using the Copy tool along with a **relative polar** key-in (⌨DL=△*distance,angle*). This will copy the features from the front view at different distances along the receding axis. Using a negative distance with a positive angle can also bring features "forward" on the receding axis.

MOVE

It doesn't really matter if you click on the Copy tool or the Move tool! Even if you start by clicking on the Copy tool, if Make Copy is unchecked (OFF) then you'll get the Move Element dialog box as shown here. That's because the Move tool works like the Copy tool <u>except that new elements are not made</u>. The existing elements are just moved by grabbing onto them (either by where you select a single element or by defining the origin when using a fence) and then defining the distance and direction to move them. However, when using a fence, **the fence also moves**. Once again you need to be careful of the fence contents changing.

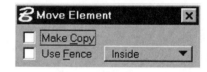

If all you want to accomplish is a **relative** translation, such as moving the element(s) 5 master units in the X-direction and 2 master units in the Y-direction, then it really doesn't matter where you select the element (or define the origin, if using the fence contents). However, if you want to move the element so that a precise point on the element is moved to a precise coordinate, then you'll need to pay close attention and snap to where you want to "grab" onto the element(s). This concept has been noted before, but this is an appropriate time to see an example of it in action.

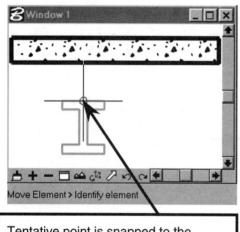

Tentative point is snapped to the midpoint of the closed shape's upper line segment.

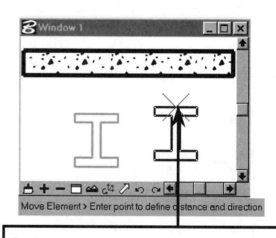

Tentative point was accepted (with a data point) and so the cursor is now attached at that precise midpoint. Closed shape moves with the cursor.

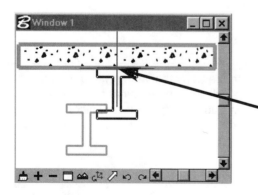

Now you can move the shape to a precise location by snapping to the midpoint of the block.

MOVE PARALLEL

 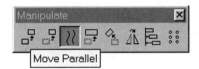

The Move Parallel tool allows you to move an element so that its orientation stays the same; it is just shifted parallel. It can be applied to numerous situations. From making parallel traffic lanes to offsetting concentric circles, it is a very useful tool. The Move Parallel tool requires identifying the element and then accepting the move on one side or the other of the original. It has two settings that can be toggled on or off. The Distance setting will allow you to specify a value for the perpendicular distance between the two parallel elements. The Make Copy setting checked ON does just that—it makes a copy of the selected element. The original element will be left where it is and a **copy** of it will be moved parallel.

Example of Move Parallel with Distance Checked ON

In this case, the line will be moved parallel by a set distance.

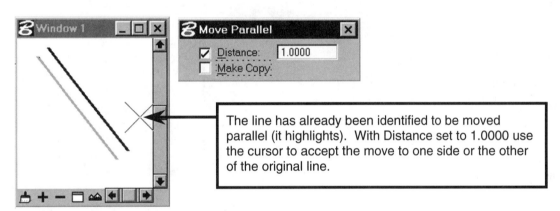

The line has already been identified to be moved parallel (it highlights). With Distance set to 1.0000 use the cursor to accept the move to one side or the other of the original line.

Example of Move Parallel with Make Copy Checked ON

This closed shape is all one element, and it will be moved parallel by a set distance, but a copy will be made. **The copy has the same attributes as the original.** It doesn't come in at the Active settings. Again the element has been identified with a data point and then it needs to know which side—a data point outside will make a larger shape and a data point inside will make a smaller one. You can keep going and make one copy after another. When you're done, use the right mouse button to finish.

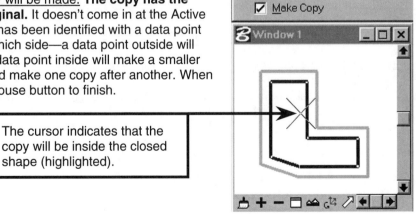

The cursor indicates that the copy will be inside the closed shape (highlighted).

SCALE

The Scale tool has a few more settings than the others do. To manipulate the scale of the element means to change its size. There are two different Method options to choose from. With the Active Scale Method, you can change the XScale and YScale settings independently (if unlocked), so you can make some really radical manipulations. The Active Scale Method allows for just a numerical scale factor. The other Method of 3 points is much more difficult to use. Instead of scaling by a set multiplier, the sizes are adjusted according to distances determined by graphically specifying points. This is useful when trying to scale the element in relation to another, but it will be left for you to try on your own when you become more advanced.

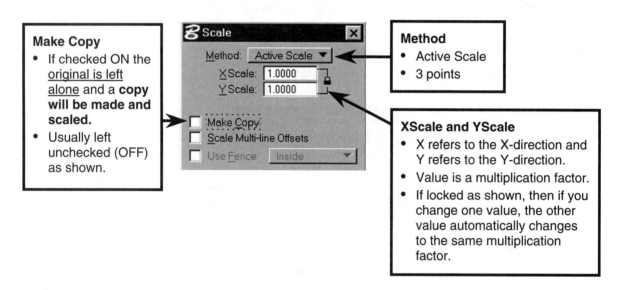

Make Copy
- If checked ON the original is left alone and a copy will be made and scaled.
- Usually left unchecked (OFF) as shown.

Method
- Active Scale
- 3 points

XScale and YScale
- X refers to the X-direction and Y refers to the Y-direction.
- Value is a multiplication factor.
- If locked as shown, then if you change one value, the other value automatically changes to the same multiplication factor.

When using the Scale tool, you will be prompted for point to scale about. This is the coordinate that will remain the same and everything else will scale around it. It doesn't have to be on the element, but it is good practice to specify a distinct point to scale about.

After an object is scaled using either method, it doesn't remember its original size.

For Example: Using the Active Scale Method on a 2" x 4" block.

With XScale and YScale values at 0.75, the block would be scaled to 1.5" x 3.0".

Using the Scale tool again, with XScale and YScale set to 2.00, the block would then be 3.0" x 6.0". It wouldn't be 4.0" x 8.0" (2 times the original block size).

Active Scale

The Active Scale Method is the easiest one to use. As discussed in the previous example, the XScale and YScale values are multiplication factors—not percentages. A factor of 1.00 will leave the element at its original size. A factor greater than 1 will make the element larger, less than 1 will make it smaller.

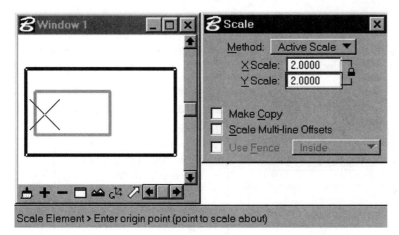

This highlighted block has already been identified with a data point. **It doesn't really matter where it was identified at.** Since XScale and YScale are both 2 it will proportionally double its size. Now the cursor can be moved around and the larger block can be seen moving dynamically. You still need to Enter origin point (point to scale about). Since you can still "see" the original block you can still snap to its corner (if you want to keep the large block position relative to its original one).

Here is the result of snapping to the original block's lower left corner. The larger block is now located with its lower left corner in the same spot. When this tentative point is accepted, the original size block will no longer appear. Instead all that is left is the block at the larger scale.

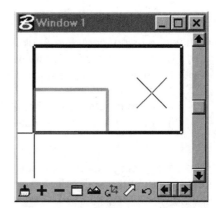

 Do **not** use the Scale tool to get items to "fit on the paper." That should be done by adjusting the Print scale. Remember—CAD drawings are usually done with everything ACTUAL size.

ROTATE

Use this tool if you need to rotate elements in the X-Y plane. The Active Angle Method is the easiest to use, so it is covered here. You can also rotate using point methods, either 2 or 3 points. These allow you to set up the angle of rotation graphically.

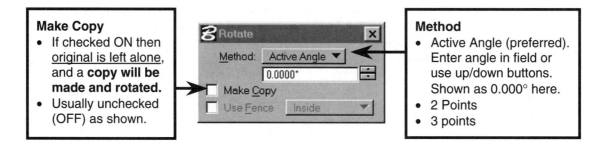

Make Copy
- If checked ON then <u>original is left alone</u>, and a **copy will be made and rotated**.
- Usually unchecked (OFF) as shown.

Method
- Active Angle (preferred). Enter angle in field or use up/down buttons. Shown as 0.000° here.
- 2 Points
- 3 points

Active Angle

When you're using the Active Angle Method of the Rotate tool, there are a few things to remember:
- 0° is oriented to the right and a positive angle is counterclockwise.
- Changing the Active Angle here will change the main Active Angle setting. This setting affects other tools.
- A key-in for setting the Active Angle is ⌨AA=.
- You can quickly set the Active Angle back to 0° by keying in ⌨AA=0 in the Key-in field.

When doing a single element, first you will select the element by data pointing anywhere on it. It isn't grabbed onto there, so location of data point doesn't matter. The element will highlight as shown here (light gray). When prompted to Enter pivot point (point to rotate about), the middle of the original line was snapped to as shown at the left. By accepting the tentative point, the line will be rotated to the position previewed here.

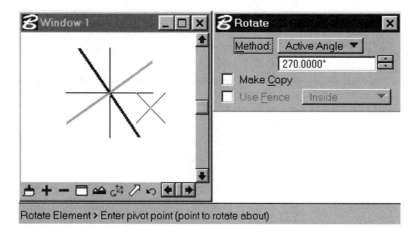

Example of Rotate Using Fence

In this example, there is an existing fence (Type-Block, Mode-Inside) that is to be rotated. The status bar shows that you need to Enter pivot point (point to rotate about). As in the scale procedure, this is the point that will not change and the fence will rotate about it. Usually, you will want to specify a distinct point (with a snap or key-in). The existing fence outline is seen rotating dynamically, but again you don't get to see its contents.

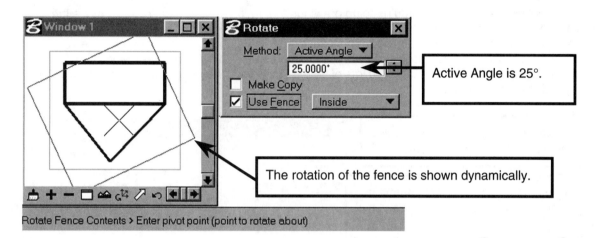

Active Angle is 25°.

The rotation of the fence is shown dynamically.

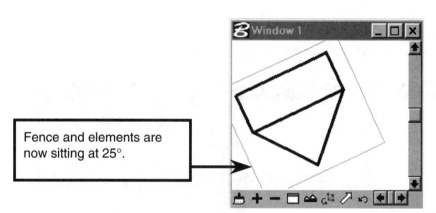

Once the pivot point has been specified, then the contents of the fence are rotated (because Make Copy is unchecked (OFF) there is no duplication). The result is shown on the left. Notice that the fence has also been rotated! Once again, be aware that the **fence contents can change** since it's been moved.

Fence and elements are now sitting at 25°.

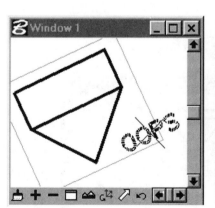
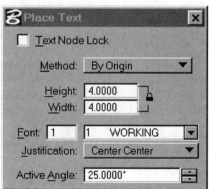

If you then go to use the Place Text tool—the Active Angle has been changed to 25° as shown on the left. You can change it to zero here or with the key-in of ⌨AA=0.

MIRROR

The Mirror tool will be useful when you're doing drawings that involve symmetry. You can move the element(s) by flopping them over the symmetry line. By using the Make Copy setting, you can easily create a symmetric object. This is a great tool to use for reversing house plans, in which case you wouldn't want to make a copy. The line of symmetry can be specified three different ways to allow you better control. The Mirror tool can save you lots of time if you realize when symmetry is involved in the drawing and take advantage of it.

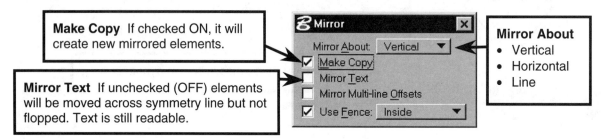

Make Copy If checked ON, it will create new mirrored elements.

Mirror Text If unchecked (OFF) elements will be moved across symmetry line but not flopped. Text is still readable.

Mirror About
- Vertical
- Horizontal
- Line

Mirror About Vertical

This uses a vertical line of symmetry. The element needs to be identified (or a fence placed and Use Fence checked ON). Then you will just need to locate the line of symmetry since its orientation has already been set.

Here a fence has been placed and is used to specify the elements to mirror. The new placement appears, and accepting the fence contents will also locate the vertical line of symmetry.

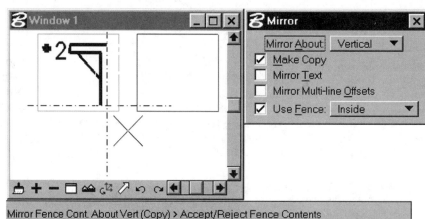

To accept the fence contents Mirror manipulation, the vertical centerline was tentatively snapped to and was accepted. The Reset button was used to finish the command (otherwise it would keep letting you Mirror the fence contents). The result is as shown here. Once again the fence has moved too.

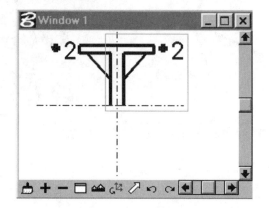

Mirror About Horizontal

The Mirror About Horizontal works the same as Mirror About Vertical except it is a horizontal line of symmetry. The elements go from top to bottom or vice versa. All that is needed is the location of the line of symmetry.

Now the items from the previous Mirror manipulation have already been selected using PowerSelector (that's why they're highlighted.) The Mirror tool is started with the Mirror About Horizontal option and with Make Copy checked ON. The existing horizontal centerline has been snapped to as shown here. The mirrored elements are shown below the original location and are waiting to be accepted. Since Mirror Text was checked ON for this Mirror manipulation, you'll end up with upside-down text. Not necessarily what you'd want, but interesting to see nonetheless.

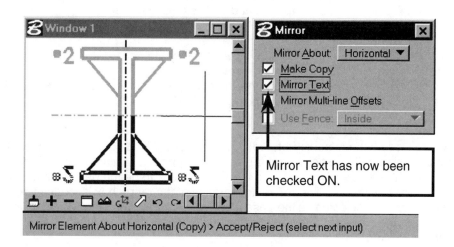

Now it should be noted that the result that looks like an I-beam is **not** a closed shape. It is still made up of individual elements even though it appears continuous. Using Mirror just took advantage of the symmetry of the final drawing to speed up the process.

Mirror About Line

This setting doesn't require an existing line to mirror about. Mirror About Line allows a symmetry line that doesn't have to be horizontal or vertical. Instead you can specify two points that set up the location and orientation of the symmetry line (it's an **imaginary one**—a line is not actually created). You can snap to an existing line if you want, but it is not necessary. Shown in this illustration is the selected triangle shape. Point 1 was already entered as 1st point on mirror line, and Point 2 is being tentatively snapped to for the 2nd point on mirror line. You can preview the element that will be the mirrored copy. By accepting the tentative, the mirrored copy will be placed and you can keep specifying another 2nd point if needed (the 1st point stays the same).

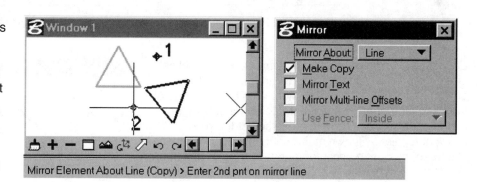

ALIGN EDGES

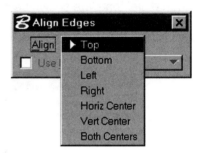

The tool tip reads Align Elements By Edge, but the dialog box calls this the Align Edges tool so we'll call it Align Edges. Actually it can also align elements based on their centers, not just their edges. The Align options available are shown in the Align Edges dialog box illustrated at the right. You select a base element and what elements you want to align. The tool moves the elements into alignment leaving the base element at its original position. The use of the "center" options does not mean the centroid; rather it means the center of the element's range block. A range block is the smallest orthogonal rectangular boundary that can surround an element, line string, or closed shape. Let's look at a couple of different examples of using Align Edges.

Let's start by aligning these four elements (rotated block, line, circle, and polygon) using the Align Top option. First you are prompted to Select a base element for alignment as illustrated below on the left. Shown below on the right is the result of selecting the rotated block to be the base element. It highlights and its range box appears as dashed lines. The appearance of the range box is to give you a visual clue as to which element has been selected as the base element. For a regular block element you won't see this clue because the range box and the element coincide. Once a base element has been selected, you can select an element to align.

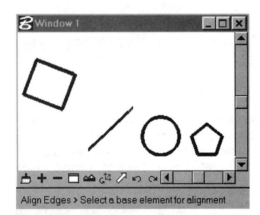 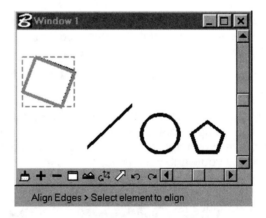

Each time you data point on an element, it will highlight and be repositioned (in this case, straight up in the Y-direction) so that its top edge aligns with the top edge of the base element. The next data point will accept the alignment and can also be used to select the next element to align. Shown at the right, the line and circle alignment has been accepted and the polygon has been selected. A right mouse button action would cause the alignment of the polygon to be rejected. The line and circle would still remain aligned. Whenever you are done aligning elements just use Reset (right mouse button).

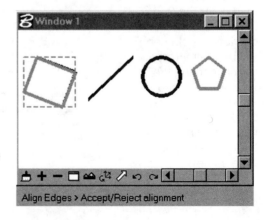

Manipulation 217

Now let's look at using the Align Edges tool to align three different closed shape elements to their horizontal centers. The Align Edges dialog box is shown at the right with the Align option set to Horiz Center (which is short for horizontal centers).

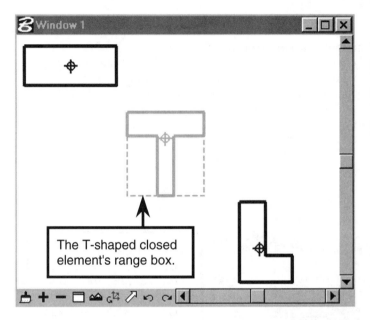

The three closed elements have their centroids marked with a ⊕ symbol. The T-shaped one has been selected as the base element. Note that the center of the T-shaped closed element's range box does not correspond to the centroid of the T-shaped closed element.

The T-shaped closed element's range box.

Here is the end result of aligning the rectangle and the L-shaped closed elements with the T-shaped one.

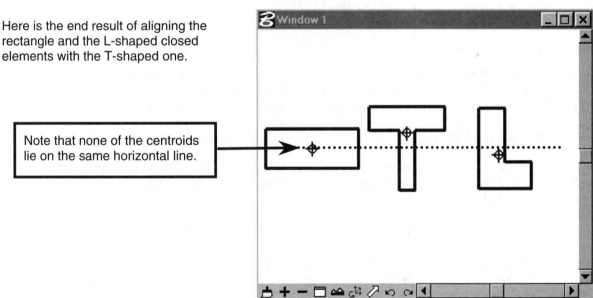

Note that none of the centroids lie on the same horizontal line.

 The Align Edges tool is great to use when trying to align separate text elements. Just remember that it uses the range block of the text too!

CONSTRUCT ARRAY

There are two types of arrays, Polar and Rectangular. An array is an organized repetition of the same elements over and over. Rectangular Array Type is based upon a system of rows and columns. Polar Array Type is based on a circular system, with the items located relative to a radius point. Both types of arrays are used in situations where copying could be used but there are so many items that it becomes extremely tedious (and there is more room for mistakes). Each Array Type option results in a different Construct Array dialog box with different settings. They both can use a fence to select the items to array. Be careful using PowerSelector to select the items to array: You may get more arrayed elements than you bargained for!

Rectangular Array Type
Here is the Construct Array dialog box for the Rectangular type of array.

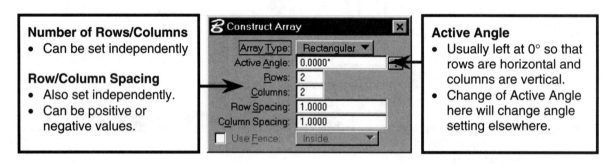

Number of Rows/Columns
- Can be set independently

Row/Column Spacing
- Also set independently.
- Can be positive or negative values.

Active Angle
- Usually left at 0° so that rows are horizontal and columns are vertical.
- Change of Active Angle here will change angle setting elsewhere.

It is important to understand what constitutes a row or a column and how they are counted.
- The original item to array is included in the count—so there must be at least 1 row and 1 column.
- The spacing of the rows is actually a Y-distance and the spacing of the columns is an X-distance. These values can be positive or negative.
- The spacing is from **a point** on one item in the array **to the same point** on the next item.

Shown here is a 2 row and 4 column Rectangular Array, with both Row and Column Spacing settings as **positive** values. If they both were negative, the new items would have been to the left and below the original. Try it and see for yourself.

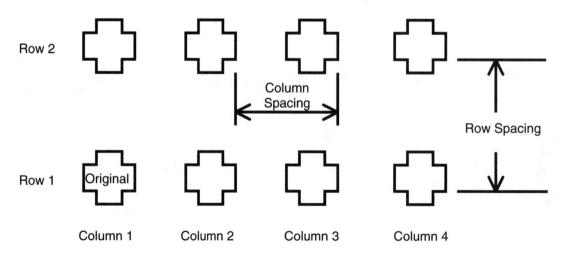

Once the Construct Array dialog box has been set correctly, then the process is extremely simple. If you aren't using a fence, the element needs to be selected (with a data point). Then accept the array with another data point in the view. If you are using a fence, it will automatically prompt you to Accept/Reject. Often you will need to fit the view in order to see the results of the Construct Array manipulation.

Polar Array Type

The Construct Array dialog box for a Polar Array Type is shown below. Notice that the radius of the array was not specified in the dialog box. That is because the data point that accepts the array will also specify the center of the array. Once again, it's time to remind you to read the prompts.

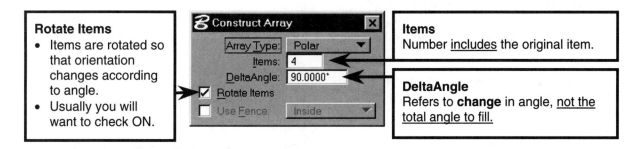

Rotate Items
- Items are rotated so that orientation changes according to angle.
- Usually you will want to check ON.

Items
Number <u>includes</u> the original item.

DeltaAngle
Refers to **change** in angle, <u>not the total angle to fill</u>.

The most common mistake here is to give it a DeltaAngle of 360° (mistaking it for angle to fill). The element will array itself on top of itself—so you think that there is only one and that the array didn't work. You try it again (and usually again and again) and you will have LOTS of elements without knowing it! OOPS—Delete Fence Contents to the rescue.

When filling 360°, the computer can calculate the DeltaAngle for you. In the DeltaAngle field, enter 360/number of items, and then hit ENTER. It will do the math for you!

As an example of using Construct Array (Polar Type), this line will be arrayed around the center of the circle to form a light symbol. The Use Fence option is unchecked (OFF) so it needs an element identified.

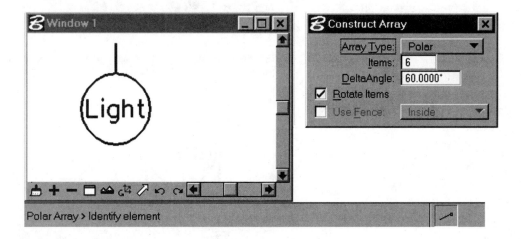

Be careful not to array a circle about its own center—you will have circles directly on top of circles. This can make the size of your design file grow rather quickly. Plus, if you're using a pen plotter, you can end up cutting holes in the paper!

Continuing on with our Construct Array (Polar Type) example, the line element is selected with a data point and it highlights. Notice the prompt to Accept, select center/Reject. The next data point will do two things: accept the polar array <u>and</u> also specify the center to array about.

In order to be precise, the center of the circle will be snapped to using a tentative point and then accepted.

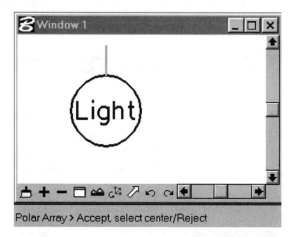

Here is the result of the Polar Array. Remember that the setting to **Rotate Items was checked ON.**

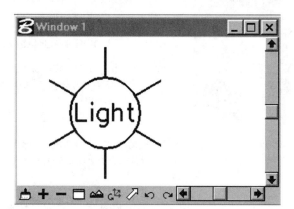

Here is the result of a similar Polar Array but the option to **Rotate Items was unchecked (OFF).**

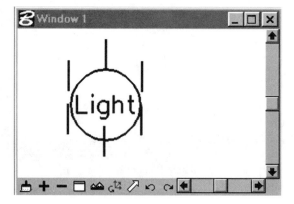

QUESTIONS

① The object below on the right is to be drawn using the object on the left (and using the gray fence shown). Give the settings of the fields (A–E) specified in this Construct Array dialog box.

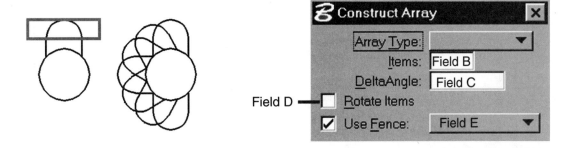

② Name the **three** Mirror About options available in the Mirror dialog box. For each method, sketch the result of using the Mirror tool **(Make Copy checked ON) on this triangular closed shape**. The first data point was at Location 1. If another data point is required, assume that it is at Location 2.

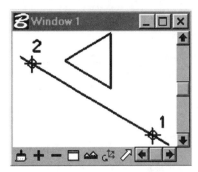

③ When you're copying elements with the Use Fence setting checked ON, does the fence move with the copied elements, or does it remain in its original location? How does the behavior of the fence location affect the contents of the fence?

Questions 4 and 5 refer to using the Move Parallel tool (with the settings as shown on the right) on the block shape.

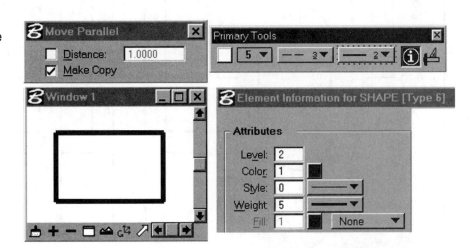

④ Is there a specific distance at which the copy will be made? If so, what is it?

⑤ What will the listed attributes of the new shape be?
  a) Level
  b) Style
  c) Weight

⑥ Assume you start with only the I-beam shown in the gray rectangular fence. State the values of fields A–D in this Construct Array dialog box that will result in the I-beam layout shown below.

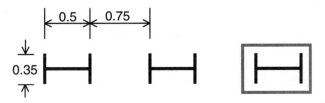
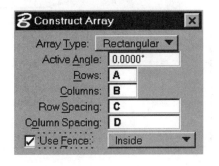

EXERCISE 10-1 Fan Cover

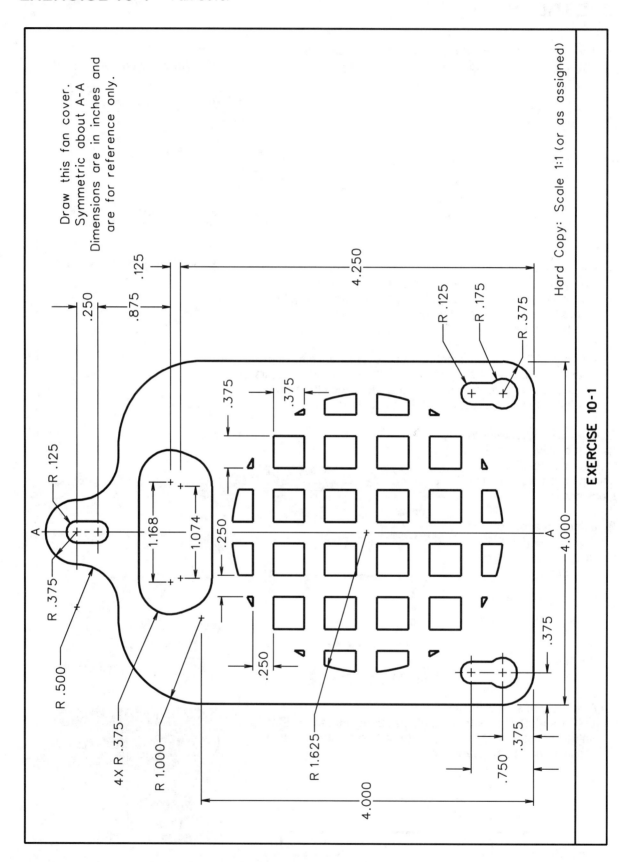

EXERCISE 10-2 Flume Plan

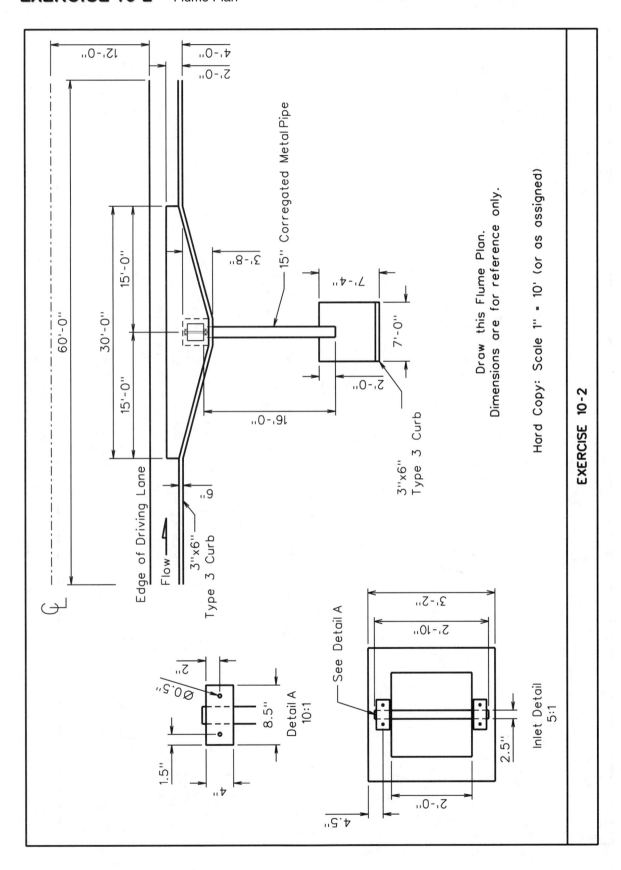

EXERCISE 10-3 Two Simple Gears

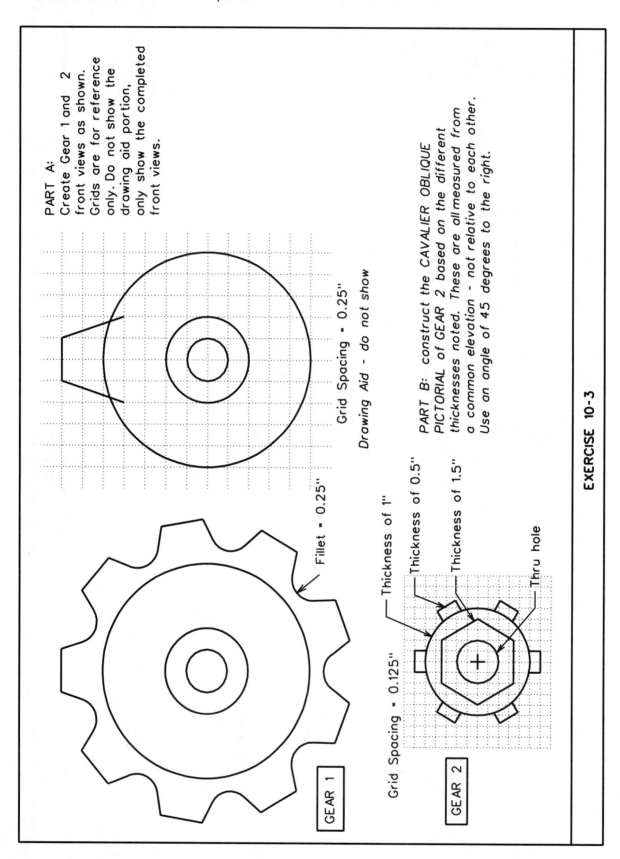

EXERCISE 10-4   Grate Inlet

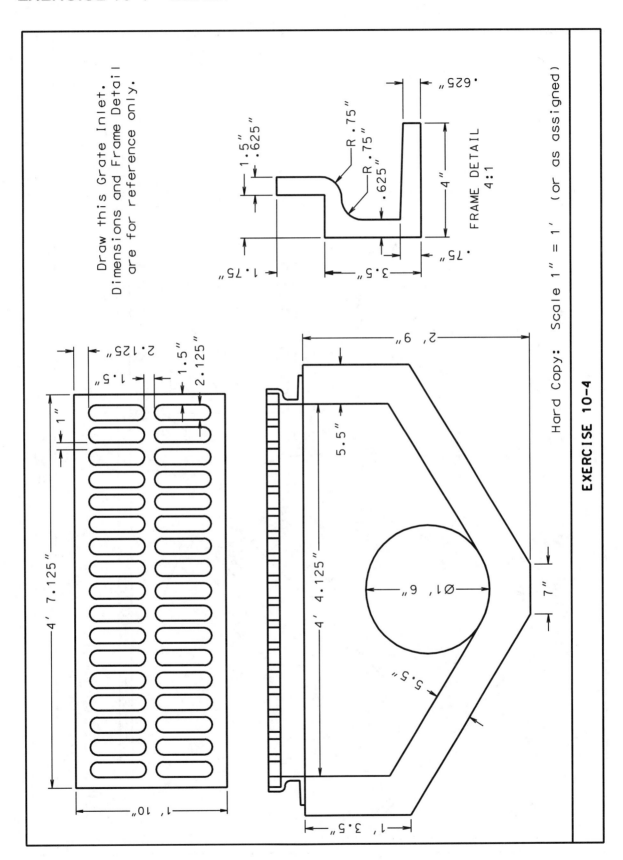

EXERCISE 10-4

EXERCISE 10-5 Proposed Park Layout

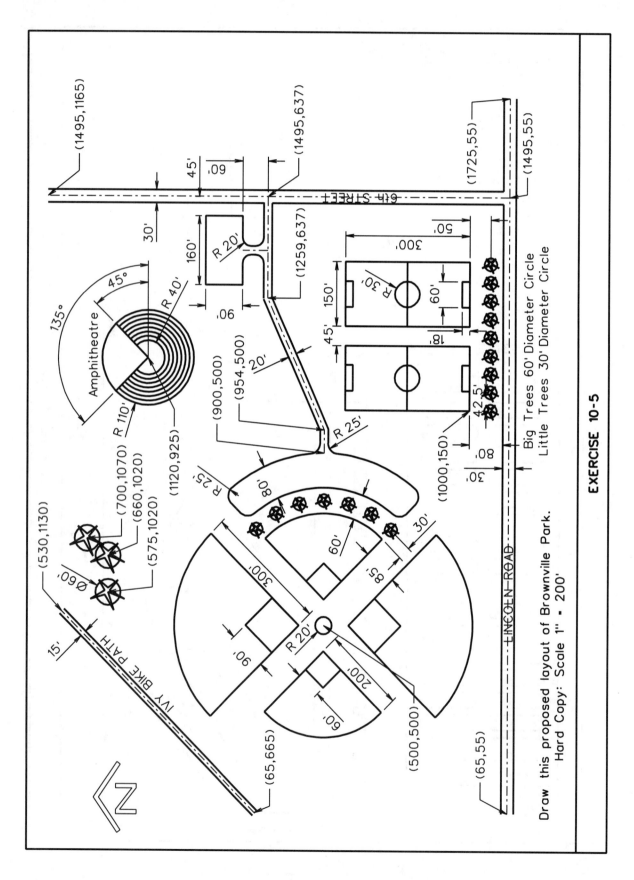

Chapter 11: AccuDraw and SmartLine

SUBJECTS COVERED:

- **Starting and Quitting AccuDraw**
- **AccuDraw Compass and Coordinate System**
- **AccuDraw Keyboard Shortcuts**
- **AccuDraw Settings**
- **Applications of AccuDraw**
- **SmartLine**
- **Place SmartLine**
- **Groups Tool Box**
- **Drop Element**
- **Create Complex Shape**
- **Create Region**
- **Popup Calculator**
- **Table of AccuDraw Keyboard Shortcuts**

AccuDraw is an advanced drawing aid. It isn't just a single tool; instead it works in conjunction with other tools and input by using lots of shortcuts and intuitive settings. The basic use of AccuDraw is for **dynamic relative precision input** rather than using key-ins for coordinates—and that's what is covered here. But there are many other uses for AccuDraw, such as using it for customizing keyboard shortcuts of commands for advanced users who can zip right along. By now you should be familiar enough with the regular tools that AccuDraw shouldn't be overwhelming. Since it is an advanced drawing aid, we will just scratch the surface of its capabilities but it will be enough to get you started.

Smart Tools are those with the light bulbs. The light bulb is to let you know that the tool has some special characteristics. They don't have to be used with AccuDraw but often are. Likewise, AccuDraw can be active and used with any tools, not just "smart" ones. However, using both at the same time is usually the norm. SmartLine® and Smart Drop Element are the two "smart" tools that show up in the Main tool box and will be covered here.

The use of AccuDraw and SmartLines often depends on personal preference. You find AccuDraw either extremely frustrating or extremely handy (especially when you become a power user). There are some applications and techniques in which using AccuDraw and SmartLine is very advantageous. Those items are covered here and include text and isometric lines using AccuDraw and using SmartLine for line strings and closed elements. AccuDraw especially takes a bit of getting used to, but it is well worth at least experimenting with. We'll see some more discussion on AccuDraw in Chapter 16 since AccuDraw really shines in 3D.

STARTING AND QUITTING ACCUDRAW

AccuDraw is actually a tool that is found in the Primary Tools tool box (which is usually docked and has level, color, style, and weight with it). It looks like a T-square and triangle, standard hand-drafting equipment. AccuDraw can be left on all the time, but often it is turned off and on as needed.
By clicking on the Start AccuDraw tool in the Primary Tools tool box, you will start AccuDraw and the AccuDraw dialog box (sometimes referred to as the AccuDraw window) will open. It is a unique dialog box because like a tool box it can be docked. Shown below is the floating AccuDraw dialog box. AccuDraw stays on until you turn it off. **In order to quit AccuDraw, you just close the AccuDraw dialog box by clicking on the X in the upper right corner.** If it is docked, you'll need to drag it out to float it and then close it. You can also quit AccuDraw by using one of its own keyboard shortcuts (discussed later).

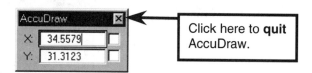

Another indication that AccuDraw is on is that a compass appears when you go to place an element or modify/manipulate one. More on that later—but here is what it can look like when you're placing a line.

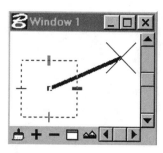

AccuDraw Dialog Box

The AccuDraw dialog box will dynamically change to show different fields of input. In addition to the precision input of X and Y values, the window can also display polar coordinates. AccuDraw uses a variety of keyboard shortcuts to toggle and lock different settings **but they work only if the AccuDraw dialog box is the <u>active</u> one.** Otherwise you'll be typing in the Key-in field (which isn't appropriate) or just typing to nowhere (which makes you feel particularly silly). When referring to an AccuDraw keyboard shortcut, we will use the following notation: <*keyboard*>. You don't type the < or the > but just the characters or key noted inside them.

You can tell if the AccuDraw dialog box is **active**:
- Docked—look for the | cursor to be blinking in one of its fields.
- Floating—look for the title bar to be darkened as well as the blinking | cursor. Sometimes it is preferred to leave the AccuDraw dialog box floating since it's easier to see if it's active.

If it isn't active, just click in one of its fields to put the focus there and get a blinking | cursor.

ACCUDRAW COMPASS AND COORDINATE SYSTEM

AccuDraw's main premise is that it makes using precision input especially quick and easy. It is supposed to be intuitive and easier to use than keying in ⌨DL= or ⌨DI= for relative coordinates. Since these deal with rectangular and polar coordinates, AccuDraw also has these types of coordinate system entry, and its compass reflects that. Shown below are the two compasses for rectangular (square one) and polar (circular one). The compasses show the 0°, 90°, 180°, and 270° with dashes. The X and Y positive axes are indicated by thicker dashes. +X is red and +Y is green. These are all in accordance with the coordinate axes of the AccuDraw compass. As its orientation changes, so do the location and rotation of its axes. It is a bit hard to tell in this illustration but they show up prominently in the actual window. **You can toggle between the rectangular and polar coordinate inputs by using the <*spacebar*>.**

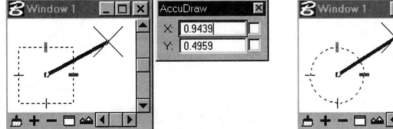
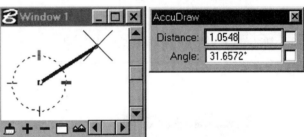

An important thing to remember is that the values that show in the AccuDraw dialog box are precision inputs **relative** to the dynamic coordinate system of the **AccuDraw compass**. If you see a Distance of 2, that is two master units relative to the origin of the compass. The compass is dynamic, and it can change relative to the elements being placed. You can also quickly change its orientation and location with keyboard shortcuts.

With AccuDraw, when the cursor is in the vicinity of the axis of the compass, it will snap to that axis. This is a strong point of AccuDraw since Axis Lock doesn't need to be used to get horizontal and vertical orientations. The Smart Lock keyboard shortcut <*Enter*> can actually lock the X or Y value to 0. Which value is locked depends on the location of the cursor when the shortcut is used. If the cursor is close to the X-axis, it will lock the Y value to 0. If the cursor is close to the Y-axis, it will lock the X value to 0. Using <*Enter*> again will disengage Smart Lock. A locked value is indicated by a checkmark in the setting's check box. Specific values can also be locked by typing the number in the input field and then hitting the ENTER key. The value will stay locked until you place a data point, then it will automatically toggle off.

If the focus is already in one of AccuDraw's input fields, just type the number you want. It isn't necessary to backspace over the numbers already there or move the cursor to the field in order to highlight it before typing. Once you type, all of the existing numbers are removed and the new value is seen.

For Example: In the illustration below on the left, the focus is in the Y input field. The blinking | cursor is at the end of the 6.7639 value. If you type a 4, the 6.7639 will be removed and the 4 will show up (as in the middle illustration). It will not make the value 6.76394! If after typing the 4, you hit the ENTER key, the Y value will be locked at 4.0000 as shown below on the right.

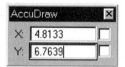  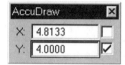

AccuDraw is very difficult to demonstrate with illustrations since it is so dynamic. But let's look at an example of the rectangular compass and how AccuDraw behaves when you're drawing lines.

Example of the Dynamic AccuDraw Compass and Coordinate System

Shown below is an example of how the AccuDraw compass dynamically moves and how AccuDraw allows you to lock values. The following aren't really steps but rather snapshots of changes occurring.

Using the regular Place Line tool, a horizontal line has been placed. The polar compass moves to the end of this line. The X-axis red dash (appearing dark here) is still in the horizontal position and the Y-axis green dash (appearing light gray here) is still vertical. The AccuDraw dialog box shows Distance locked to 1.0000.

The cursor was moved close to the Y-axis and the line (1 master unit in length) snapped into the vertical orientation (Angle was NOT locked). The line was placed with a data point. The compass then reoriented itself to the new endpoint as shown below. Notice the new orientation of the X and Y axes. The Distance and Angle values will dynamically reflect the cursor's position **relative** to the AccuDraw compass.

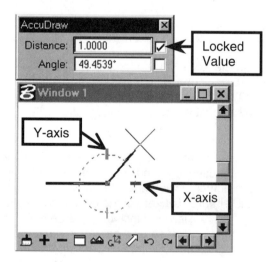 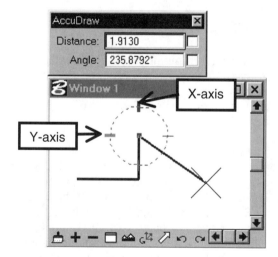

Example (Continued):

Now the angled line will be placed using values entered in the AccuDraw dialog box's fields.

The Distance and Angle values are locked. However, even though Angle is 45° the cursor specifies the direction of the 45°. A data point actually places the line's endpoint.

After the data point establishes the line, again the compass moves its origin and reorients itself relative to the element.

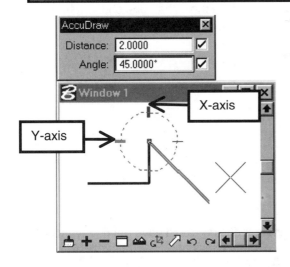
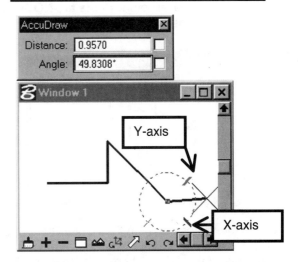

If the next line is to be perpendicular relative to the angled line, it is very simple to do.

The cursor is moved near one of the colored dashes of the compass. This time it is the one at 270°. The line actually snaps to that orientation as long as the cursor is somewhat close. Notice that 270° appears in the Angle field but it is not locked.

What if you wanted to specify rectangular coordinates instead? Simply use the <spacebar> and the compass immediately changes its appearance. Its location and orientation remain the same. The cursor can be moved close to the Y-axis and the X value will go to 0.

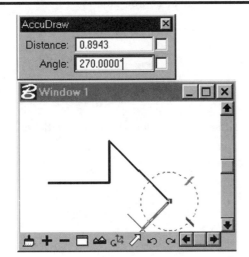
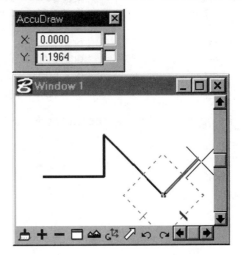

ACCUDRAW KEYBOARD SHORTCUTS

As mentioned previously, AccuDraw is capable of many operations. There are AccuDraw keyboard shortcuts that are one or two characters that you type on the keyboard to perform these operations. In order for these keyboard shortcuts to work, the AccuDraw dialog box must be the one that is active.

For instance, AccuDraw has an operation that takes the compass and reorients it. Imagine (in continuing the previous example) that the line continues to be drawn, and then you decided to rotate the compass. Type <*R*> (or <*r*> because **shortcuts are NOT case-sensitive**). Since more than one shortcut begins with R a list box pops up as illustrated below.

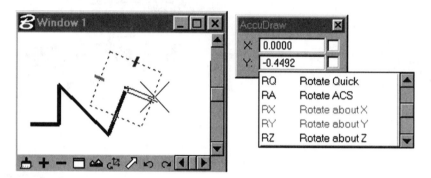

Here you can choose and run the Rotate Quick operation. Another choice would have been to type <*RQ*> (while the AccuDraw dialog box is active) which would run the Rotate Quick operation immediately without going through the list box. For example, by typing <*D*> the Lock Distance shortcut is done immediately because there is only one shortcut that begins with D. A quick box pops up to visually indicate the shortcut being done as illustrated at the right.

Some more commonly used AccuDraw keyboard shortcuts are:

Keyboard Shortcut	What it does
<*spacebar*>	Toggles between rectangular and polar coordinates
<*q*>	Quits AccuDraw
<*x*>	Lock X toggle
<*y*>	Lock Y toggle
<*d*>	Lock Distance toggle
<*a*>	Lock Angle toggle
<*o*>	Set Origin

A larger listing of AccuDraw keyboard shortcuts that are available can be found at the end of this chapter.

ACCUDRAW SETTINGS

One of the advantages of AccuDraw is that you can make it do what you want by controlling its settings. If you don't like that the AccuDraw compass reorients itself automatically, you can change it so that its orientation is always orthogonal (horizontal or vertical). If you don't like the colors that indicate the X and Y axes you can change those too!

The AccuDraw Settings dialog box is accessed from the main pull-down menu SETTINGS>ACCUDRAW. There are quite a few settings, but they won't all be discussed. The ones that seem to be the most useful for beginners to modify are Unit Roundoff and Context Sensitivity. The others are best left set as shown.

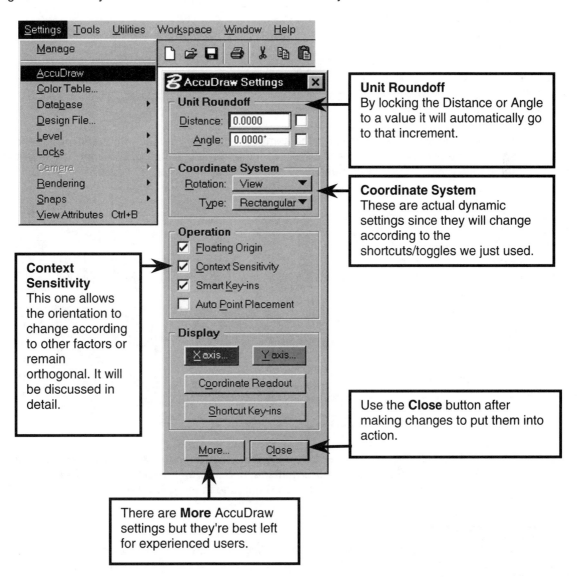

Unit Roundoff
By locking the Distance or Angle to a value it will automatically go to that increment.

Coordinate System
These are actual dynamic settings since they will change according to the shortcuts/toggles we just used.

Context Sensitivity
This one allows the orientation to change according to other factors or remain orthogonal. It will be discussed in detail.

Use the **Close** button after making changes to put them into action.

There are **More** AccuDraw settings but they're best left for experienced users.

Unit Roundoff

This is a very handy way to move in incremental distances. For instance, you can lock the Distance Unit Roundoff to 0.25 and it will go in increments of 0.25 master units. This is often a better approach than Grid Lock!

Key-in values and snapped tentative points still take precedence over the Unit Roundoff setting.

Context Sensitivity

When the Context Sensitivity is checked ON (the default) then the compass orients itself to existing factors. The most obvious result is that the rotation of the compass will automatically change to be relatively orthogonal to the last line placed. This is how it behaved in the previous example.

If Context Sensitivity is unchecked (OFF) then the compass remains oriented orthogonal no matter what it is being used for. The origin of the compass will still move and relocate itself. The coordinates are still relative to the origin of the AccuDraw compass. This can affect any tool using AccuDraw such as Move.

Here is an example of a line being placed with AccuDraw's two settings as shown. Context Sensitivity is unchecked (OFF) and Distance Roundoff is locked to 0.25 master units.

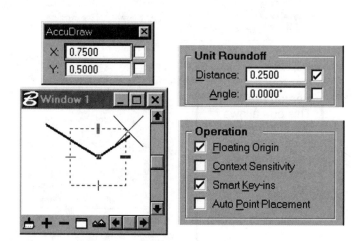

An option called Rotate AccuDraw to segments in the Place SmartLine tool allows you to avoid having to use the Context Sensitivity setting when using SmartLine. It only works with SmartLine whereas Context Sensitivity affects other tools. See the discussion on SmartLine in the following section for more details.

Sometimes the changes made in the AccuDraw Settings dialog box aren't immediately apparent. You may need to restart a tool after making changes to AccuDraw Settings.

APPLICATIONS OF ACCUDRAW

In addition to using AccuDraw and its Unit Roundoff in place of Grid Lock, there are other times when AccuDraw seems to be extremely handy even for beginners. One of those is when placing text and the other is when doing isometric pictorial drawings. Those specific applications are illustrated here. But as always, you're encouraged to try AccuDraw with other tools as well. The ones in the Modify and Manipulate tool boxes would be a good place to start.

Place Text

One of the irritating things about placing text (by origin) is that it is difficult to get each subsequent text element to be in a straight line with the previous text placed. AccuDraw takes care of that with its compass location/orientation.

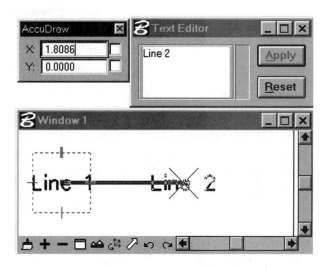

In this illustration on the right, Line 1 is text that was just placed. The next text to place is typed into the text editor as Line 2. The cursor shows its dynamic position. Since AccuDraw's compass moved to the origin of the last text placed, the next text can easily be lined up by moving the cursor until it just "snaps" into the direction of AccuDraw's X-axis. As long as you still keep placing text, it will line up nice and easy.

Isometrics with AccuDraw

Isometric refers to making a 2D drawing that appears three-dimensional. An isometric pictorial is drawn using an isometric axis, which has three axes 120° apart from each other. The 120° on paper actually represent 90° in space. Measurements in the X, Y, and Z axes of orthographics must be made on the X, Y, and Z axes of the isometric axis. Go back and review an engineering graphics book regarding pictorials if you need to. **Just keep in mind that this orientation actually refers to 90°, 210°, and 330° in standard angle convention.**

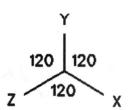

MicroStation does come with tools such as Isometric Grid, Isometric Axis, and Isometric Circle and Block. Generally speaking, they work okay if exact distances are not needed. However, to draw a true-size isometric using these tools alone is difficult, especially for beginners. AccuDraw and its Unit Roundoff setting work much better for producing isometric pictorials. You'll want to use the following specific isometric tools: Isometric Axis, Isometric Circle, and then either Place Line or SmartLine. Leave the Isometric Block and Isometric Grid (especially with Grid Lock on) out of it—they aren't very handy.

Check ON Isometric Axis Lock

The Isometric Lock setting can be found in the full Locks dialog box. Remember, this is accessed under the main pull-down menu of SETTINGS>LOCKS>FULL. The Isometric Plane set to All is the best choice. This locks the angles so that lines, etc., are oriented along the isometric axes.

Use AccuDraw's Unit Roundoff

If the true measurements are in certain increments such as 1/8th of an inch, then set the Distance Unit Roundoff of AccuDraw to this value and have it checked ON. Remember that you can get to the AccuDraw Settings dialog box from the main pull-down menu of SETTINGS>ACCUDRAW. Now since the axis is locked to isometric (done previously), the measurements will be along this isometric axis also and will jump to known increments. This saves having to key-in relative polar coordinates and coming up with the correct angle all the time.

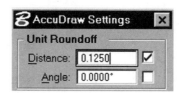

Toggle OFF AccuDraw's Context Sensitivity

In the same AccuDraw Settings dialog box, under Operations, toggle OFF the Context Sensitivity setting. It is easier to have the compass stay oriented orthogonal to the view instead of to the elements. That way it isn't trying to "snap" into an orientation other than the isometric axis.

For Example: Once these settings have been established, then by using AccuDraw and Place Line, you can draw a rectangular plane in the Top Isometric Plane that is 1" x 1.625". It's the right distance measured along the correct axis just by watching the value in the AccuDraw Distance field. Move the cursor until it shows the right distance and then use a data point.

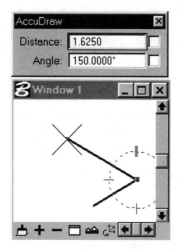 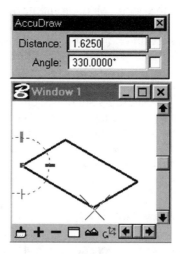

Isometric Circle

A circle in an orthographic will appear as an ellipse in an isometric pictorial. The Isometric Circle tool is located in the Isometric tool box—which is **not** found in the Main tool box. To open the Isometric tool box, you must use the main pull-down menu and check it on.

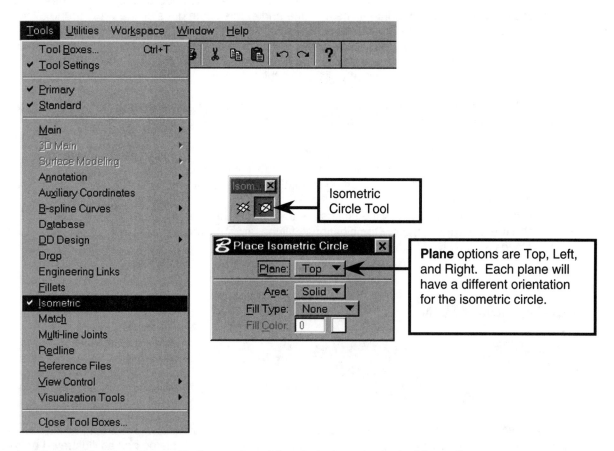

The actual tool requires you to specify the **center** of the circle. In order to do this easily, you may want to place a line that can be snapped to. Then once the center has been specified, it will want an edge point. To get the correct size of isometric circle, you'll want to be sure that you are specifying the edge at an isometric axis location.

Shown here is the middle of a Place Isometric Circle (Top Plane). A line was drawn so that its midpoint could be snapped to (it will be deleted later). Then with AccuDraw settings and Isometric Lock checked ON the isometric circle that represents a 0.5-radius circle can easily be drawn correctly to size.

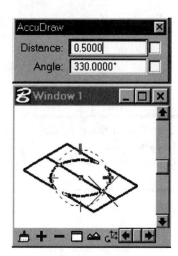

This gives you the general idea of a fairly easy way to accomplish simple "true-size" isometric pictorials with the advantage of AccuDraw.

SMARTLINE

SmartLine allows for lots of lines and arc options to be combined. You can place individual lines, line strings, rounded vertices, and closed shapes all with one tool. Pretty "smart" tool, huh? A SmartLine involves segments that can be lines or arcs. There is a vertex at each end of a segment. These segments can be joined together into a chain. If the end of this chain goes all the way around to its beginning, then it can become a closed element. The important aspect of a SmartLine is the vertex, where one segment ends and another segment begins. A vertex can be assigned a type that can be sharp, an arc, or a chamfer. The type is easily modified even after it has been placed. This all sounds fairly complicated but actually works pretty slick. Just think vertex.

Elements that are joined together are actually called different things and are assigned a "type." Often, the graphic type is listed in the status bar. So that you won't get confused, here are some simplified definitions:
- **Line String:** Lines joined together but not closed.
- **Complex Chain:** Lines, arcs, and other elements joined together but not closed.
- **Closed Element:** Where the starting point of the line string/complex chain and its ending point are joined together.
- **Closed Shape:** Only lines/line strings are involved in the closed element.
- **Closed Complex Shape:** Combinations of lines, arcs, other elements, complex chains, and line strings are involved in the closed element.

PLACE SMARTLINE

The Place SmartLine tool is defaulted as the tool on top in the Linear Elements tool box, which can be found in the Main tool box. Clicking on it results in the Place SmartLine dialog box as illustrated below.

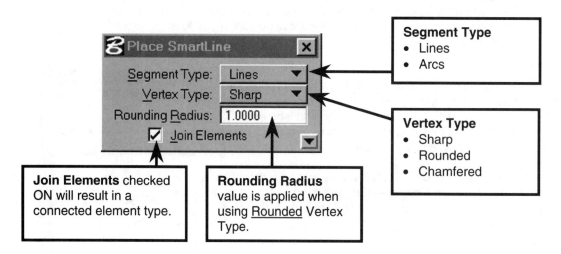

SmartLine Placement Settings

There are two options found in the SmartLine Placement Settings that appear when you click on the ▼ found in the Place SmartLine dialog box as illustrated below. The Rotate AccuDraw to segments option is useful even to beginners. Previously, preventing the AccuDraw compass from rotating to align with the last line segment placed would require that AccuDraw's Context Sensitivity setting be unchecked (OFF). With the introduction of the Rotate AccuDraw to segments setting, you can just toggle it on or off, and AccuDraw's compass will behave accordingly when using SmartLine. It is recommended that the other option of Always start in line mode be left checked ON as illustrated.

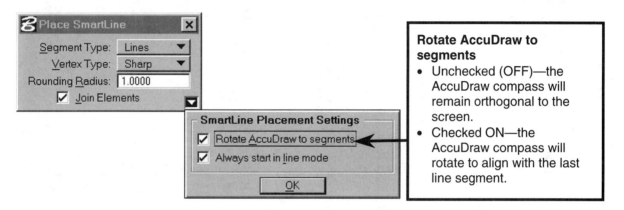

Rotate AccuDraw to segments
- Unchecked (OFF)—the AccuDraw compass will remain orthogonal to the screen.
- Checked ON—the AccuDraw compass will rotate to align with the last line segment.

Segment Type: Lines

With this option, the Place SmartLine tool creates straight lines. The endpoints are specified with data points (tentative snap is allowed). Each endpoint is a vertex, which can be any of the three types (if the size is valid). By joining the elements, the segments will form one element. With Join Elements checked ON, in order for SmartLine to create line strings or closed shapes the options need to be Lines (Segment Type) and Sharp (Vertex Type). With the other options, SmartLine will create "complex " elements.

Vertex Type: Sharp

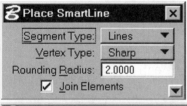

Shown below is a line string being placed using SmartLine. Settings are as shown on the left. Since Join Elements was checked ON then all the segments were joined, and it is one element. Selecting any segment would then get the entire line string as indicated with the highlighted single element. Its 7 vertices are labeled for your information.

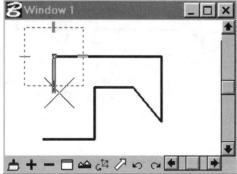
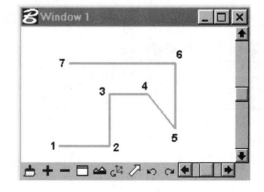

Vertex Type: Rounded

In this case the vertex will not be Sharp but instead will be rounded with an arc. It is similar to constructing a fillet. With this Vertex Type, **you need to have a segment on each side of the vertex in order to have the rounding arc**. This means that you won't see the rounding arc until the third vertex (second segment) is being entered (and the rounding radius fits). If the chain goes around so that the first vertex also becomes the last vertex, then a closed element can be made. In this case, the first vertex will then have segments on each side of it, so then its rounding radius will show up. All of this sounds complicated, but seeing is believing so let's look at an example.

For Example: Let's look at creating a rectangular shape with Vertex Type Rounded for its four vertices.

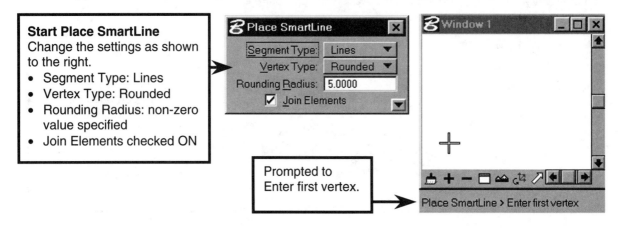

Start Place SmartLine
Change the settings as shown to the right.
- Segment Type: Lines
- Vertex Type: Rounded
- Rounding Radius: non-zero value specified
- Join Elements checked ON

Prompted to Enter first vertex.

1st Vertex was entered (Rounded Type) and now moving to 2nd Vertex. The first segment will be visible and still dynamic until the 2nd Vertex has been entered.

2nd Vertex has been entered and now moving to specify 3rd Vertex. The first segment has been placed. The second segment is now dynamic, waiting for the 3rd Vertex.

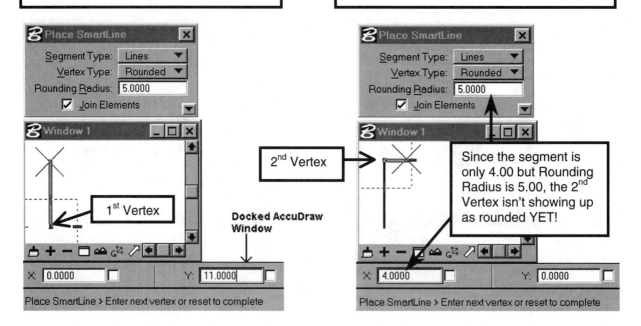

Since the segment is only 4.00 but Rounding Radius is 5.00, the 2nd Vertex isn't showing up as rounded YET!

AccuDraw and SmartLine 241

Vertices will keep being entered (with the same Type and Rounded Radius settings) and going around to make a closed shape.

As the length of the second segment is increased (to 15.00) the Rounded Type of the 2nd Vertex now appears since the Rounding Radius value of 5.00 now fits.

The 3rd Vertex has been entered and now going for the 4th Vertex (third segment).

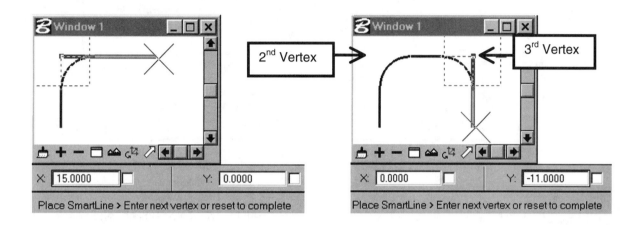

Now when the 4th Vertex has been placed (the lower right-hand corner of the rectangle), the cursor will be moved near the very 1st Vertex. Now the dialog box will change because a closed element is possible, and the last segment can use the 1st Vertex rather than have its own 5th Vertex.

Since the 1st Vertex was Rounded Type, now that it has a segment on each side of it, it remembers and shows up rounded!

This section appears and Closed Element is checked ON.

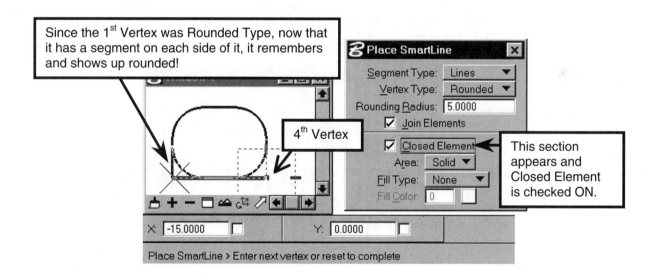

If Closed Element is unchecked (OFF) it will be a complex **chain** and there will be 5 vertices. **It can still look like a closed element** (since the 1st and 5th vertices are at the same location)—**but it won't be.**

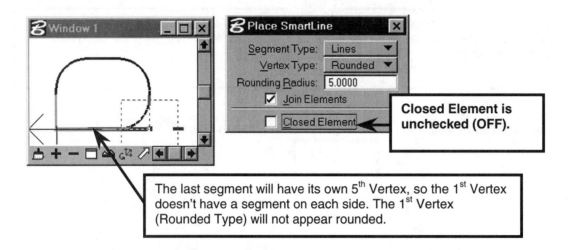

The last segment will have its own 5th Vertex, so the 1st Vertex doesn't have a segment on each side. The 1st Vertex (Rounded Type) will not appear rounded.

 It is a good idea to snap to the endpoint to be sure of getting a closed element. Also it is a bit tricky to toggle off the Closed Element setting. The dialog box changes back small when you move the cursor away from the 1st Vertex! Snap and that won't happen; the larger box with the option stays open.

Vertex type: Chamfered

This uses a straight line instead of an arc. Each chamfer must have the same distance from the vertex specified. **NO NEW VERTICES ARE ADDED FOR THE LINE.** As long as Join Elements is checked ON, the chamfer line is associated to the vertex and is not a stand-alone element.

Shown here is a **3-segment, 4-vertex** SmartLine. The vertices are all of Chamfered Type with a Chamfer Offset of 5. Notice the 4 vertices are labeled and their actual locations indicated.

This is a bit tricky for beginners to get used to. A common mistake is to try to establish where the chamfering line starts rather than giving the vertex.

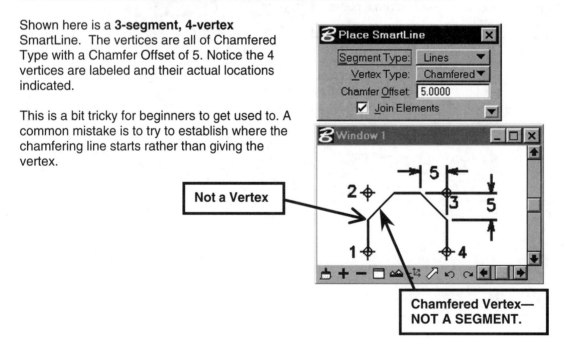

Mixing Different Vertex Types

Placing a SmartLine with all the vertices having the same Vertex Type is fairly straightforward. Set the type and enter the vertices. However, when you want a line string with different types of vertices then the sequence of events needs to be considered. **First the vertex's position must be specified and then its type can be set/changed**. The result of the Vertex Type setting won't show up until it gets segments on each side of it.

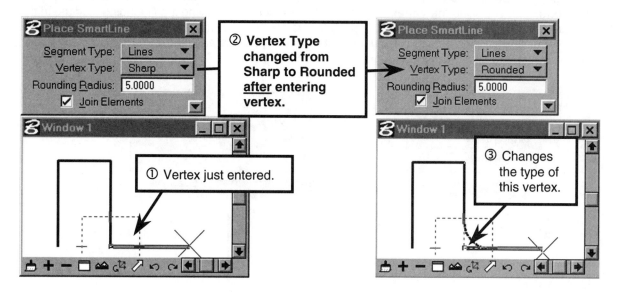

Segment Type: Arcs

The Arcs Segment Type isn't as complicated as the Lines option, especially since we're sticking to only one Vertex Type—Sharp. If Join Elements is checked ON, you can now easily create a complex chain with different-size arcs. With the settings of Arcs (Segment Type) and Sharp (Vertex Type) the Place SmartLine tool works like the Place Arc (Center Method) tool. A point on the arc is the first vertex, then the center of the arc is specified. The next vertex is also a point on the arc. One of the main advantages of using Place SmartLine for arcs is that the direction of the sweep angle is easily changed just by moving the cursor around—it isn't limited to counterclockwise.

Shown here is Place SmartLine in action with Arcs Segment Type and Sharp Vertex Type. The AccuDraw dialog box is docked at the bottom with the Distance/Angle precision input fields there. Using AccuDraw to set up the arc's radius is a good technique.

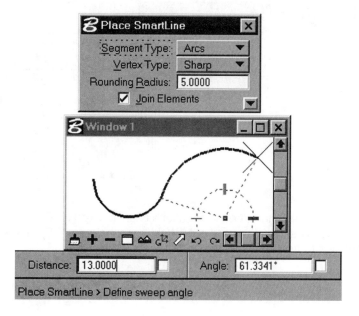

Modify Element and SmartLine Modifications

It is important to go back and look at the Modify Element tool and how it works with SmartLine. We saw a preview of this when modifying a shape, but it makes much more sense now that you're familiar with a vertex. If you select a joined element (closed or not) to modify, you can make changes to each of its vertices' Vertex Type setting too! Shown below on the left is a vertex (Vertex Type Sharp) selected for modification. The Modify Element dialog box has the Vertex Type and Rounding Radius settings for that specific vertex. By changing the Vertex Type setting to Rounded (and giving it a Rounding Radius) in the Modify Element dialog box, you can not only move the vertex but can change how it appears too! The modified vertex is illustrated below on the right. **Often it is much easier to ignore any mistakes when placing a SmartLine and then go back and modify them later.**

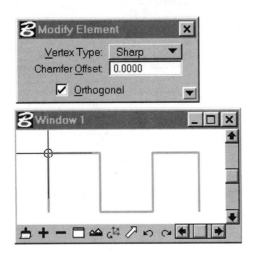 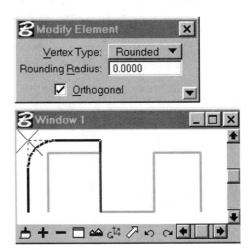

The SmartLine Modification Settings are available by clicking on the ▼ in the Modify Element dialog box as shown below. If the Enable SmartLine modifications is checked ON, then the Modify Element tool has the settings that were explained above. It is recommended that you leave Enable SmartLine modifications checked ON.

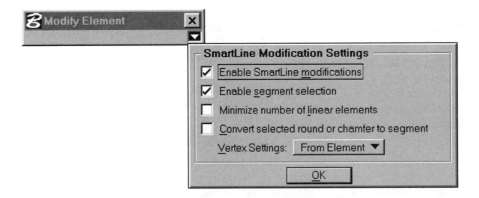

If the Enable SmartLine modifications is unchecked (OFF), then the other settings (which are for the advanced user) are grayed out. With it unchecked (OFF) then the Modify Element tool won't have any settings to use to modify the vertices, etc.

GROUPS TOOL BOX

The Groups tool box contains tools that can join elements together to make closed shapes. Create Complex Shape will take existing open elements such as lines, line strings, and arcs and join them together to create a closed complex shape. The Drop Element works in reverse—it can take elements that are joined together and "explode" them into individual elements. The Create Region tool allows you to use Boolean operations and other methods to generate a new closed element based on existing elements.

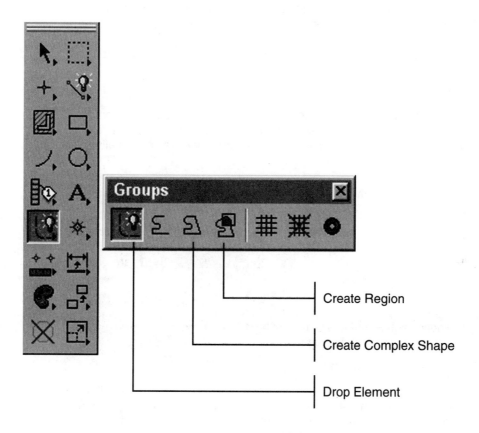

DROP ELEMENT

This is another tool that is "smart" and so has a light bulb next to it. It is the Drop Element tool and is the default on top of the Groups tool box. This tool can be used for other things, but it's mainly for dropping the status of a line string or closed element and making them just individual elements. If you used SmartLine and joined elements, this tool will take them apart and leave separate elements. Be careful—they are no longer "smart" after that! The Drop Element tool also has the option to work on dimensions, but that isn't recommended.

Shown below is the Drop Element dialog box. The check boxes are for different types of things that can have their status dropped and be separated into individual elements. As an example, the SmartLine illustrated below (with 7 segments and 8 vertices) has been selected and is all highlighted so that you can tell that it is 1 element.

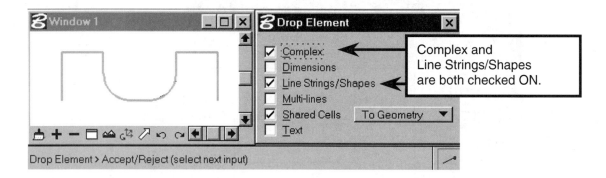

The element has been dropped (the status bar will say so). The SmartLine will be dropped to its simple elements, including lines, **line strings,** and arcs. If you would also like to drop the line strings into individual lines, you would have to execute the command again, this time selecting the line string portion shown in gray on the left below. Since it is no longer "smart," the vertices become just regular endpoints and the arcs turn into their own elements too. However, everything will still look as if they are all one element until you try to modify or delete it.

Below on the left, the line string (with Sharp vertices) that remains is highlighted for deleting. The illustration on the right shows what modifying the arc element will look like. Notice that it now has endpoints. There won't be a Vertex Type setting available in the Modify Element dialog box because the arc is no longer "smart."

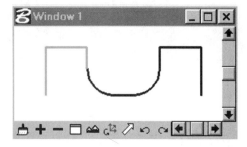 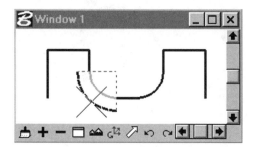

CREATE COMPLEX SHAPE

This Create Complex Shape tool allows you to identify elements, line strings, and/or complex chains that will be joined together to form a closed shape. The elements identified must be contiguous, so it's best if they have originally been placed very precisely (by snapping to keypoints). Start the tool, and you'll be prompted to identify the elements. When you identify the elements with a data point, be sure to select them **in order** around the perimeter of the complex shape-to-be. After identifying the last item, then data point in an empty space in the view to accept. Now **read the prompt to see if it actually closed the shape.** If it didn't, usually the original element's endpoints weren't close enough to be considered contiguous. You can increase the Max Gap value and try again.

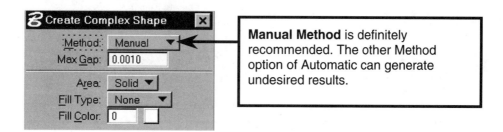

Manual Method is definitely recommended. The other Method option of Automatic can generate undesired results.

CREATE REGION

The Create Region tool also creates a closed shape, but with this tool, the elements don't need to be contiguous. It uses the concept of Boolean operations that were discussed in Chapter 8 and the methods available for the Measure Area tool. With the Create Region tool, a new closed shape is created (with the Active attributes) from the region defined by the Boolean operation. The Flood Method will have a Max Gap setting available and will require a data point inside the area to flood. It too works similarly to the Flood Method of the Measure Area.

Keep Original
- If unchecked (OFF) the original elements selected to define the region will be deleted.
- If checked ON it will create the new closed shape without deleting the original elements.

Method
- Intersection
- Union
- Difference
- Flood

POPUP CALCULATOR

There are times when being able to calculate a value "on the fly" would be handy. MicroStation SE comes with a new popup calculator that does just that. Some fields that support the popup calculator are AccuDraw's X, Y, and Z fields and the Active Angle and Active Scale fields that show up in various dialog boxes. You can use the mathematical operators of +, –, *, /, =, or go to the advanced features and define variables and other functions. As the name "popup" implies, the calculator pops up when these operators are typed into a field that supports it. The calculation is shown and is accepted by using the ENTER key, doing a data point, or taking focus out of the field. If you don't want the calculated value, use the ESC key to reject it.

Let's look at the Scale tool and using the Active Scale field to illustrate the popup calculator. We'll use it to do the simple calculation of 3 divided by 2.

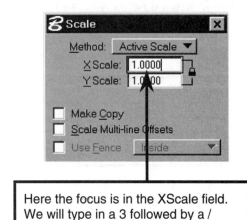
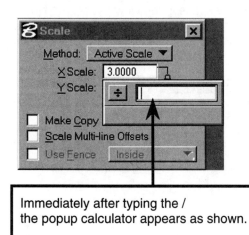

Here the focus is in the XScale field. We will type in a 3 followed by a /

Immediately after typing the / the popup calculator appears as shown.

Now to use the popup calculator to do that tough math problem for us!

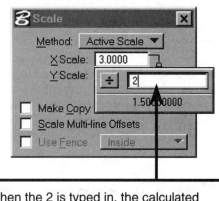
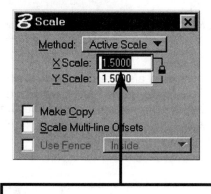

When the 2 is typed in, the calculated value will be shown below the field. You can use the ENTER key to accept or the ESC key to reject.

Here's the result of accepting the popup calculator's value. The XScale has been changed.

Now this was a simple example. You can try out using the popup calculator in conjunction with AccuDraw. Please refer to Help for a thorough explanation of the accuracy and sign conventions that pertain to the popup calculations.

TABLE OF ACCUDRAW KEYBOARD SHORTCUTS

Key:	Effect:
<*?*>	Opens the AccuDraw Shortcuts window.
<*Return*>	Smart Lock • In Rectangular coordinates, locks X to 0 if the pointer is on the drawing plane y-axis or Y to 0 if the pointer is on the X-axis. • In Polar coordinates, locks Angle to 0°, 90°, –90°, or 180° if the pointer is on a drawing plane axis or otherwise locks Distance to its last entered value.
<*spacebar*>	Switches between Rectangular and Polar coordinates.
<*O*>	Moves the drawing plane origin to the current pointer position.
<*X*>	Toggles the lock status for the X value.
<*Y*>	Toggles the lock status for the Y value.
<*Z*>	Toggles the lock status for the Z value.
<*D*>	Toggles the lock status for the Distance value.
<*A*>	Toggles the lock status for the Angle value.
<*N*>	Activates Nearest snap mode.
<*C*>	Activates Center snap mode.
<*I*>	Activates Intersect snap mode.
<*K*>	Opens the Keypoint Snap Divisor settings box, which is used to set the Snap Divisor for keypoint snapping.
<*R*>,<*Q*>	Used to quickly and temporarily rotate the drawing plane.
<*R*>,<*A*>	Used to permanently rotate the drawing plane. Because it rotates the current ACS, this rotation will still be active after the current command terminates.
<*R*>,<*X*>	Rotates the drawing plane 90° about its X-axis.
<*R*>,<*Y*>	Rotates the drawing plane 90° about its Y-axis.
<*R*>,<*Z*>	Rotates the drawing plane 90° about its Z-axis.
<*F*>	Rotates the drawing plane to align with the axes in a standard Front view.
<*S*>	Rotates the drawing plane to align with the axes in a standard Right view
<*T*>	Rotates the drawing plane to align with the axes in a standard Top view.
<*V*>	Rotates the drawing plane to align with the view axes.
<*W*>,<*A*>	Saves the drawing plane alignment as an ACS.
<*G*>,<*A*>	Retrieve a saved ACS.
<*P*>	Opens the Data Point Key-in settings box for entering a single data point.
<*M*>	Opens the Data Point Key-in settings box for entering multiple data points.
<*G*>,<*K*>	Opens (or moves focus to) the Key-in window (same as choosing Key-in from the Utilities menu).
<*G*>,<*S*>	Opens (or moves focus to) the AccuDraw Settings box (same as choosing AccuDraw from the Settings menu).
<*G*>,<*T*>	Moves focus to the Tool Settings window.
<*Q*>	Deactivate AccuDraw.

Table courtesy of Bentley Systems, Incorporated

QUESTIONS

① Explain the result of having the Context Sensitivity setting unchecked (OFF) in the AccuDraw Settings dialog box. Also describe how the compass in AccuDraw will behave.   *orientate to existing factor*   *or thogonally*

② What key on the keyboard will toggle between the polar and rectangular coordinate systems?

Space bar.

③ State whether the following statements dealing with AccuDraw are True or False:
- *false* a) The keyboard shortcuts are case-sensitive.
- *false* b) Key-in values and snapped tentative points will take precedence over the Unit Roundoff setting.
- ✓ c) The values shown in the AccuDraw dialog box are relative to the origin of AccuDraw compass.
- ✓ d) The X-axis and Y-axis of AccuDraw's compass can be reoriented.

④ What is the coordinate system indicated by each of the compasses shown below?

Coordinate System **A**   Coordinate System **B**

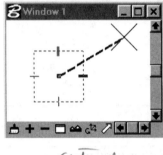   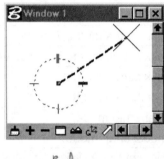

rectangular   *polar.*

⑤ What are the three different Vertex Types available in Place SmartLine?

Sharp, round, chamfered.

⑥ Is the element on the right a closed element? Explain your answer.

YES IF YOU SNAP TO END POINT OTHERWISE NO

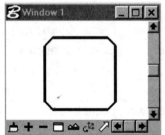

⑦ How many vertices would be involved in the SmartLine (done with Vertex Type Chamfered) shown on the right?   *5*

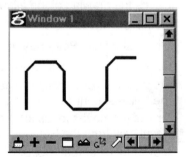

⑧ What tool will take a Line String or Shape done with SmartLine and break it up into smaller individual elements?

— *drop element*

EXERCISE 11-1 Belt Drive

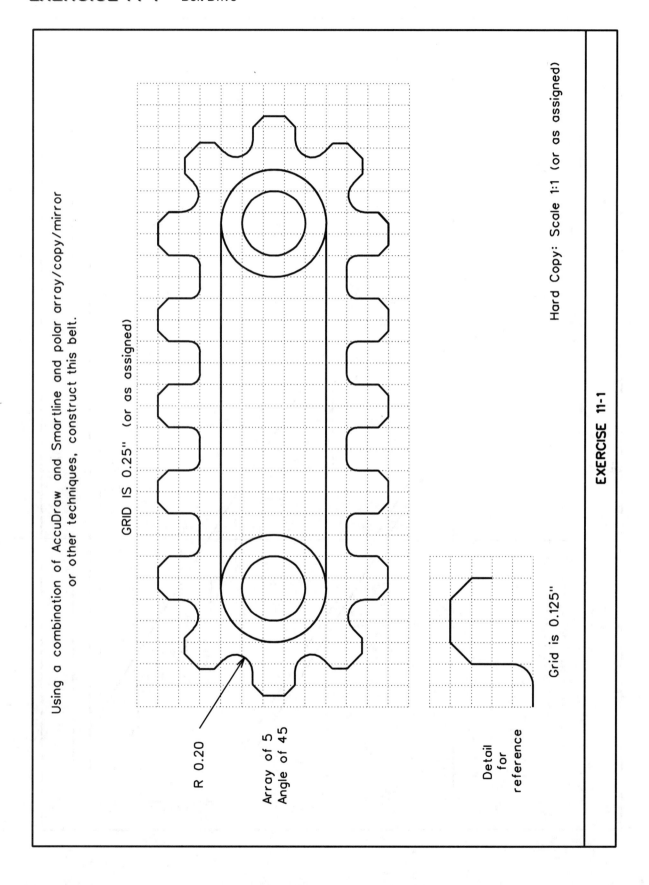

EXERCISE 11-2   Stairs and Railing

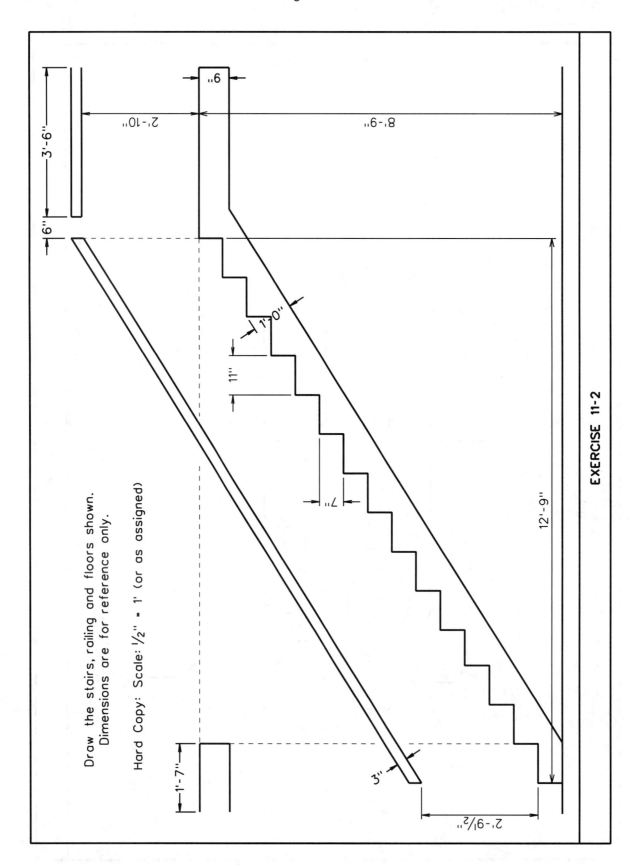

EXERCISE 11-2

EXERCISE 11-3   Dangerous Intersection

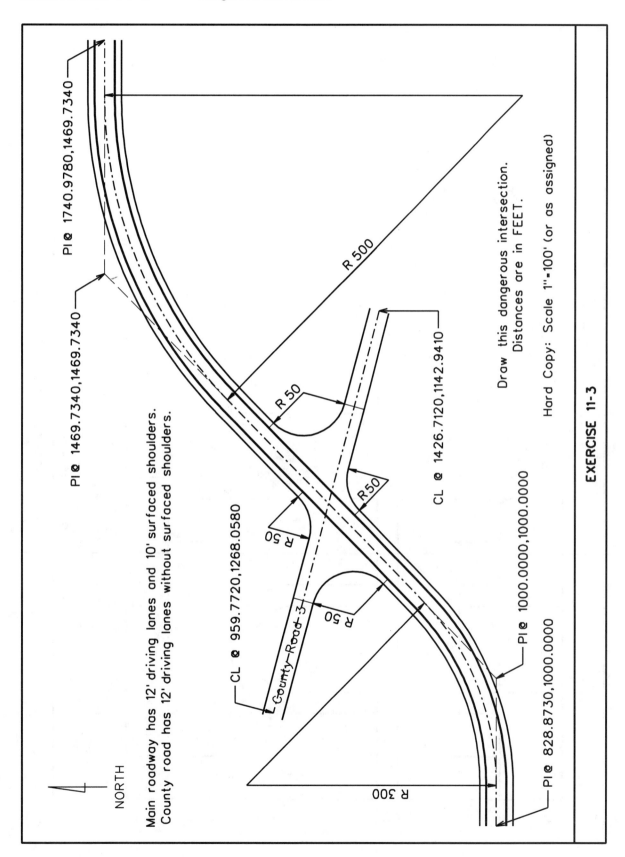

EXERCISE 11-3

EXERCISE 11-4 Isometric Pictorial of Bracket

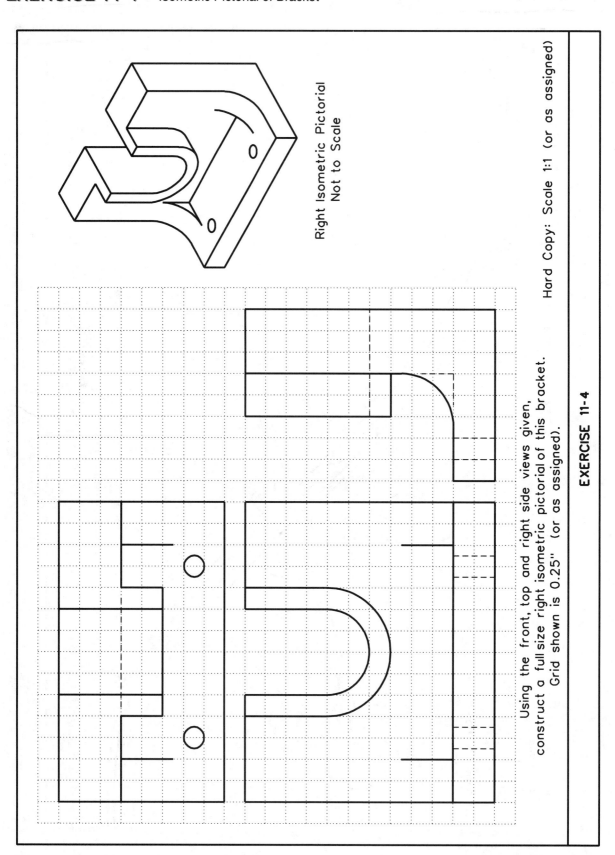

Chapter 12:
Cells and Cell Libraries

SUBJECTS COVERED:

- Cell Types
- Cell Library
- Cells Tool Box
- Creating Cells
- Different Uses of Cells
- Place Active Cell
- Replace Cells
- Select and Place Cell
- Cell Selector Utility
- Other Buttons in the Cell Library Dialog Box

When you find yourself creating the same items over and over again for each drawing then you may want to make a cell out of the elements. A cell is a group of elements that have been tied together to form a new entity. A cell can be placed over and over again at various scales, attributes, etc., in **ANY** drawing. Cells allow for a standardization of many drawings. To make the organization easier, cells are stored in libraries. **You can store numerous cells in a single cell library.** MicroStation comes with default libraries and their cells that you can use. It is a simple process to create your own cells either in your own new cell library or in an existing library.

CELL TYPES

There are two different types of cells: Point and Graphic. How the cell acts when it is placed in a drawing depends on what type it is. There are advantages and limitations for each type. When you are using the libraries available, you must pay attention to what type of cell you have selected.

Graphic Cell Type
- It maintains the settings/attributes that it had when the cell was created.
- It is good for maintaining standards for a company.
- The keypoints of the individual elements that make up the cell are still recognized.

Point Cell Type
- It takes on the Active attributes when it is placed.
- You can snap only to the origin of the cell.

CELL LIBRARY

You must attach a library to your drawing file before you can use the cells that it contains. This doesn't mean that you are copying all of its contents into your drawing, instead it specifies the path to the library file. These cell library files have a .cel extension. You can open the Cell Library dialog box with the main pull-down menu under ELEMENT>CELLS.

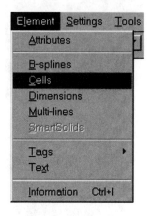
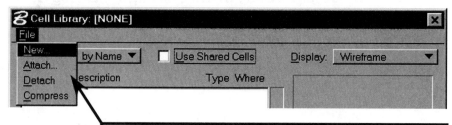

New
You can create your own cell library.

Attach
- **There can be only one** library attached at a time.
- It automatically detaches any other library.

Detach
- You may detach a library when you are done with it.

Cells and Cell Libraries 257

Attaching an Existing Cell Library

To open the Attach Cell Library dialog box, use the Cell Library pull-down menu FILE>ATTACH as illustrated at the right.

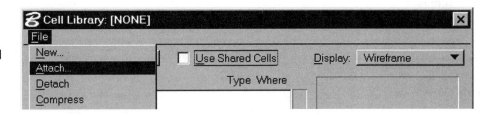

Numerous cell libraries are available. Some contain 2D (two-dimensional) cells and others 3D (three-dimensional) cells. A warning will be given if you try to attach a 3D cell library to a 2D file. Mixing of 2D and 3D is not recommended.

Some of the cell libraries already available are found under the **examples** folder in the path illustrated at the right. It contains the various sub-folders as listed. In most of these folders there is a **cell** folder that contains the cell libraries available for that type of example project.

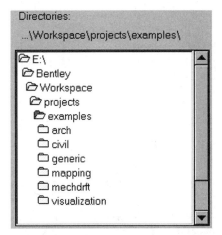

Other standard cell libraries that you may want to attach are stored in the system directory as illustrated below. The areapat.cel library includes some ANSI standard material patterns, which will be used later for patterning.

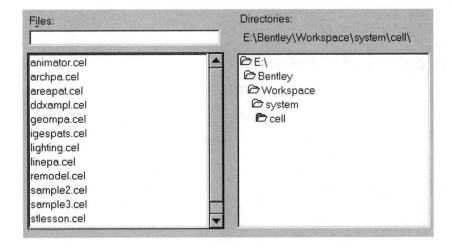

 Remember that only one library can be attached at a time. If there is already a file attached, it will be detached automatically and the selected cell library will be attached.

Illustrated below is the cell folder found under the mapping folder. There are four cell libraries found in the mapping discipline. They appear in the list box. Each of the cell libraries can contain numerous cells available for use.

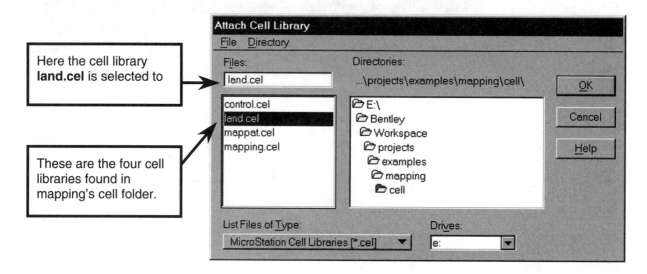

Here the cell library **land.cel** is selected to

These are the four cell libraries found in mapping's cell folder.

Here are the land.cel library and its available cells.

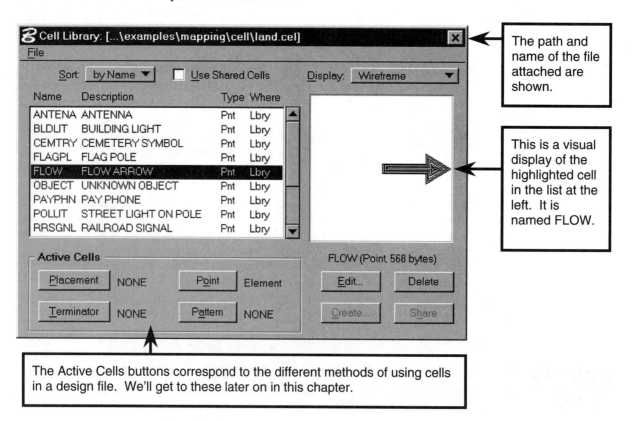

The path and name of the file attached are shown.

This is a visual display of the highlighted cell in the list at the left. It is named FLOW.

The Active Cells buttons correspond to the different methods of using cells in a design file. We'll get to these later on in this chapter.

Cells and Cell Libraries   259

Creating a New Cell Library

A cell must belong to a cell library. If you are not adding a cell to an existing library file, then you must create your own **NEW** file using the Create Cell Library dialog box. This dialog box is accessed from the **Cell Library** pull-down menu FILE>NEW as shown below.

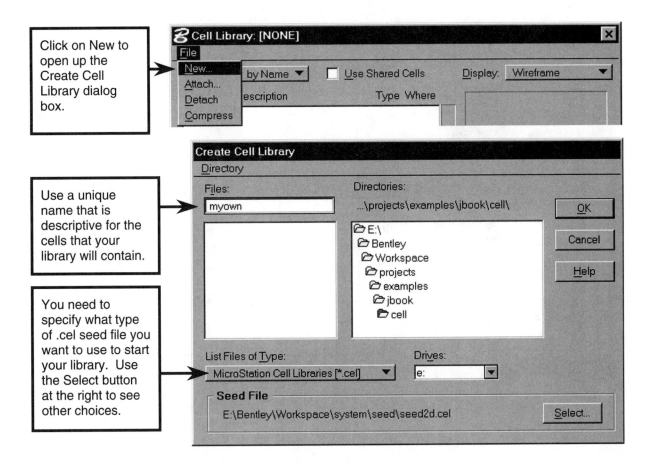

Detaching a Cell Library

You may Detach a cell library. The option is found in the Cell Library pull-down menu FILE>DETACH as illustrated below. Remember that only one attachment can be made at a time. Therefore you are not required to specify which cell library to Detach—since there can be only one attached, there's only one to detach! If a library is still attached upon exiting a drawing, then MicroStation will make that same attachment when the design file is opened. If it can't find the cell library file, it will still open the design file (without the attachment). Please note that closing the Cell Library dialog box does not Detach the cell library.

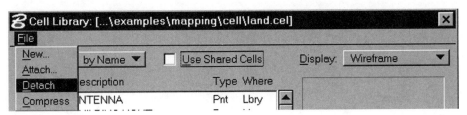

CELLS TOOL BOX

The Cells tool box is found in the Main tool box. We're going to look at the one dealing with placing a cell, the tool that defines the origin for the cell, and the ones that let you select an existing cell and place it or replace it. The others will be left for you to experiment with on your own. First let's look at how to actually create a cell, then we'll figure out how to place it.

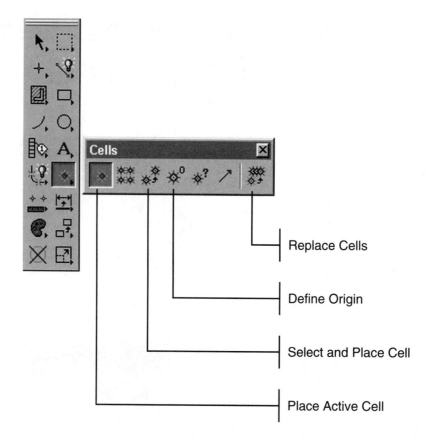

CREATING CELLS

First there must be elements that you want to make up your cell. Some special consideration must be given to creating a Graphic type cell because the attributes of the elements used to make a Graphic cell will be retained as the attributes when the cell is placed. Also the use of cells included in other cells is not recommended. This is called nesting cells. A better procedure would be to drop status on the cell that would have been nested. This will break it up into its basic elements and these can be included in the new cell. The Drop Element tool discussed in Chapter 11 can be used to drop status of the cell.

Basic Steps for Creating Cells

❶ **Be sure that you have a cell library attached.** If not, do so.

❷ **Select the elements to be included.** You can use a fence or an element selection tool.

❸ **Define an origin for the cell** using the Define Origin tool in the Cells tool box.

 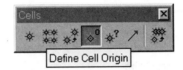

You will be prompted for a data point. The origin is the point that will be used to position the cell later on when it is placed. Once again, in simple terms this is where the cell will be "grabbed onto." It is very important to use an origin that makes sense for later placement. Keypoints of an element make very good positions for defining an origin. For example, the center of a circle is a good origin point. An arbitrary point off the object is not.

❹ **Click on the Create button in the bottom right corner of the Cell Library dialog box.** If you've done these steps correctly, this button will be available. If the Create button is dimmed, be sure that an origin has been defined **and** elements are selected.

❺ **Complete the Create New Cell dialog box that appears.** Complete the information for the name of the cell, a brief description, and what type of cell it will be. When you've entered at least the name information, the Create button will be active. You must click on it to actually create the cell.

Here are some elements forming a T-shape that are all ready for cell creation. An origin has been defined and a fence exists to select the elements.

Origin Defined (midpoint of the top line).

- The letter O is placed near the origin point—it is NOT a circle whose center marks the origin.
- **Only one origin can be defined at a time**, but the letter stays there until you update the view.
- If you are creating more than one cell, be sure that you define a new origin for the new cell.

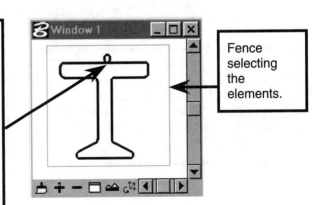

Fence selecting the elements.

Here's the Create New Cell dialog box that appears when you click on the Create button of the Cell Library dialog box.

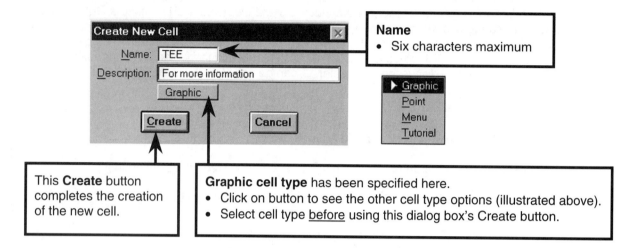

Name
- Six characters maximum

This Create button completes the creation of the new cell.

Graphic cell type has been specified here.
- Click on button to see the other cell type options (illustrated above).
- Select cell type <u>before</u> using this dialog box's Create button.

Things to Remember When Creating Cells

Be sure that you've defined the origin for the cell correctly. **The letter O can show up all over the place, but there is only one actual origin defined at a time**. Illustrated below are some of the other common stumbling blocks.

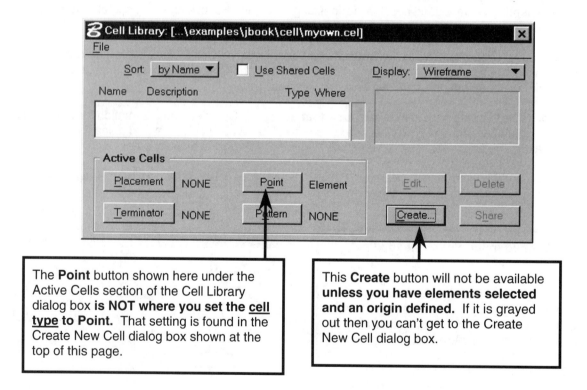

The **Point** button shown here under the Active Cells section of the Cell Library dialog box **is NOT where you set the cell type to Point.** That setting is found in the Create New Cell dialog box shown at the top of this page.

This **Create** button will not be available **unless you have elements selected and an origin defined.** If it is grayed out then you can't get to the Create New Cell dialog box.

DIFFERENT USES OF CELLS

There are four different ways to use cells in your drawings.
- **PLACEMENT** is used with Place Active Cell. This is the most practical application.
- **TERMINATOR** is used for putting arrowheads, etc., on elements.
- **POINT** is used for placing unique-looking "points" but has limited settings.
- **PATTERN** is used to fill in a closed shape with the cell repeated over and over. It also specifies the cell to use for linear patterning. This will be discussed in the next chapter.

These correspond to the **buttons** in the Active Cells section of the Cell Library dialog box. Highlight a cell name in the cell library and click on one of the four buttons. That cell will become the **Active Cell for that individual command**. Each button can have its own Active Cell. The name of the cell will appear to the right of the button. As illustrated, the Active <u>Placement</u> Cell is TEE but the Active <u>Pattern</u> Cell is TIN. The NONE indicates that Terminator doesn't have an Active Cell specified for it. If there is an Active Cell specified for a command then it will automatically be used in the command. These buttons in the Cell Library dialog box only set their Active Cell—**they do not perform the command**. Tool buttons for the commands are found in different tool boxes. We'll look at the **Place** Active Cell tool that uses the Active **Placement** Cell.

PLACE ACTIVE CELL

The Place Active Cell tool is the most common use of cells. It allows you the most control of your cell. Along with the location, you can also set the angle and scale of the cell. The tool is found in the Cells tool box and is the default tool on top. Once you've specified the Active Cell, then you are prompted to Enter cell origin. This refers to specifying where you want to place the cell in your design file. The origin point that was defined for the cell will be placed at the coordinates specified here (graphically or by key-in).

Illustrated on the right is the Place Active Cell dialog box. Its settings will be discussed in detail.

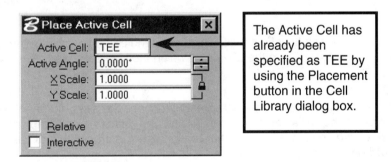

The Active Cell has already been specified as TEE by using the Placement button in the Cell Library dialog box.

Active Placement Cell

The Active Cell specifies **which** cell from the attached cell library is to be placed using the Place Active Cell tool. In this illustration, it had already been specified as TEE from the Cell Library dialog box. There are other ways to specify the Active Cell setting for the Place Active Cell tool.
- If you know the name, you can use the key-in ⌨AC=*cellname*. This key-in will automatically open the Place Active Cell dialog box without having to click on its tool.
- You can type the cell's name in the Active Cell field of the Place Active Cell dialog box.

Active Angle

The Active Angle setting allows you to rotate the cell when it is placed. **This setting is relative to the orientation of the cell at the time of its creation.** Assume the original cell was rotated 30° at creation. If Active Angle is set at 10° then the Placed cell will be rotated at 40°. The Active Angle setting can affect other tools. It will remain at the value specified until you change it. The key-in for setting the Active Angle is ⌨AA=.

XScale and YScale Settings

The most important thing to remember here is that the **XScale** value is a multiplication factor that will be applied to the cell's size in the **original X-direction**. The **YScale** value is also a multiplication factor but it will be applied to the cell's size in the **original Y-direction**. The little padlock symbol indicates whether or not the settings values must be the same. If the padlock appears closed, then the XScale and YScale settings must have the same value. If you change one, the other will automatically change to the same value. If the padlock appears open as shown below, then the XScale and Yscale settings can have different values. Changes to the XScale and YScale settings here also change the same settings in the Scale tool (Active Scale Method) found in the Modify tool box. There is also an Active Scale key-in that would set the one value for both the XScale and the YScale. The key-in is ⌨AS=*scalefactor*.

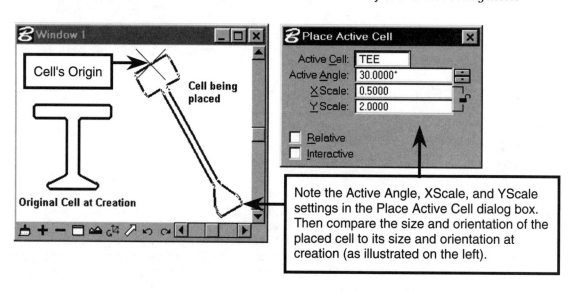

Note the Active Angle, XScale, and YScale settings in the Place Active Cell dialog box. Then compare the size and orientation of the placed cell to its size and orientation at creation (as illustrated on the left).

Relative

This option deals with Level settings, which are handled for Graphic cell types. If unchecked (OFF), the elements will be assigned to the Levels on which they were created. This is recommended for beginners.

Interactive

With Interactive checked ON, you **graphically** choose the size and orientation of the cell. After specifying where to place the cell's origin, you can stretch and rotate the cell graphically. The cell origin remains anchored where you placed it—another reason why it is crucial that at cell creation time you define a logical cell origin.

 A placed cell is considered an element. The Delete Element tool will delete the **entire** cell, but it will not delete the cell defined in the cell library.

Example of Active Cell Placement

Shown below are two different ways of using Place Active Cell. Notice the more important parts indicated.

TEE is a Graphic cell type, so even though the Active Line Type is 3, the actual cell placed has a 0 Line Type (what it was when the cell was created).

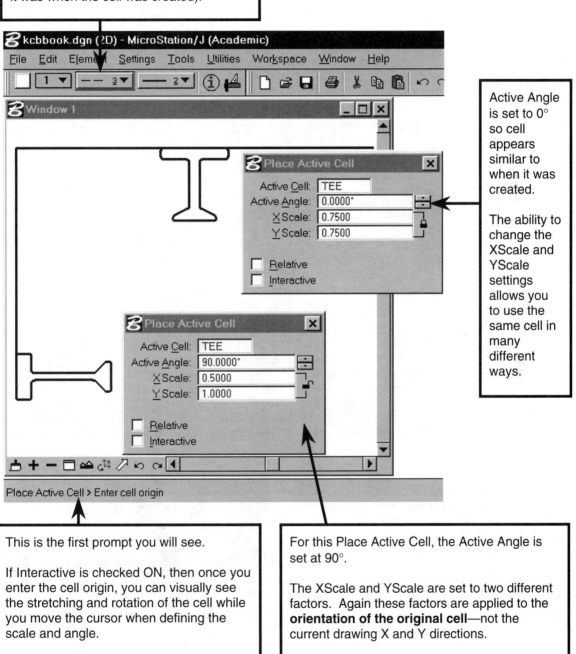

Active Angle is set to 0° so cell appears similar to when it was created.

The ability to change the XScale and YScale settings allows you to use the same cell in many different ways.

This is the first prompt you will see.

If Interactive is checked ON, then once you enter the cell origin, you can visually see the stretching and rotation of the cell while you move the cursor when defining the scale and angle.

For this Place Active Cell, the Active Angle is set at 90°.

The XScale and YScale are set to two different factors. Again these factors are applied to the **orientation of the original cell**—not the current drawing X and Y directions.

REPLACE CELLS

 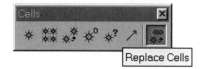

This Replace Cells tool is wonderful for those of us who either make mistakes or have to do revisions—covers about everybody, doesn't it? It has two different Method options, Update and Replace. Update utilizes a cell that exists in a different cell library but has the same name. Replace uses one cell to replace another cell (or cells) that don't have the same name.

Update Method

Let's assume that we've created a second cell library and called it secondmyown.cel. In this second cell library there is also a cell named TEE but its base is different from the cell TEE in the myown.cel library. It is set as the Active Placement Cell. This is illustrated on the right.

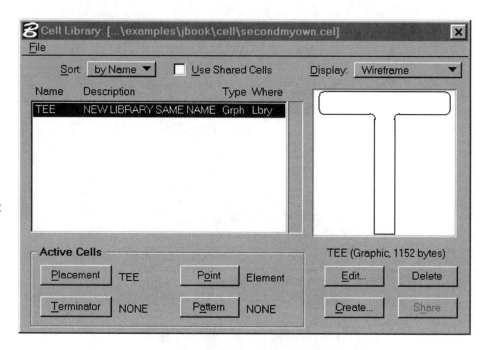

The original TEE cell was placed three times in a row as illustrated at the right. A fence surrounds two of them and the Replace Cells tool has been selected with the settings as shown. The prompt reads Accept/Reject Fence.

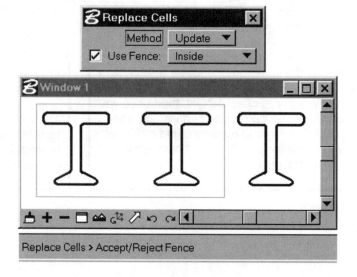

Cells and Cell Libraries 267

If the replacement is accepted, the cells will be replaced with the TEE from the secondmyown.cel library as shown here on the right.

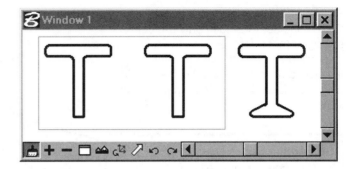

Replace Method

The Replace Method allows you to replace cells with a differently named cell. It has two different Mode options: Global or Single. Single allows you to specify individual cells that need to be replaced. Global replaces all instances of the specified cell(s).

For the following examples of both modes, we will use the layout of cells as illustrated on the right. There are four cells of our original TEE, two triangular-shaped cells called ANTENA, and three cells called TOWER. The ANTENA and TOWER cells can be found in the land.cel library referred to earlier in this chapter.

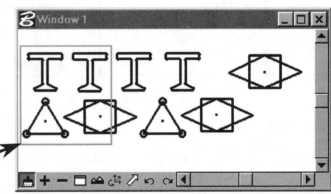

A fence is placed, it surrounds two TEE cells and one ANTENA cell. The fence overlaps one TOWER cell.

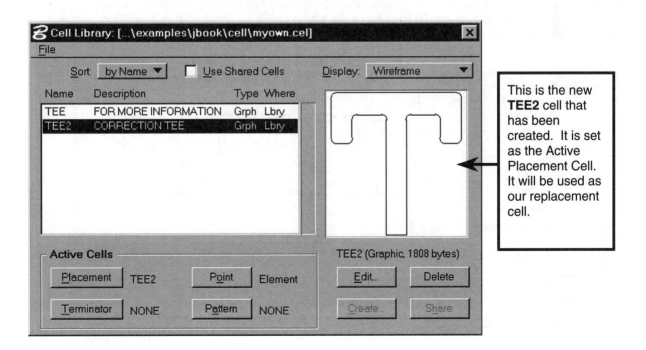

This is the new **TEE2** cell that has been created. It is set as the Active Placement Cell. It will be used as our replacement cell.

Single Mode

With Mode set to Single, only the cells specified (not necessarily all instances of the cell) will be replaced with the new cell. Illustrated here on the right is the result of using Replace Cells with the settings as indicated.

Use Active Cell is checked ON.
- The replacement cell is TEE2.

Use Fence is checked ON.
- **All** cells (two TEE cells and one ANTENA cell) <u>inside</u> the fence are replaced with TEE2.
- If an Overlap fence had been specified, the one TOWER cell would have been replaced also.

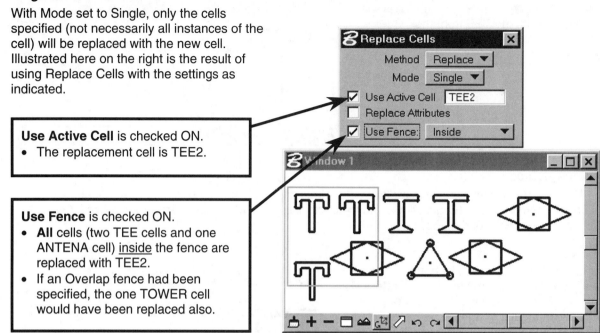

Global Mode

With the Global Mode, the cell to be replaced is identified and then **all instances of cells with the same name** will be replaced. If the Use Fence option is unchecked (OFF) then you will identify the cell to be replaced with a single data point. An Alert box will pop-up stating how many cells will be replaced and you can say Yes or No to the global replacement. It is pretty straightforward. Now if you have the Use Fence option checked ON the whole thing requires a bit of finesse. The fence is used as search criteria (by cell name) for which cells will be replaced. **Therefore even cells that aren't in the fence specification will still be replaced <u>if</u> they have the same name as a cell that <u>is</u> specified by the fence.** This requires an example to really make sense.

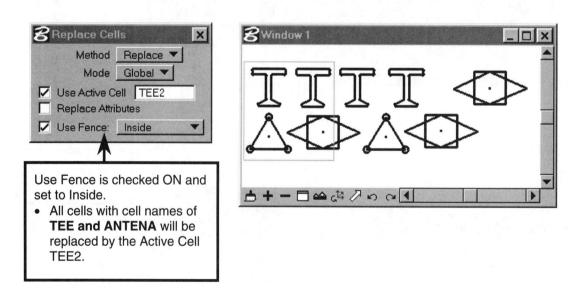

Use Fence is checked ON and set to Inside.
- All cells with cell names of **TEE and ANTENA** will be replaced by the Active Cell TEE2.

Cells and Cell Libraries 269

Upon accepting the selection with a data point, the Alert shown below will pop-up.

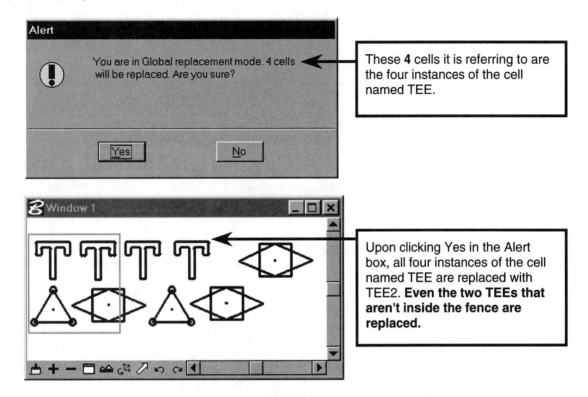

These **4** cells it is referring to are the four instances of the cell named TEE.

Upon clicking Yes in the Alert box, all four instances of the cell named TEE are replaced with TEE2. **Even the two TEEs that aren't inside the fence are replaced.**

Another Alert pops up because cells that are named ANTENA were also specified by the fence.

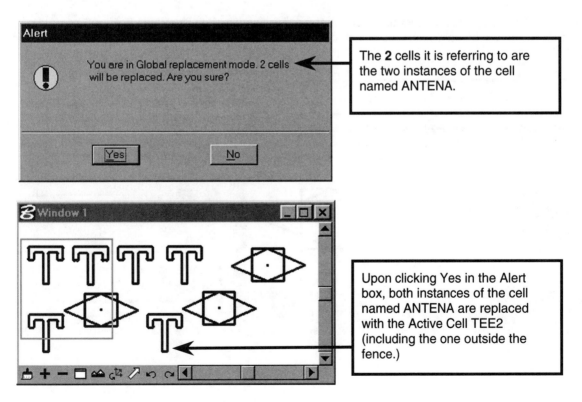

The **2** cells it is referring to are the two instances of the cell named ANTENA.

Upon clicking Yes in the Alert box, both instances of the cell named ANTENA are replaced with the Active Cell TEE2 (including the one outside the fence.)

SELECT AND PLACE CELL

 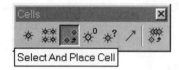

This tool is handy if there are cells already used in the design file and you want to select a cell and immediately place another instance of it. This is useful to those of us who forget what they named the cell but can see it and want to just use it again.

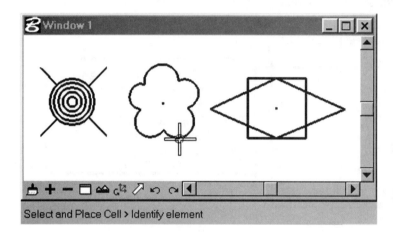

Here are three cells from the land.cel library that have been placed. You are prompted to Identify element. This is done with a data point on the edge of the shrub-like shape as illustrated.

Once the data point has been done, the element (cell) will highlight as illustrated below. The data point used to Accept will actually place the cell at that position with the settings active in the Select and Place Cell dialog box.

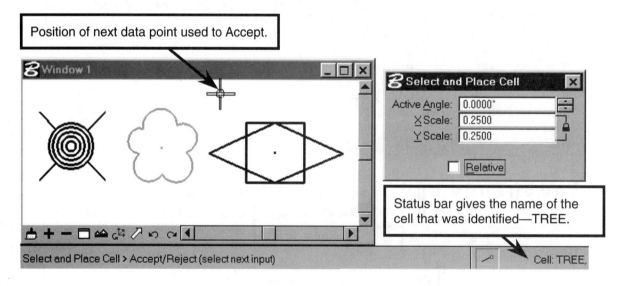

Position of next data point used to Accept.

Status bar gives the name of the cell that was identified—TREE.

Cells and Cell Libraries 271

You can continue to place more cells by just using a data point to enter the cell orgin.

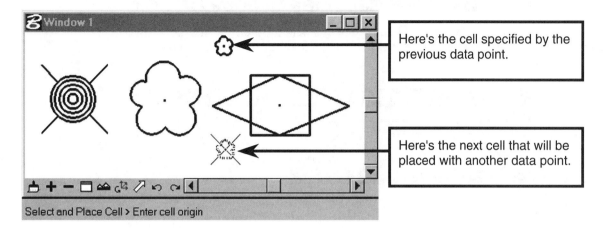

Here's the cell specified by the previous data point.

Here's the next cell that will be placed with another data point.

CELL SELECTOR UTILITY

This is a handy utility that lets you view, select, and place a cell with a button tool. It is found in the main pull-down menu under UTILITIES>CELL SELECTOR. The buttons represent the cells available. By using the button, the Place Active Cell dialog box will open and the Active Cell will be set to the button's corresponding cell.

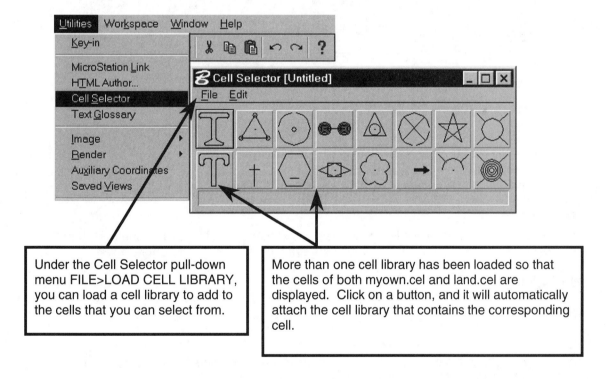

Under the Cell Selector pull-down menu FILE>LOAD CELL LIBRARY, you can load a cell library to add to the cells that you can select from.

More than one cell library has been loaded so that the cells of both myown.cel and land.cel are displayed. Click on a button, and it will automatically attach the cell library that contains the corresponding cell.

OTHER BUTTONS IN THE CELL LIBRARY DIALOG BOX

There were a few buttons in the Cell Library dialog box that weren't discussed. The ones in the lower right-hand corner deal with making corrections to the information saved with the cell.

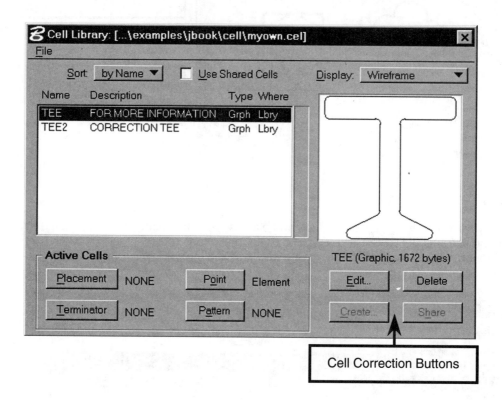

Cell Correction Buttons

Edit Button
This button **affects just the name and description** of the cell, not the elements making up the cell. So it is useful for typos but that is about it.

Delete Button
This deletes the cell definition in the library and lets you start over. This is usually the easiest option for beginners. Do not try to create another cell with the same name—be sure to Delete it first.

Create Button
This is available if elements have been selected and a cell origin has been defined. This was discussed in detail earlier on in this chapter.

Share Button
Shared cells save space because only one copy exists in the design file. It is used when you're placing a large number of the same cell. However, it is not recommended for beginners.
- The benefit is that if you replace one shared cell then all of the same shared cells are also updated.

QUESTIONS

① What **type** of cell will take on the Active attributes and settings when it is placed?
  Point Cell

② What must be defined for a selected set of elements in order for the Create button of the Cell Library dialog box to be available? *Must have elements selected and an origin defined*

③ How many cell libraries can be attached at one time?
  only one

④ What are the four different buttons that set their own Active Cell for different uses? *Placement, Point, Terminator, Pattern*

⑤ What does checking ON the Interactive setting in the Place Active Cell dialog box allow you to do?
  you can graphically choose size and orientation

⑥ What orientation are the XScale and YScale settings in the Place Active Cell based on?
  on the original x & y directions

⑦ If the Active Angle is changed in the Place Active Cell does it have any effect on other tools? If so, what?

⑧ What does this Delete button in the Cell Library dialog box do?
  This deletes the cell definition lists

⑨ Describe what the Cell Selector utility does.
  place a cell with button tool

⑩ Why is the definition of a cell's origin important—how does it affect a cell's placement?

This will be the point the cell will be placed
10. keypoint & origin good

— importantly, especially if you choose a active origin then ✓
interactive to stretch settings and rotate a cell
therefore the origin stays the same

274 Chapter 12

EXERCISE 12-1   Chess Pieces

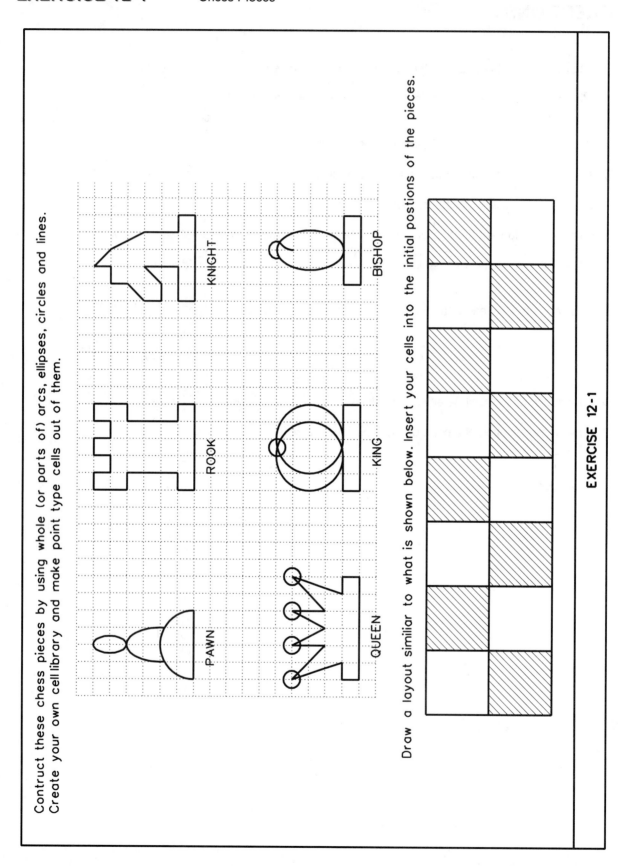

Cells and Cell Libraries

EXERCISE 12-2 Part A: Piping Symbols

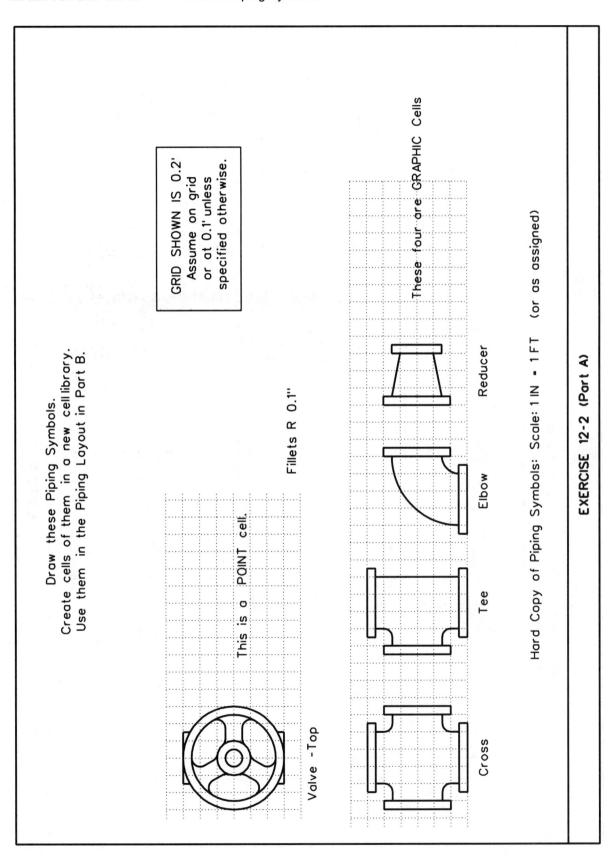

EXERCISE 12-2 (Continued)   Part B: Piping Layout

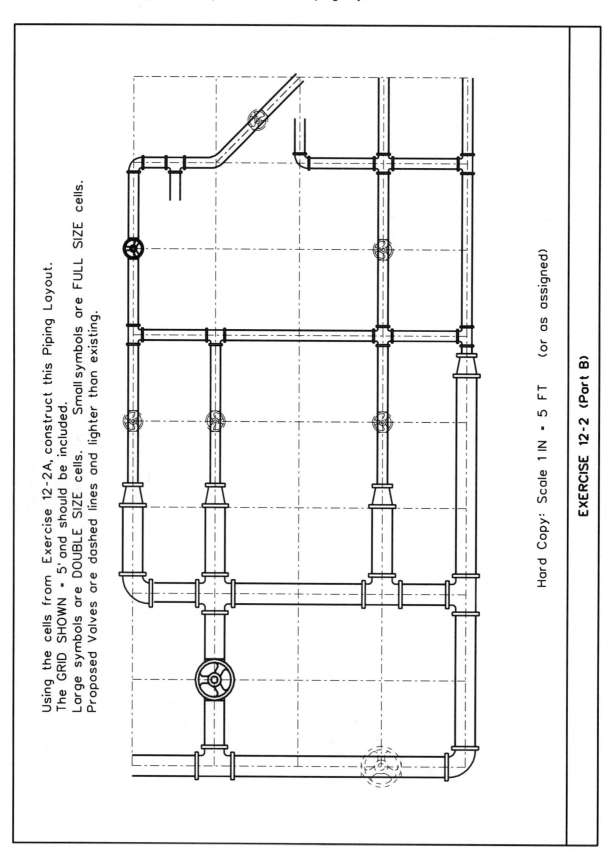

Cells and Cell Libraries 277

EXERCISE 12-3   Part A: Park Symbols

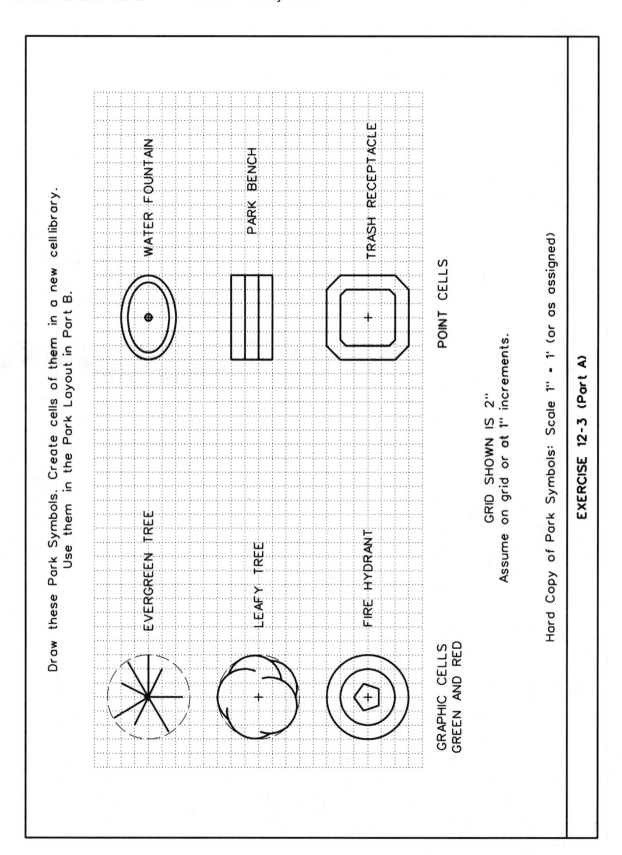

EXERCISE 12-3 (Continued)   Part B: Mini-Park Layout

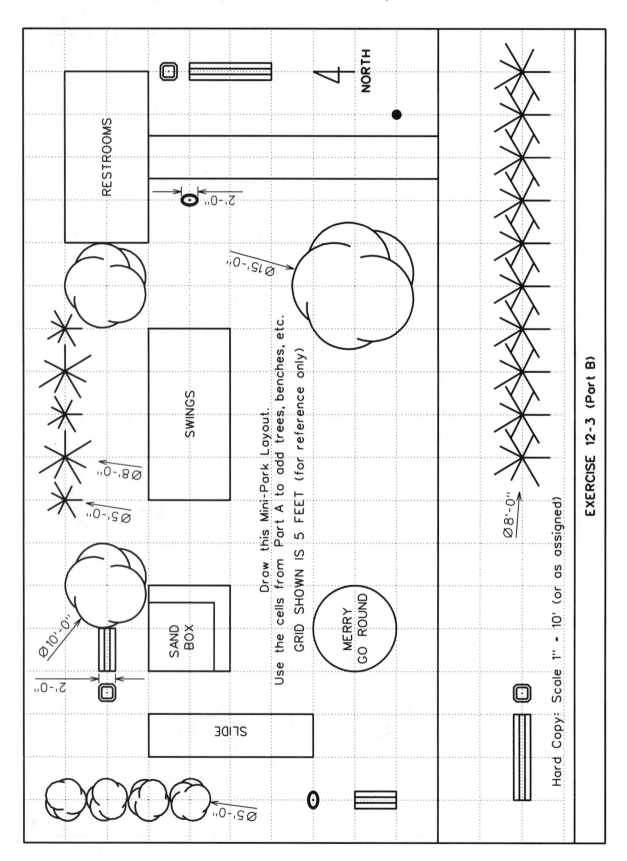

EXERCISE 12-4   Part A: Sign Symbols

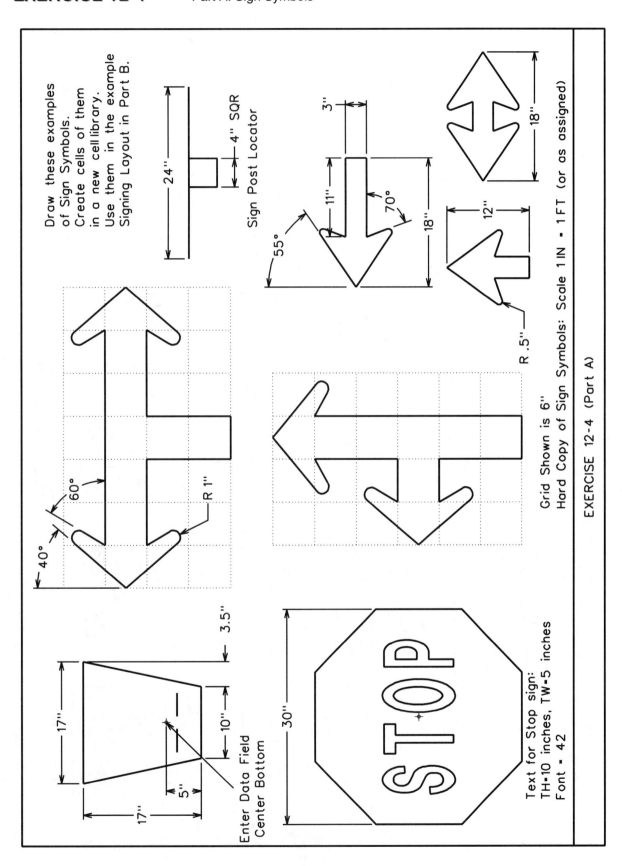

EXERCISE 12-4 (Part A)

EXERCISE 12-4 (Continued)   Part B: Intersection Signing Layout

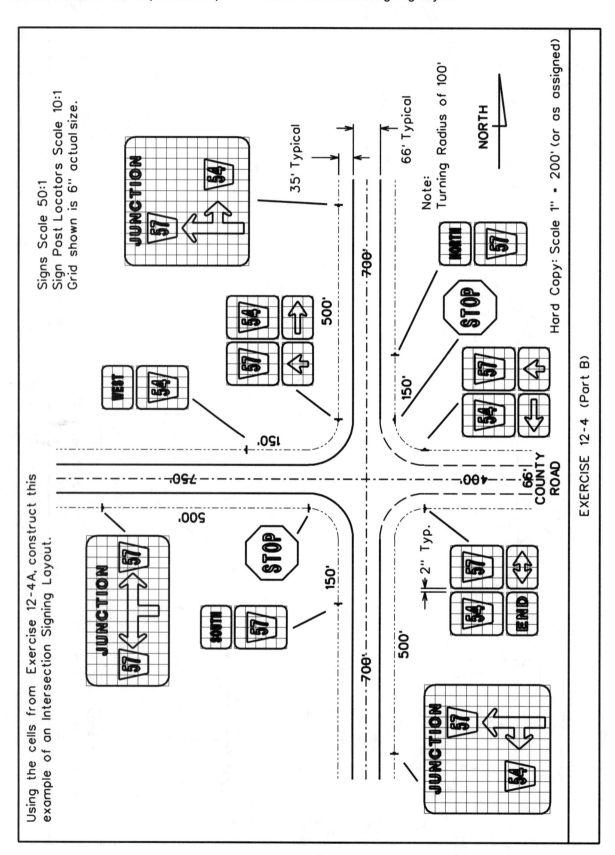

EXERCISE 12-4 (Part B)

Chapter 13: Patterning

SUBJECTS COVERED:

- Area Type
- Change Element to Active Area
- Displaying Patterns
- Patterns Tool Box
- Delete Pattern
- Hatch Area
- Element Method
- Flood Method
- Boolean Methods
- Crosshatch Area
- Pattern Area
- Linear Pattern
- Show Pattern Attributes
- Match Pattern Attributes

Patterning refers to filling areas with a repeated pattern over and over. It is sometimes referred to as section lining when used in engineering graphics. In a sectioned view, part of the object is cut away and its material is exposed—section lining is used to represent that material. Patterning is also used to indicate different materials in a cross-section. If elements were placed over and over again individually rather than using the Patterns tools, the time involved would be extremely extensive. With patterning, the boundaries of the area to fill are specified and the type of pattern. Then the software can figure the rest out on its own—slick and fast.

AREA TYPE

In previous chapters, when you were placing a shape or closed element, there was an Area setting noted as "will be covered later." Well, now is later. There are two different types of areas: Solid or Hole. Patterning will only show up in **Solid** Area type. If the Area type is set to **Hole**, then the patterning will not appear. Usually, the Area setting is defaulted to Solid. This Area setting is not only given to closed elements but also applies to other elements such as text (it just isn't as obvious).

CHANGE ELEMENT TO ACTIVE AREA

There is a tool found in the Change Attributes tool box that will change the Area type of an element. It is called the Change Element to Active Area tool (although the tool tip drops the word "element"). The tool is very simple to use. Using the dialog box, change the Area option to the type wanted (Solid or Hole) and then select the element. There isn't a Use Fence setting. The Area type set here will remain the Active Area type until you change it in this or some other dialog box such as Place Block.

DISPLAYING PATTERNS

Each view has the option of displaying patterns or not. Often, if a lot of patterning is being done, the patterns will not be displayed to speed things up. However, beginners are highly encouraged to check the View Attributes settings to be sure that the Patterns is checked ON (default is toggled off). Do this before starting to experiment with patterning. That way you can actually see what you're doing! Remember, CTRL-B will get you to the View Attributes dialog box quickly.

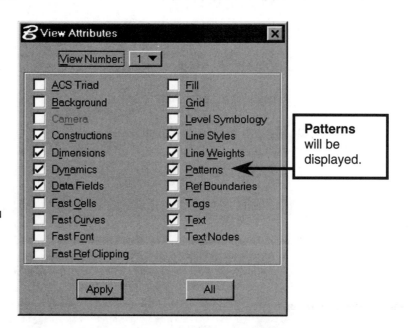

Patterns will be displayed.

PATTERNS TOOL BOX

Patterning tools are found in the Patterns tool box. Standard patterns that use lines at different spacing and angles for the section lining have their own tools: Hatch Area and Crosshatch Area. Pattern Area uses cells for the repeating elements. Cells allow you to have different materials symbols such as dictated by ANSI standards. This tool box has its very own Delete Pattern tool. There are also tools for showing information about and matching existing patterns.

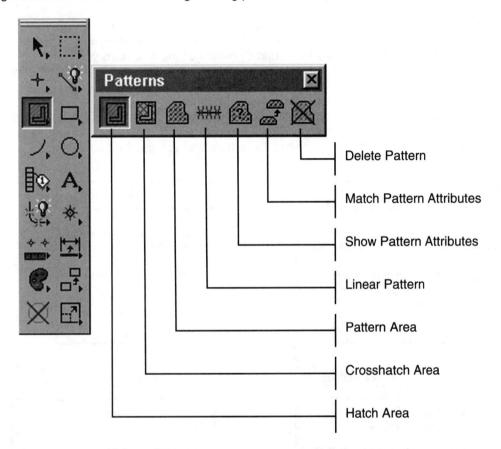

- Delete Pattern
- Match Pattern Attributes
- Show Pattern Attributes
- Linear Pattern
- Pattern Area
- Crosshatch Area
- Hatch Area

Associative Pattern

An associative pattern means that the patterning itself is associated to the closed element that defined the patterning boundary. This means that if the element is modified or manipulated, its associated pattern will automatically update to reflect the change. The Hatch Area, Crosshatch Area, and Pattern Area tools have an Associative Pattern setting. If it is checked ON, then the pattern placed will be referred to as an associative pattern and it will be associated to an element. We'll look at some examples of associative patterning later.

 Dimensions can also be associative. The dimension tools have an Association Lock setting that is comparable to the Associative Pattern setting in the patterning tools. This associative terminology will be seen again in Chapter 14.

DELETE PATTERN

 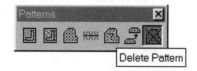

Let's cover how to delete a pattern before going through the different patterning tools. That way, when you make mistakes while experimenting with patterning tools, you'll know an easy way to recover. **The easiest way by far is to do an UNDO immediately after getting a pattern you don't want.** However, if you don't catch your mistake right away, this doesn't work too well. Then it is time for Delete Pattern, the last tool in the tool box.

Since the pattern is repeated over and over, there are a large number of "elements." Delete Element will work on each individual piece but would take forever. It is much quicker to use the Delete Pattern tool. Then all you have to do is data point on one part of the pattern and you get the whole thing.

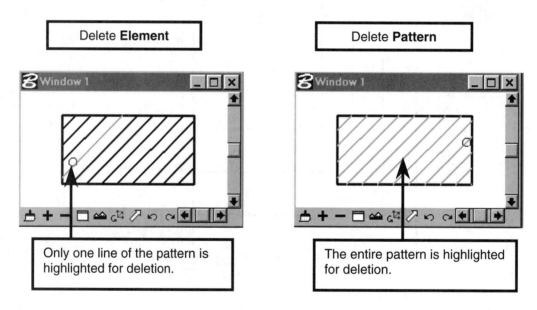

You'll need to select on the boundary of an associative pattern in order to get it to select it! Associative patterns will be discussed later on—just be aware that the selection data point location will be dependent on whether or not the pattern is associated to an element.

HATCH AREA

 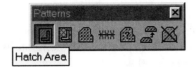

The Hatch Area is the simplest of the pattern tools. It places parallel lines (Active Attributes settings) at the specified spacing and angle. Since it is the simplest of the tools, it is also the best for illustrating some of the settings that are common to all three of these tools: Hatch Area, Crosshatch Area, and Pattern Area. These settings include Associative Pattern and Snappable Pattern as well as numerous methods of patterning.

Spacing (in master units) and **Angle** of parallel lines to use for pattern.
- Adjust these according to your needs.

Tolerance deals with how big a gap is allowed when approximations are done for pattern lines inside a curved boundary. Leave at 0.00 for good results.

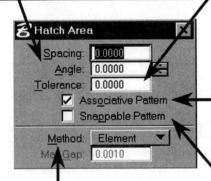

Associative Pattern
If checked ON, then the pattern is associated to an element. This means that if the element is modified then the pattern will automatically modify to reflect the change.

Snappable Pattern
If unchecked (OFF), then the lines that make up the pattern will not have keypoints. This is recommended because then it doesn't interfere with snapping to main elements.

Method
- Element and Flood are the most common ones.
- Intersection, Union, and Difference are Boolean operations (similar to the concepts covered with the Measure Area tool).
- Fence and Points are left for the advanced user.

ELEMENT METHOD

This method uses a **closed** element to specify the boundary for the patterning. A line string or complex chain is not necessarily closed. An easy way to get a closed element is to use the SmartLine tool. Its settings of Join Elements and Closed Element would need to be checked ON. You can also use the Create Complex Shape tool or the Create Region tool to create the closed element. All of these items were discussed in Chapter 11; go back and review if you need to!

When patterning using the Element Method, you'll be required to identify the element. Once the closed element has been identified, the prompt will be Accept @pattern intersection point. A data point is needed for this. The data point will accept the element **and** the location of this data point will specify a pattern intersection point. If the data point is located **inside** the closed element, one of the lines that make up the pattern will pass through this pattern intersection point. If the data point is located **outside** the closed element, then an imaginary extended pattern line would intersect that point. The location of the data point determines the position of the pattern.

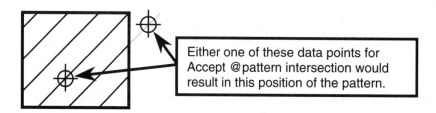

With Associative Pattern unchecked (OFF), once the data point for Accept @pattern intersection point has been placed (either inside or outside the closed element), **the patterning will automatically be completed.** The same is true for the case when Associative Pattern is checked ON and the data point for Accept @pattern intersection point is **outside** the closed element—the patterning will automatically be completed.

There is one situation in which a preview of the patterning will appear. This occurs when Associative Pattern is checked ON <u>and</u> the data point for Accept @pattern intersection point is **inside** the closed element. The pattern will be highlighted so that you can see its position. In this situation, you can keep placing a data point inside the element until the pattern position is acceptable. **A Reset (right mouse button) will complete the patterning.** If you data point outside the closed element, this will reposition the pattern so that it intersects that point and automatically completes the patterning.

For simplicity's sake, we will assume that the data point to Accept @pattern intersection point will be outside the closed element (unless stated otherwise). In that case it won't matter if Associative Pattern is toggled on or off. The data point will automatically complete the patterning.

Patterning 287

Let's look at using the Element Method with the Hatch Area tool. Associative Pattern is checked ON. The block on the left has a Solid Area type. The block on the right has a Hole Area type. We need to see the effect of the Area Type setting on the patterning.

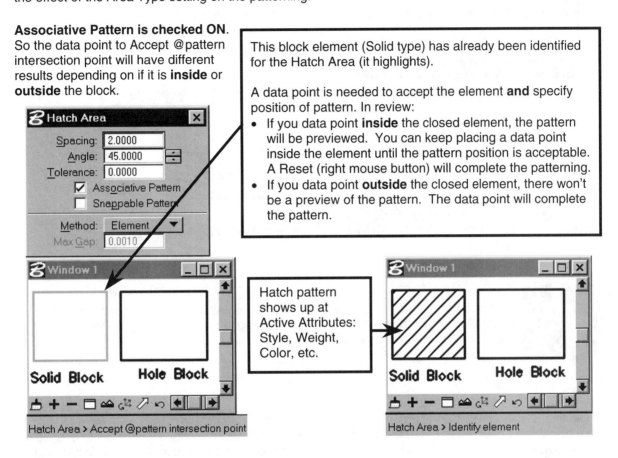

Associative Pattern is checked ON. So the data point to Accept @pattern intersection point will have different results depending on if it is **inside** or **outside** the block.

This block element (Solid type) has already been identified for the Hatch Area (it highlights).

A data point is needed to accept the element **and** specify position of pattern. In review:
- If you data point **inside** the closed element, the pattern will be previewed. You can keep placing a data point inside the element until the pattern position is acceptable. A Reset (right mouse button) will complete the patterning.
- If you data point **outside** the closed element, there won't be a preview of the pattern. The data point will complete the pattern.

Hatch pattern shows up at Active Attributes: Style, Weight, Color, etc.

Now the next element identified will be the block on the right (Hole type). This is trickier because when you've identified the element and then data point (inside or outside the closed element) to Accept @ pattern intersection—NOTHING SHOWS UP. It doesn't matter if the data point was inside or outside the closed element, the pattern is completed and the tool immediately prompts Identify element. This is because **a Hole Area type will not show patterning—but a Solid Area type will.**

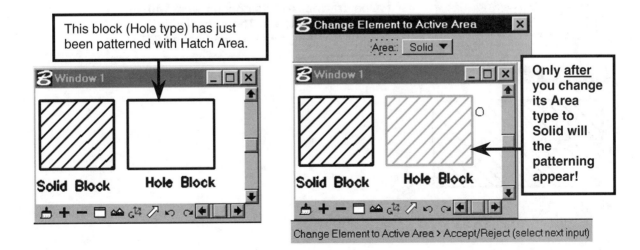

This block (Hole type) has just been patterned with Hatch Area.

Only **after** you change its Area type to Solid will the patterning appear!

Now let's look at how the associative pattern behaves when the element is modified.

Modify Element is used to dynamically show the changing size.

Once the element has been modified, the pattern associated to it is still there and changes to match the new area boundary.

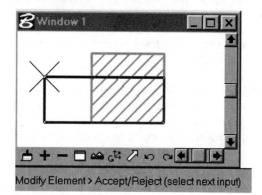
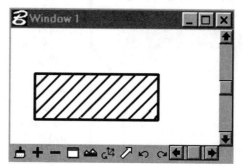

 Section lining is usually a smaller Weight than the visible edges. You can achieve this by drawing the element at a heavier line weight, and then changing the Active Weight setting before doing patterning.

Hole and Solid Area Types Together

The fact that a Hole area will not show patterning can be used to your benefit. Often an area to be patterned will need to be labeled with text. If the text has been placed when the Active Area type is Hole then it won't be covered up by the pattern. Now there isn't an Area type setting in any of the text tools, but you still can use Change Element to Active Area to your advantage. Either place text and then change it to make sure it's the correct Area type or use the Change Element to Active Area to adjust the Active Area setting before placing text. This technique works with other elements too, but it is most handy with text.

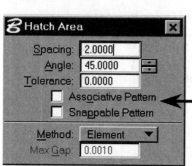

One drawback—this doesn't work with an associative pattern. The Associative Pattern setting must be unchecked (OFF).

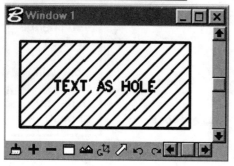

Patterning

Uses of Element Method

Using the Element Method for patterning usually requires some planning ahead. You need to determine the shape of the closed element since patterning will fill the area that it bounds. The perimeter of the closed shape should not include extended surface limits.

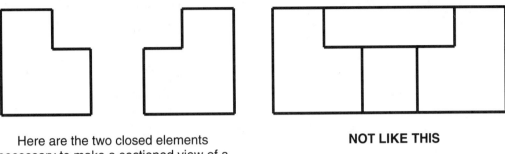

Here are the two closed elements necessary to make a sectioned view of a counterbored hole.

NOT LIKE THIS

Because this preplanning is necessary, a different method of hatching may be more suitable. Often the **Flood** Method is the next choice.

FLOOD METHOD

The Flood Method involves selecting inside an open space and filling the area until an element is hit to stop it. It's very simple, no planning ahead required, just element(s) enclosing an area. Shown below are individual lines that enclose an area. They overlap but are not line strings, nor is there a closed element involved.

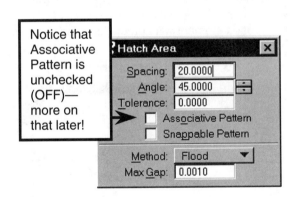

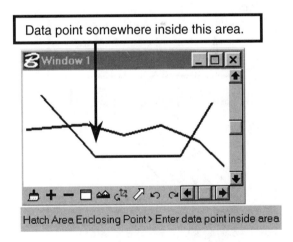

Once the data point inside the area has been made, then the enclosing area will be highlighted as shown below on the left. By accepting at the pattern intersection, the Hatch Area (Flood Method) will be completed as shown below on the right. Very simple and quick—isn't that great!

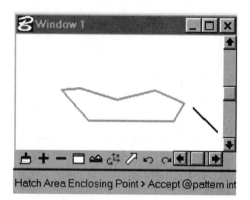
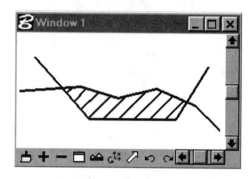

Non-Element Methods and Associative Patterns

When using the Element Method of patterning, there was already a closed element that could be associated to (there had to be—it's the Element Method). However, the other methods don't have a single closed element that the pattern can be associated to. Therefore, when using methods such as Flood and the Boolean operation ones, if Associative Pattern is checked ON then a **closed element will be automatically made!** That way the pattern can be associated to something and it's happy.

In the illustration below, the original lines that formed the perimeter of the enclosed area have been deleted so you can tell the difference between having Associative Pattern toggled on or off.

This was the original setting with Associative Pattern unchecked (OFF). A new closed element was not made because it wasn't needed.

The Flood Method was done again, but this time Associative Pattern was checked ON. A closed element will automatically be made for you.

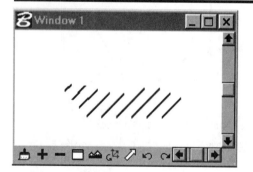
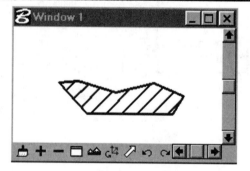

If you use Delete Pattern on an associative pattern done in these "non-element" methods, the automatically created closed element will also be deleted.

BOOLEAN METHODS

The other methods of Intersection, Union and Difference use elements and their relationships to form the boundaries of the patterns. Those boundaries are similar to using the Measure Area tool. Go back and review that material to see which areas will be patterned. As with the Flood Method, if the pattern is associative, then a new closed element will be created just so that the pattern has a single element to be associated to.

CROSSHATCH AREA

This tool works like Hatch Area, but you get two different sets of parallel lines. Each set has its own settings independent of each other (but both at the same Active Attributes). This is how a block is done with the Crosshatch Area tool as shown.

Once again, when using the Element Method and the Crosshatch Area tool, the Associative Patterning setting and the location of the data point to Accept @pattern intersection (inside or outside the element) will determine the sequence of events.

If Associative Pattern is unchecked (OFF):
1) Identify the element.
2) Use a data point to Accept @pattern intersection point (inside or outside the element, it doesn't matter).
3) The patterning will automatically be completed.

If Associative Pattern is checked ON:
1) Identify the element.
2) Use a data point to Accept @pattern intersection point.
   - If you data point **inside** the closed element, the pattern will be previewed. You can keep placing a data point inside the element until the pattern position is acceptable. A Reset (right mouse button) will complete the patterning.
   - If you data point **outside** the closed element, there won't be a preview of the pattern. The data point will complete the pattern.

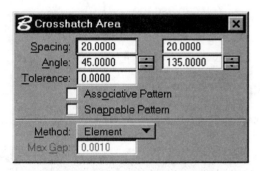

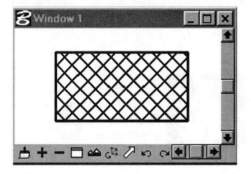

PATTERN AREA

The Pattern Area tool requires a cell for the pattern. This means that you need to **attach a cell library** and select the pattern cell you want to use. Go back and review Chapter 12 on cells if you need to. The other settings are similar to doing an array except that it will figure out how many rows and columns are needed in order to fill the area.

The sequence of identifying the element works like what was just discussed with the Crosshatch Area tool. It depends on if the pattern is associative or not. We won't cover that concept again.

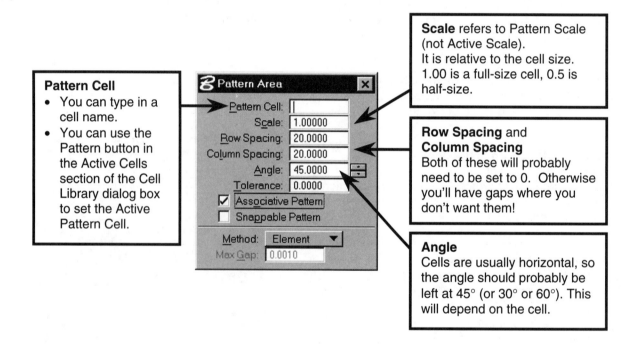

Pattern Cell
- You can type in a cell name.
- You can use the Pattern button in the Active Cells section of the Cell Library dialog box to set the Active Pattern Cell.

Scale refers to Pattern Scale (not Active Scale). It is relative to the cell size. 1.00 is a full-size cell, 0.5 is half-size.

Row Spacing and **Column Spacing**
Both of these will probably need to be set to 0. Otherwise you'll have gaps where you don't want them!

Angle
Cells are usually horizontal, so the angle should probably be left at 45° (or 30° or 60°). This will depend on the cell.

 The Pattern Scale is based on the cell size. However, the size of the elements in the cell are based on their positional units at their creation. However, when determining the Pattern Scale to use, you will need to take the final hard copy scale into consideration too.

Patterning   293

This illustration shows the use of the ANSI33 cell as the Active Pattern Cell. Notice that the Row/Column Spacing is set to 0 and the Scale to 2. Sometimes you'll have to experiment with different scale factors to find the one that works well. The pattern may look a bit wavy here, but it will print out fine.

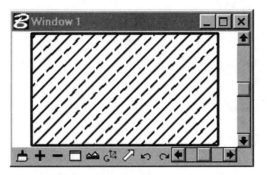
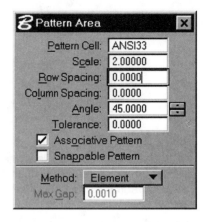

Here's the Cell Library dialog box to remind you of how it looks! There are a lot of cells available. Check under the **...\Workspace\system\cell** directory for some existing ones.

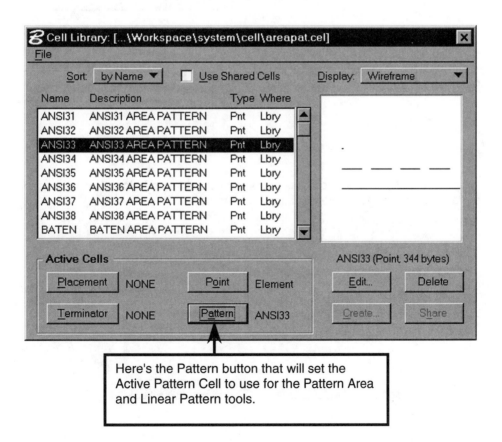

Here's the Pattern button that will set the Active Pattern Cell to use for the Pattern Area and Linear Pattern tools.

LINEAR PATTERN

The Linear Pattern tool also uses a cell for its pattern. This tool doesn't fill an area; instead it patterns linear elements. This tool is sometimes used instead of Custom Line Styles. Once the line (or line segment) has been patterned, the element itself is not visible. If you use Delete Pattern, then the element that had been selected to undergo linear patterning will return.

Here is the Linear Pattern dialog box and a line that has been identified near its right endpoint. The Cycle was left as the default of Truncated. You can experiment with the other options on your own.

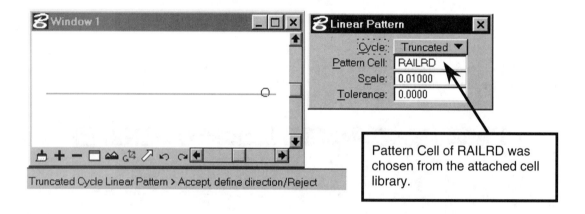

Another data point was specified, and the line is patterned as illustrated below. This is much easier than trying to use a manipulation tool to create all of these elements.

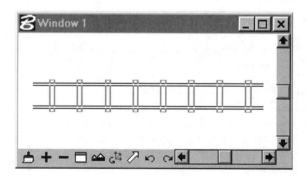

SHOW PATTERN ATTRIBUTES

The Show Pattern Attributes tool will let you see what the Pattern Cell, Scale, and Angle settings of an existing pattern are. If the Hatch or Crosshatch tools were used to place the existing pattern, then it shows the Spacing and Angle of the lines. It is for information purposes only—you can't edit the pattern—just see its settings. The information shows up in the status bar.

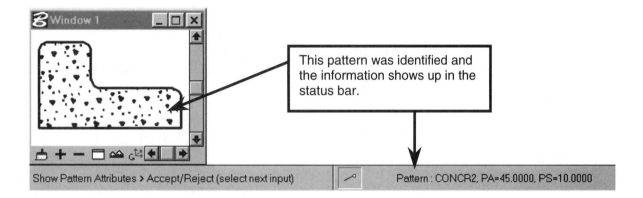

This pattern was identified and the information shows up in the status bar.

MATCH PATTERN ATTRIBUTES

The Match Pattern Attributes tool is bit more useful than the Show Pattern one. It will actually match the settings of an existing pattern. This is really nice if you're working with others and need to match some existing patterning. All you do is identify the pattern to match and then go to the Pattern tool—it will now have the settings of the existing pattern. Now when you use the tool, the resulting pattern will match the existing pattern. Sorry, but there isn't a "Change Pattern Attributes" tool—you have to delete the pattern and do it again. This tool at least saves you some of the headaches.

Pattern Area settings BEFORE using Match Pattern Attributes.

Pattern Area settings AFTER using Match Pattern Attributes.

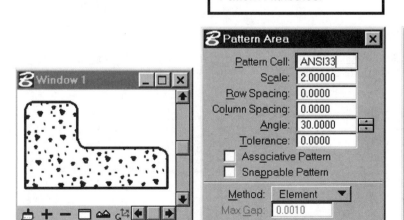
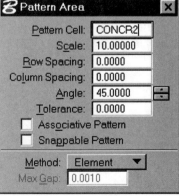

QUESTIONS

① Under what tool box do you find the Change Element to Active Area tool? What is it used for?

② What does it mean to have an **associative** pattern?

③ When you're using the Element Method of patterning:
   a) What type of element does it require?
   b) What settings and actions will result in a preview of the pattern showing up? Be specific.

④ What is the difference between Hatch Area and Crosshatch Area?

⑤ What happens if you use the Hatch Area tool with the Element Method, and the element is a Hole Area Type?

⑥ In order to use the Pattern Area tool, what must be attached first? Why?

⑦ For the two elements shown, illustrate the differences between these Boolean operations in patterning.
   a) Intersection
   b) Union
   c) Difference (A–B)
   d) Difference (B–A)

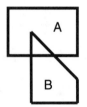

⑧ What is this tool found in the Patterning tool box called? What is the advantage of using it?

⑨ Sketch the result of patterning on these two closed elements with the Element Method (selecting the rectangle) and
   a) Associative Pattern unchecked (OFF).
   b) Associative Pattern checked ON.

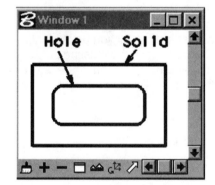

⑩ What Method of Pattern Area would <u>quickly</u> pattern the areas bounded by these lines and line strings as shown?

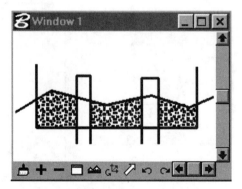

EXERCISE 13-1   Counterbored Object

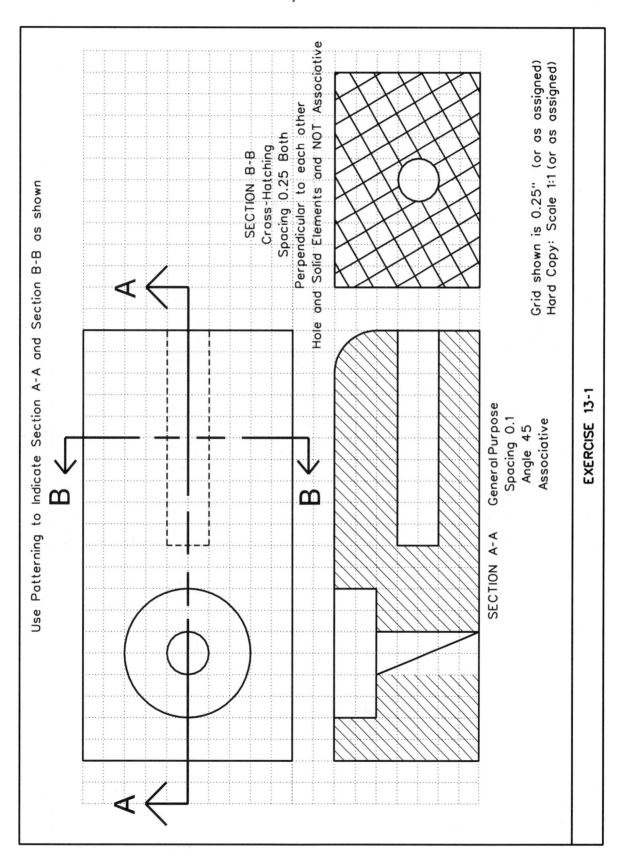

EXERCISE 13-2 Retaining Wall

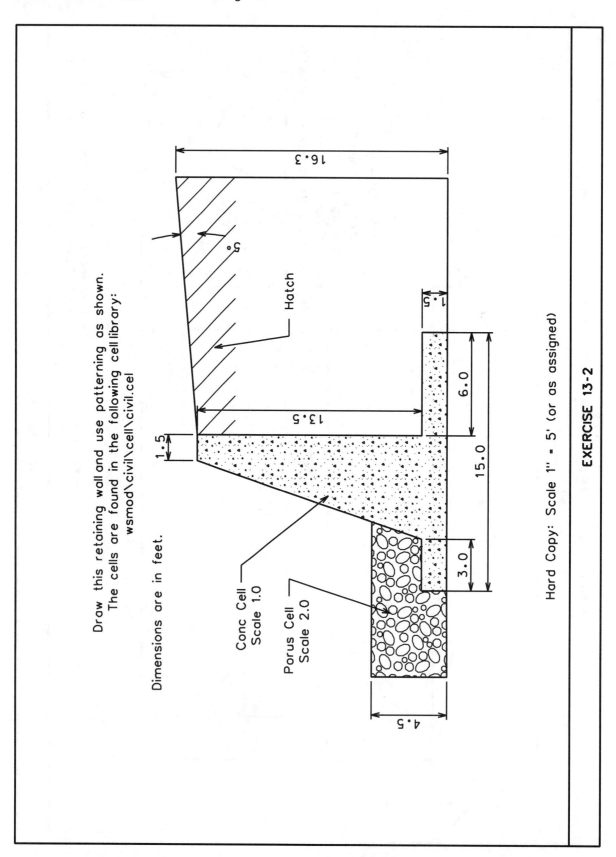

EXERCISE 13-2

EXERCISE 13-3   Sectioning the Grate Inlet

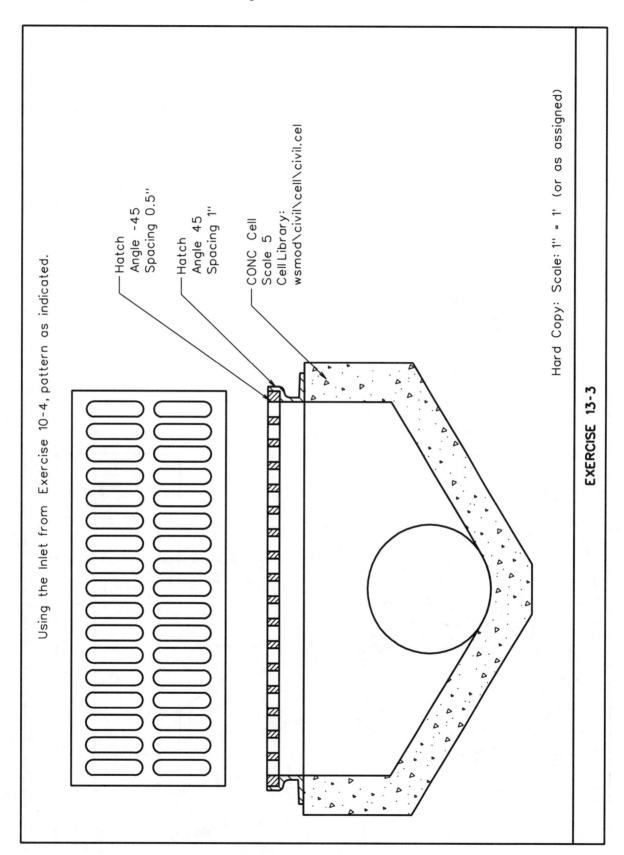

EXERCISE 13-4 Half Section of Radially Symmetric Object

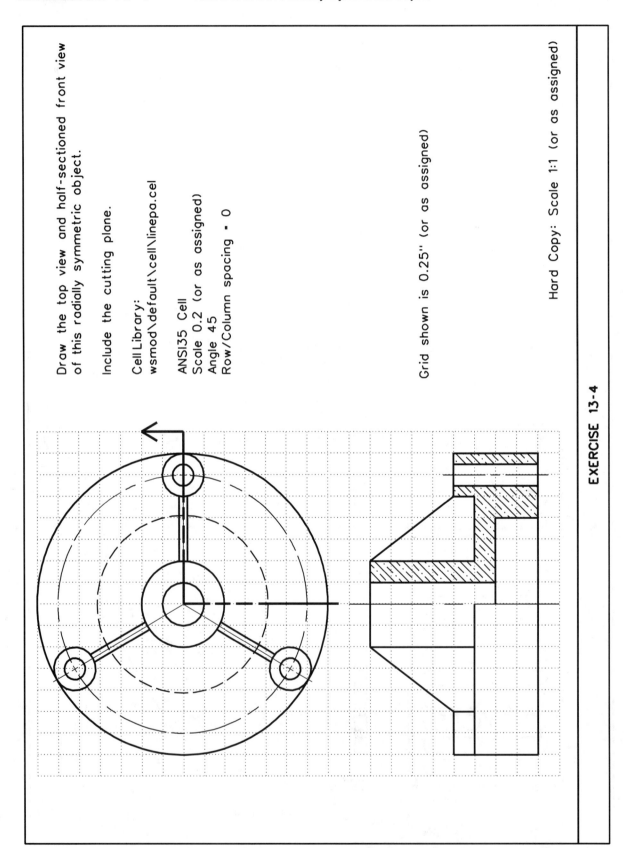

EXERCISE 13-4

EXERCISE 13-5   Offset Section of Tool Holder

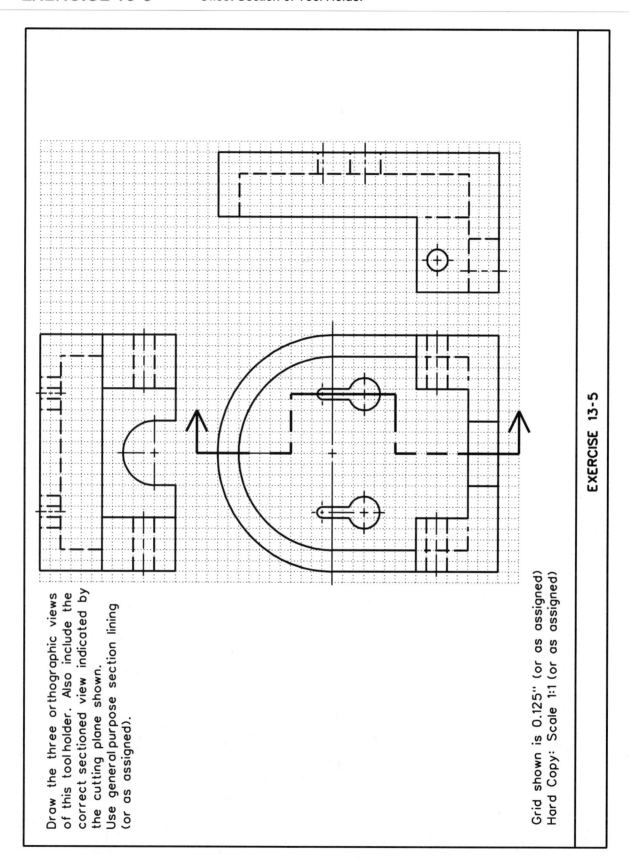

Draw the three orthographic views of this toolholder. Also include the correct sectioned view indicated by the cutting plane shown. Use general purpose section lining (or as assigned).

Grid shown is 0.125" (or as assigned)
Hard Copy: Scale 1:1 (or as assigned)

Chapter 14: Dimensioning

SUBJECTS COVERED:

- Dimensioning Terminology
- Dimension Settings
- Dimension Tool Box
- Dimension Element
- Dimension Size with Arrows
- Dimension Location (Stacked)
- Dimension Angle Size
- Dimension Angle Location
- Dimension Angle Between Lines
- Dimension Radial
- Update Dimension
- Geometric Tolerance

Making dimensioning easier is one of the strengths of any CAD product. There are many conventions controlling the look and form of dimensions. Maintaining uniformity was tough using hand-drafting techniques but is a breeze for computer-aided drafting. Dimension settings allow you to control how the parts making up a dimension element will appear. The dimensioning tools allow you to place different dimensions in a variety of ways while at the same time having consistent settings. Even though CAD makes dimensioning much easier, **it is still up to you the user to follow the conventions of dimensioning**. CAD can quickly put dimensions in any location and at any orientation even if it isn't following the rules of dimensioning—think for yourself!

DIMENSIONING TERMINOLOGY

You need to be familiar with the various terms used in dimensioning in order to know what the settings will control. If these terms are unfamiliar to you, please review the dimensioning standards found in most engineering graphics or related textbooks. Shown here is a typical dimension element that is made of different parts. You can control each individual part, but when placed, a dimension actually comes in as only **one** element.

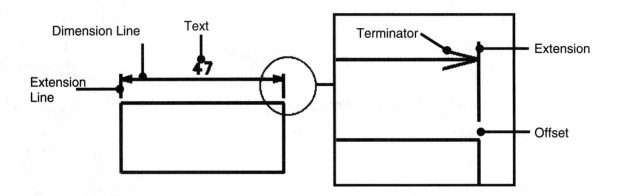

DIMENSION SETTINGS

How a dimension looks is determined by the dimension settings that are active when the dimension is created. You can edit an existing dimension element, but it is easier to get the settings how you want them and then place the dimension. These settings can also be modified at any time while placing dimensions. Changing the settings will affect only the dimensions placed after the changes. It will not go back and automatically update the previously placed dimensions unless you tell it to.

Dimension settings used can vary for different disciplines. An architecture design can have a dimension style that is different from a mechanical engineering design. Dimension settings can be stored as styles and then be used in different design files. This topic will not be covered here; see the software documentation for more information.

It is important to Save Settings for your design file before you close it. If you do, then any changes that you made to the dimension settings will be saved for that design file. When you go back and open that same design file, you won't have to make the same changes to the dimension settings again. A reminder: Save Settings on Exit can be checked ON in the main pull-down menu WORKSPACE>PREFERENCES in the Operation Category. That will automatically save the settings of a design file when you exit MicroStation. If that preference is not set, then you can use the main pull-down menu FILE>SAVE SETTINGS to save the settings of the design file at any time. You can also use CTRL-F to save the settings of the design file. Anyway you do it is fine; just be sure that all your settings changes have been saved before closing the design file.

Dimensioning

Text Height Units

Many of the settings' values are based on text height units, especially the ones dealing with the geometry (spacing and size) of the dimension parts. This is done so that when you change the scale of the hard copy, then by changing the text height, the size of the dimensions' parts will follow. Otherwise, terminators and gaps will not adjust and will be either too big or too small. By making geometry settings dependent on the text size, you don't have to go in and edit all the settings for various plot sizes.

Because of this relationship between the settings and the text height, you will want to set your text height first. That way when you place a dimension, you'll have a better perspective of how the dimension will actually appear. Remember the key-in ⌨TX= for setting the text height quickly. **If a setting is based on text height units, then the value for the setting is a multiplication factor—not a distance.** A value of 0.5 means that the actual distance is 0.5 x current text height.

This table shows the various dimension settings and what units their values are based on.

Geometry Setting		Units
EXTENSION LINES:	Extension and Offset	TEXT HEIGHT UNITS
TEXT MARGINS:	Left, Lower, and Tolerance Left	TEXT HEIGHT UNITS
TERMINATORS:	Width and Height	TEXT HEIGHT UNITS
OTHER:	Stack Offset	WORKING UNITS
	Minimum Leader	TEXT HEIGHT UNITS
	Center Size (non-zero)	WORKING UNITS

For Example: The gap between an object and the extension line is determined by the Offset geometry setting in the Extensions Line category. This setting is based on text height units. The actual distance of the gap is determined by taking the Offset setting value and multiplying it by the current text height. Shown below are two different settings and the actual distance of the gap that results.

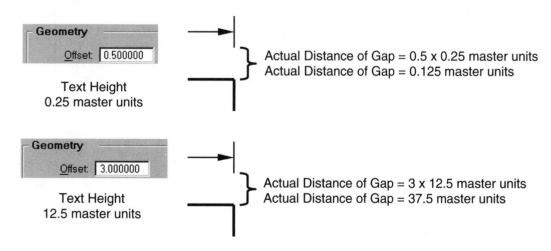

Accessing the Dimension Settings Dialog Box

Since each dimension is actually an individual element, you can access the settings under the main pull-down menu ELEMENT>DIMENSIONS. There you will find various categories of settings in the Dimension Settings dialog box. Click on the category (such as Extension Lines shown here) and the settings for that category will appear in the dialog box. You will want to be aware of most of these settings and may want to modify them. However, the Custom Symbols and Tool Settings should be left to advanced users.

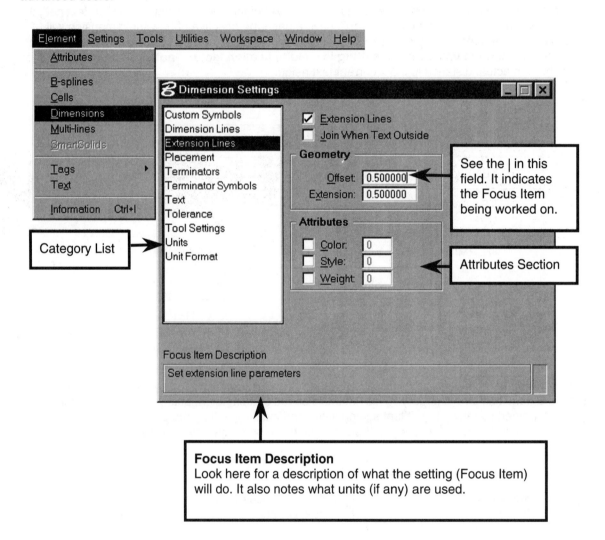

Focus Item Description
Look here for a description of what the setting (Focus Item) will do. It also notes what units (if any) are used.

The Dimension Settings dialog box is **resizable**. Not all boxes are resizable, but some are, such as this one. You change the size of the dialog box the way you change the size of a window. Move the cursor to its edge and you'll get a double-headed arrow. Click and drag to the new size. As illustrated below, a smaller dialog box doesn't shrink the entire dialog box display; instead it just clips it. This means that all of the settings available in the dialog box may not fully appear. If, for instance, the Focus Item Description isn't visible at the bottom, try resizing the dialog box.

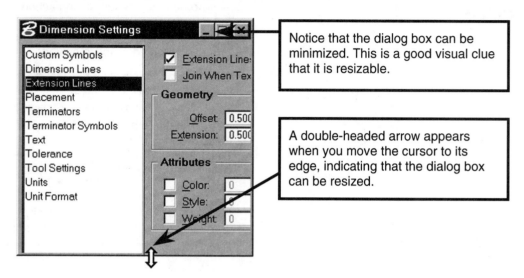

If dimensions are not being displayed, check to be sure that in View Attributes the Dimensions setting is checked ON.

Common Settings for Each Category

Most of the dialog boxes for each category will have some common settings. Most have a Geometry section with settings that deal with the geometry specific to that category. For instance, the Extension Category has the Offset and Extension settings in its Geometry section. Another common section is Attributes. These settings allow you to control the attributes of **each part** of a dimension. If checked ON for a category, then that attribute will be applied to that specific part. If all the settings in the Attributes sections are unchecked (OFF), then the entire dimension element will have the Active Attributes that are currently set for the design file.

For Example: If you want dimension lines to be green but would like extension lines to be blue, then you will want to set the attributes for each of those categories accordingly. If you want the entire dimension to appear black, then just be sure that the settings in the Attributes section of each category are unchecked (OFF) and the Active Color attribute is set to black.

The default settings for the dimensions can vary from seed file to seed file. The seed file seed2d.dgn will be used for the examples shown here.

Dimension Lines Category

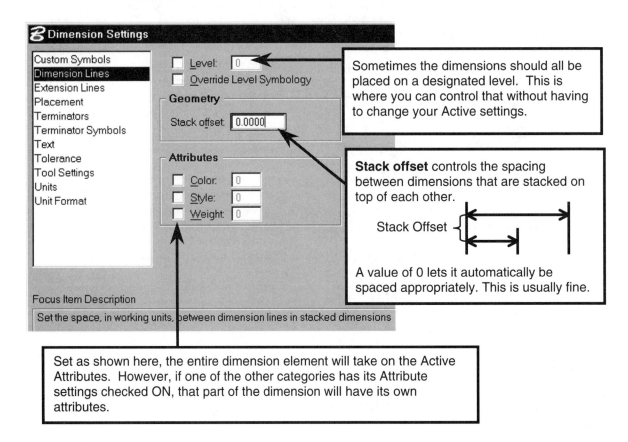

Extension Lines Category

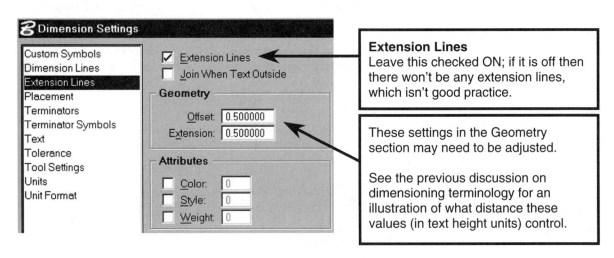

Placement Category

These settings determine how the dimension can be placed in relation to the drawing. Some need to be discussed in detail.

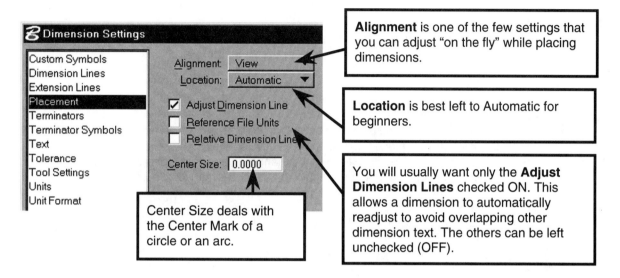

Center Size deals with the Center Mark of a circle or an arc.

Alignment is one of the few settings that you can adjust "on the fly" while placing dimensions.

Location is best left to Automatic for beginners.

You will usually want only the **Adjust Dimension Lines** checked ON. This allows a dimension to automatically readjust to avoid overlapping other dimension text. The others can be left unchecked (OFF).

Alignment

This determines the orientation of the dimension in relation to the drawing. The two most commonly used ones are View and True. A different orientation of a dimension element can also result in a different dimensioned size.

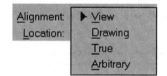

View: The dimension is placed to remain orthogonal to the view. This usually means horizontal or vertical.

Drawing: Unless the view has been rotated, Drawing will result in the same alignment as View.

True: The dimension is oriented based on the endpoints selected for the beginning and ending of a dimension. This doesn't necessarily mean orthogonal.

Arbitrary: This is one to be avoided by beginners. As the name implies, everything is arbitrary, even the convention of extension lines and dimension lines remaining perpendicular. However, it is good for dimensioning isometric drawings when Isometric Lock is on.

Here are some dimensions of a slanted line with different alignment settings.

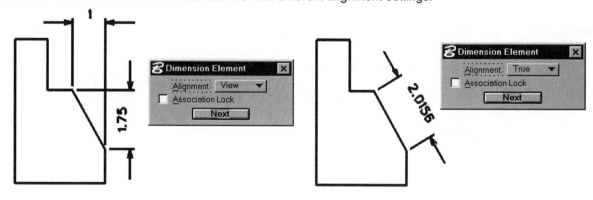

Center Size

This setting controls the + that marks the center of a circle or an arc. Its setting affects how the radial dimension Center Mark (covered later in this chapter) will appear.

Positive Center Size Value
- Only the + is shown, and its size is based on **working** units.

Negative Center Size Value
- The + is shown, and its size is based on **working** units.
- Extension lines are placed that extend past the boundary to show symmetry.
  If Associative Lock is OFF, then you dynamically specify the length of extension lines.
  If Associative Lock is ON, then the extension line length is set automatically.

Zero Center Size Value
- Only the + is shown, and its size is based on **text height** units.

Terminators Category

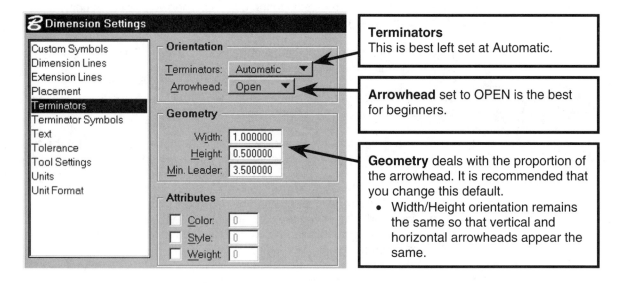

Terminators

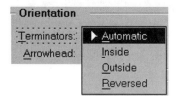

Automatic allows the program to place the terminators inside the extension lines if possible. If there isn't enough room, they are placed outside the extension lines. Changing this setting can force the terminators to be inside or outside the extension lines, but quite often that will cause overcrowding.

Dimensioning 311

Arrowhead

The Arrowhead settings control how the arrowhead terminators will appear.

Here are some examples of what Open, Closed, and Filled arrowheads look like.

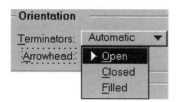

 The Filled arrowhead will not appeared darkened in unless the Fill setting is checked ON in the View Attributes dialog box. If it is unchecked (OFF), then the Filled arrowhead will appear like a Closed one. Also some printers have trouble with the Fill setting—filled arrows have been known to not print at all!

Width and Height

The Width and Height settings in the Geometry section control the size of the terminator. They are in text height units. The default is set to a 1:2 height-to-width ratio

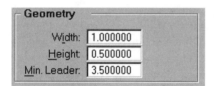
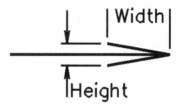

 The standard convention for arrowheads is usually a **height-to-width** ratio of 1:3. To match convention, the **Height would need to be 0.333 with the Width remaining at 1.** This also maintains the convention that the height of the arrowhead should be smaller than the text height.

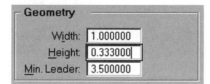

This setting shown above will match standard conventions for arrowhead geometry.

WARNING: Don't set the Height to 1 and the Width to 3, even though this matches the 1:3 ratio. The arrowhead height would be 3 times larger than the text height, which is not good practice.

Minimum Leader

The minimum leader (Min. Leader setting) is the smallest distance allowed between the tip of the arrowhead and the end of the dimension line (sometimes referred to as leader). This Min. Leader setting is based on text height units. The actual minimum leader length will be the Min. Leader value multiplied by the actual text height. If the text inside the extension line is too large to allow this minimum length, then dimension lines are put outside the extension lines (if this will allow the text to fit between the extension lines). If the text is still too large to fit between the extension lines, then it too will be placed outside the extension lines. Therefore, by making the Min. Leader value smaller, you can have more text fitting inside the extension lines. Be careful not to get the minimum leader length so small that the whole dimension line disappears or is truncated too close to the end of the arrowhead.

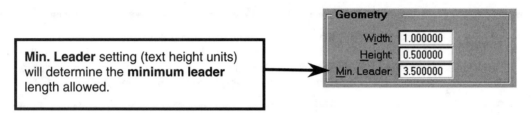

Min. Leader setting (text height units) will determine the **minimum leader** length allowed.

For Example: Let's look at some different Minimum Leaders settings and the resulting dimension line and text placement. The text height is the same in each case.

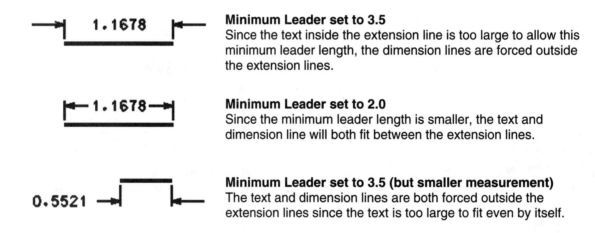

Minimum Leader set to 3.5
Since the text inside the extension line is too large to allow this minimum leader length, the dimension lines are forced outside the extension lines.

Minimum Leader set to 2.0
Since the minimum leader length is smaller, the text and dimension line will both fit between the extension lines.

Minimum Leader set to 3.5 (but smaller measurement)
The text and dimension lines are both forced outside the extension lines since the text is too large to fit even by itself.

Text Category

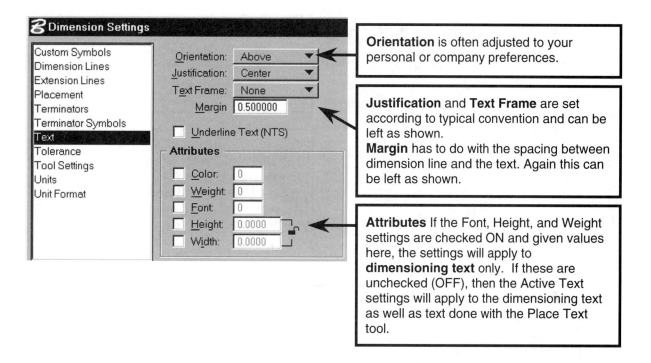

Orientation

There are three different options for the orientation of the dimension text.

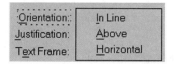

In Line will place text inside the dimension line and oriented in the same direction as the dimension line.
Above will place text above and parallel to the dimension line.
Horizontal will be the choice if you want unidirectional lettering. This places the text within the dimension, but it remains horizontal.

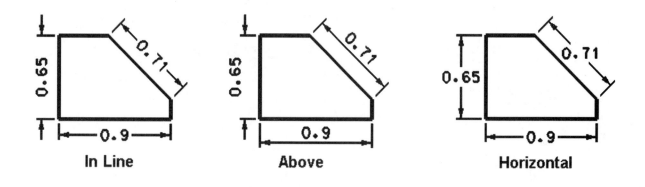

Tolerance Category

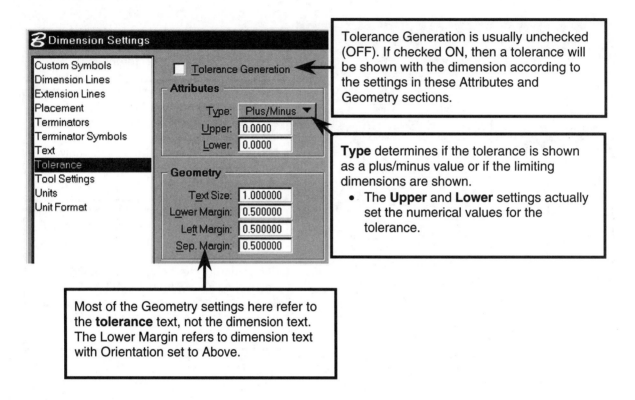

Type

The Type setting allows you to specify how the tolerance is to be displayed. There are two types of tolerances that will be shown if Tolerance Generation is checked ON. In each case a value of 0.1 master units is specified for both the Upper and Lower settings.

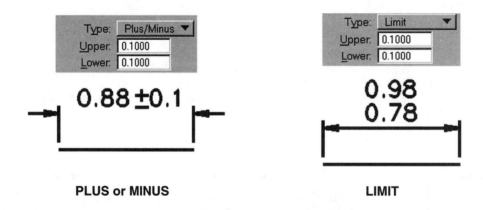

PLUS or MINUS　　　　　　　　　　**LIMIT**

Dimensioning 315

Units Category

The settings in the Units category deal with **linear** and **radial** dimensions. Units special to **angular** dimensions are controlled under the Units Format category.

Format controls how the units in the dimension value are shown.
- **Mechanical** will show the dimensions in master units **without labels**.
- **AEC** will make the Label options available to show dimensions in master units and sub units with the option of adding the Unit Names as labels.

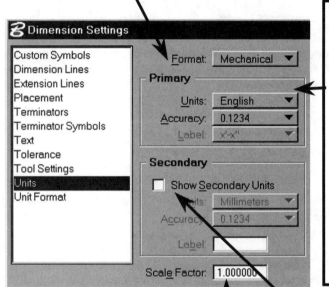

The settings in this **Primary** section refer to the main size stated in the dimension. You can't toggle off the display of the primary dimension. That would defeat the purpose of dimensioning, which is to show the size of an object.

Units allow changing from English to Metric. The Units setting should match the original working units.

Accuracy controls the number of digits displayed after the decimal point or can display fractions. The setting shown here will result in 4 digits after the decimal point. A more reasonable setting would be 1 or 2 digits after the decimal point.

The **Scale Factor** should be left at 1 when elements are drawn "true size." Dimensions should always reflect the actual size of the object even if the hard copy is not printed full size.

Show Secondary Units checked ON would allow you to display **2 sets of dimensions** (Primary and Secondary). Note that options of the Units setting are metric; this is because the Primary Units are set to English.

Shown below is an example of using the AEC option for the Format setting. Notice that the Unit Names (found under SETTINGS>DESIGNFILE>WORKING UNITS) are used as the labels in the dimension. What is typed in the Master Units and Sub Units fields will show up **character-for-character** in the dimension. If you have a typo, it will show up!

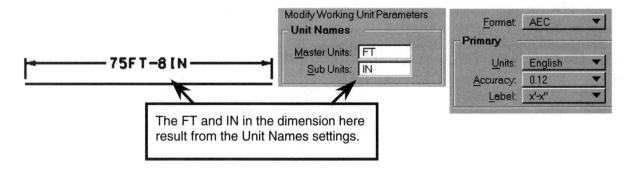

The FT and IN in the dimension here result from the Unit Names settings.

Accuracy

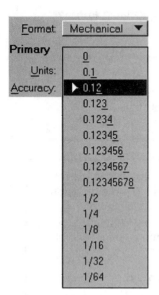

In addition to controlling the number of digits displayed, the Accuracy setting in the Primary section of the Units category also acts as a roundoff specification. The dimension value will be rounded off to reflect the Accuracy chosen. However, the actual size of the object is not changed. If the Accuracy is changed to display more digits, then the dimension can be updated to reflect the digits correctly.

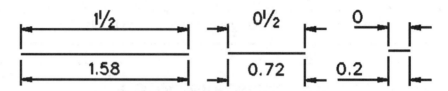

Here are three different-sized lines. Each one is dimensioned twice. The top dimension reflects an Accuracy setting of ½. The bottom dimension reflects an Accuracy setting of 0.12. Notice that the last line is small enough that it doesn't even round up to ½!

Show Secondary Units

If Show Secondary Units is checked ON, then dimensions will be shown with two values: the Primary and Secondary Units. Shown here is an example showing Secondary Units.

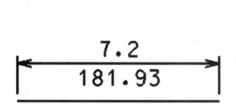
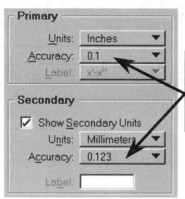

Notice that the Accuracy of the Primary and Secondary Units can be controlled independently.

Scale Factor and Details

Sometimes elements are not drawn at the actual size of the object that they represent. If you still want the dimensions to show the actual size of the **object** (not the element size), the scale factor value will need to be changed. An example of this is a detail drawn as double-size (elements drawn twice as big as the actual size of the object they represent). In order for the dimension's values to state the actual size of the object that the element represents, the scale factor setting would need to have a 0.5 value. In this case, an element that is 4 master units would be dimensioned as 2 master units. This Scale Factor setting is not related to the plotting of a hard copy.

Dimensioning 317

Unit Format Category

The Angle Format section applies only to angular dimensions. The other sections refer to linear, radial and angular dimensions.

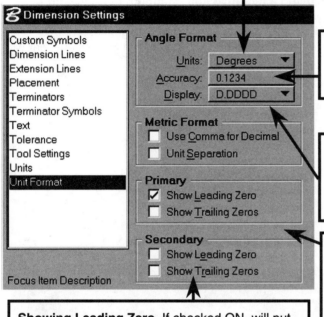

Units of angle dimension can be in
- **Degrees**, which is the standard format.
- **Length** format, which is **not** recommended.

Accuracy is again the number of digits behind the decimal point, but this time for **angular** measurements.

Display of the angle can be in:
- D.DDDD for decimal degrees.
- DD^MM'SS" for degrees, minutes, seconds.

Primary and **Secondary** dimensions each have their own settings. Be sure you are in the right section.

- Secondary settings will matter only when the Show Secondary Units setting in the previous Units category is checked ON.

Showing Leading Zero If checked ON, will put a 0 in front of the decimal point when the number is less than one.

Show Trailing Zeros If checked ON, will put a 0 at the end of a number to maintain the number of significant digits.

 The standard convention for decimal inch dimensioning is 1) Do **not** show **leading** zeros and 2) Do show **trailing** zeros. This is just the opposite of the Primary default settings shown above.

 Notice that there are settings for both the Primary and Secondary Units. If you find yourself saying, "I know I checked Show Trailing Zeros on—why aren't they displayed?" then you may need to see if you accidentally changed the Secondary settings instead of the Primary settings.

DIMENSION TOOL BOX

There are numerous dimensioning tools available. However, many of them are just specialized offshoots of some of the more basic tools. These basic tools will be explained, and you can test out the other tools on your own.

Be sure to read the prompts while doing dimensioning—it will simplify the process if you do. A dimension element is placed according to its vertices. These vertices are points that you specify. A linear dimension will have two vertices (a start point and an endpoint). An angular dimension may have three vertices. By reading the prompts you will know what data needs to be specified. **Most of the time you will be specifying the vertex points of a dimension by using a tentative and snapping to an existing element. This maintains the accuracy of dimensions to reflect the actual size of an element. Don't eyeball—close isn't good enough!** Sometimes it is necessary to select the tool again to restart a command rather than use the Reset (right mouse) button. Again, reading the prompt in the status bar will keep you informed.

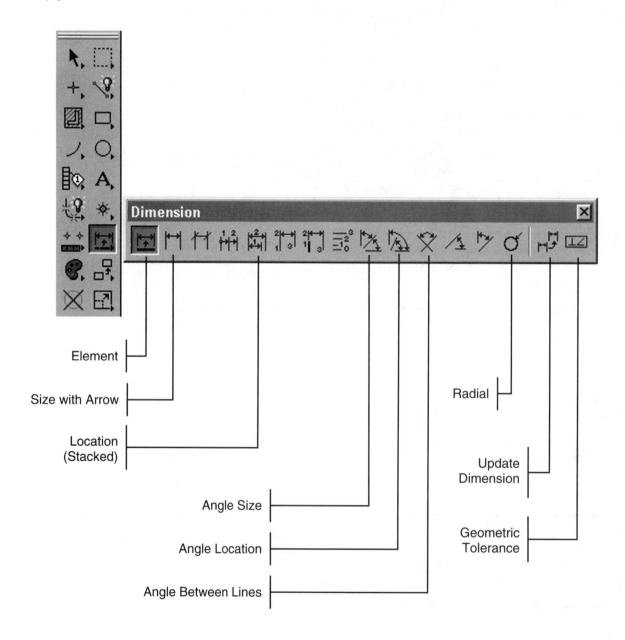

Associative Dimension

An associative dimension means that the dimension is actually linked to an element, so that if the element is modified the dimension will also be modified. Each dimension tool has an Association Lock that when checked ON makes the dimension that is placed with the tool an associative dimension. Associative dimensioning is often preferred because it can reduce the need to re-dimension a drawing.

For Example: Shown at the near right is a diameter dimension that has been associated to the circle. The circle is then modified in the middle step, and the final step (far right) shows the diameter dimension automatically reflecting the larger diameter resulting from the modification.

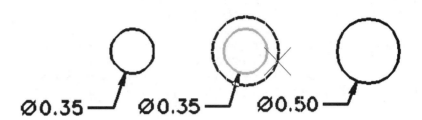

When placing an associative dimension, you need to pay attention to which elements are highlighted when you snap to specify the vertices of a dimension. The highlighted element will be the one associated with the dimension's vertex. It is possible that each vertex of a dimension can be associated to two different elements. This can cause problems when modifying one element but not another.

For Example: This start of the dimension was snapped to the end of the horizontal line, but the dimension endpoint was snapped to the end of the vertical line. This isn't going to make a difference in the size shown in the dimension. However, if either of the lines were modified then it would affect how the associated dimension would adjust.

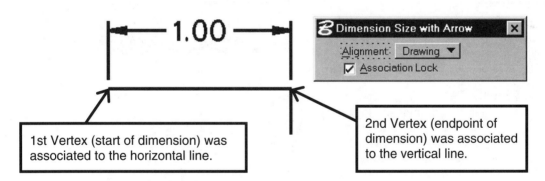

Notice that if the endpoint of the horizontal line is modified to a shorter length, **the associated dimension doesn't reflect the change.** This is because the 2nd Vertex was associated to the vertical line, not the horizontal line. This is a very simple example but you need to be aware of the ramifications of not paying attention to which element you are snapping to when specifying an associative dimension.

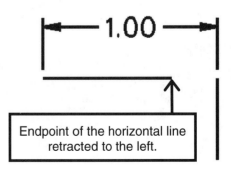

Alignment

Another common setting found in dimensioning tools is the Alignment option button illustrated at the right. The different Alignment settings available for a dimension were discussed previously in the Dimension Settings under the Placement Setting category. With this Alignment option button, the setting can be changed in the middle of a command (referred to as "on the fly"). This is handy since the alignment of the dimension will vary depending on what dimension size you want to specify, and it would be a pain to have to go back to the Dimension Settings dialog box all the time. Shown below is an example of changing the alignment "on the fly." In the middle of placing a dimension, the illustration below on the left shows the True Alignment dimension. The one on the right shows that when Alignment is changed to Drawing, the dimension being placed changes accordingly—in the middle of the command sequence.

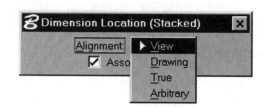

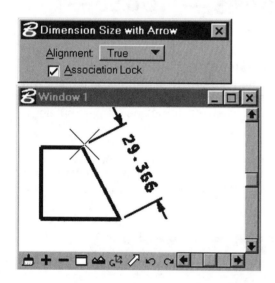
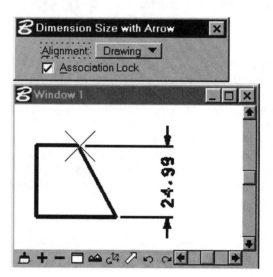

Length of Extension Line

Many of these tools will also prompt you to specify the length of extension line. This is the distance from the vertex of the dimension to the dimension line. Therefore the offset of the extension line is included in the extension line length but the extension of the extension line is not. Go back and review the section on dimensioning terminology if you need to. You can specify the length of the extension line graphically or by key-in.

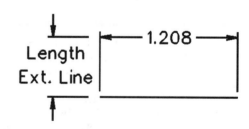

The standard minimum distance from the object to the first dimension is 3/8th of an inch (0.375). By using a relative coordinate key-in such as ▭DL=, you can follow this standard practice.

Settings Used in Examples

Taking into account that the dimension settings greatly affect how the dimension will appear, you should be aware of what settings were used for the examples that follow. Some of the settings are obvious such as accuracy, horizontal text, and arrowheads. Some that are not so obvious but very important are the settings of Placement Location and Terminators Orientation, which are both set at Automatic. If these settings aren't Automatic, the flexibility of the tool(s) is greatly increased but so is the number of variations. Therefore, I left it up to the program to automatically determine those items.

DIMENSION ELEMENT

 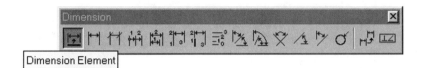

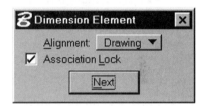

This is the easiest dimensioning tool to use. By selecting an element with a data point, the element's endpoints are used as the vertices of the dimension. The element will be highlighted and you'll be prompted to accept it. The accepting data point will do two things: 1) accept the element for dimensioning and 2) also specify how far from the element the dimension line will be placed. That's all there is to it.

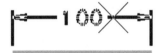

The horizontal line here has been selected for the element to dimension. The element is highlighted. The X is marking the data point placement that will accept the element and specify where the dimension line should be placed.

Since the Dimension Element tool is the easiest to use, it often becomes the tool of choice for inexperienced users. Alas, this often results in a common error of dimension placement—elimination of the standard gap between an object and its extension line.

Shown here is the common mistake that occurs when using Dimension Element. It was used for Line A and then also used for Line B. Since the extension line offset starts from the vertex of Line B, it is covered up by the vertical visible line. This results in no gap where one is needed. How to avoid this mistake? Don't use the Dimension Element tool—instead use Dimension Size with Arrows or one of the other dimensioning tools.

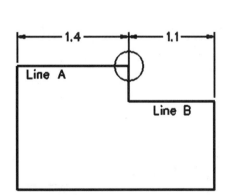

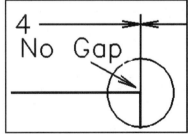

Enlarged view of mistake

Another drawback of using Dimension Element is that it is tricky to follow the standard practice of having dimensions line up. When specifying the length of the second dimension's extension line, you need to actually **snap** to the existing dimension's <u>dimension line</u>. This seems to be more work than necessary.

The Next button located at the bottom of the Dimension Element box allows you to toggle to the Next dimensioning tool available. This is a handy way to get to the dimensioning tools that are for specialized situations such as Label Line.

DIMENSION SIZE WITH ARROWS

This tool allows you much more control over the placement of the dimension. Read the prompts because you'll be asked to specify these in order:
1. Start of dimension
2. Length of extension line
3. Dimension endpoint

When you are done specifying the endpoint use the Reset button to end the command.

The points selected for the start of the dimension and the dimension endpoint will be the vertices of the dimension. The offset of the extension line will be measured from the vertex established.

 It is very important that you **snap** to the points being specified because this will determine the actual size dimensioned. In order to have a precise dimension—snap!

When placing a dimension with Alignment set to Drawing, how does the program know whether you are placing a horizontal or vertical dimension? That is determined by the **direction** of the length of extension line specification. Since the dimension line is perpendicular to the extension line, if the extension line is indicated in the horizontal direction, the dimension will be vertical, and vice versa.

Here Point 1 specified the start of the dimension. Point 2 set up the length of the extension line and also set the direction of the dimension line. Point 3 finished placing the dimension by specifying the dimension endpoint.

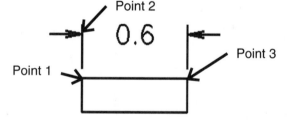

On small dimensions where the text is forced outside the extension lines, **the text will be placed on the side of the start of the dimension**. So for the vertical dimension 0.2, the start of the dimension was at the bottom and the endpoint was at the top.
Don't get stuck in a rut of left to right or top to bottom—live a little, and you may get better results.

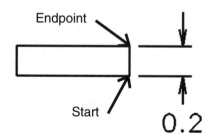

Continue to Place Dimensions

After you specify the endpoint of the dimension, the prompt still reads Select dimension endpoint. You may continue to place dimensions, with the second dimension using the first dimension's endpoint as its start of dimension. This is often known as incremental dimensioning. The ability to continue to place dimensions is a nice feature because the next dimension is actually lined up with the previous dimension according to convention (it uses the same extension line length). However, you must be aware that the dimensions are actually "joined together" so that if you erase one, they all go!

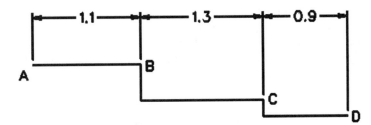

Here is an example of continuing to place dimensions with the Dimension Size with Arrows tool. The vertices of the dimensions were specified from A to D. Notice with this method there are still gaps between the object and the extension lines—good practice.

DIMENSION LOCATION (STACKED)

 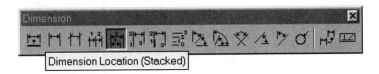

This Dimension Location (Stacked) tool allows you to do baseline dimensioning. This is where all the dimensions refer to the first extension line and are stacked one on top of the other. This tool also requires that you specify the start of the dimension, the length of extension line, and then the dimension endpoint. But it continues to prompt Select dimension endpoint. **It uses the original start of the dimension as the start of each of the dimensions.** When you want to end the dimensioning, use Reset. The dimensions placed are actually "joined together" so again, if you erase one, they all go.

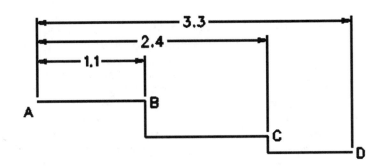

Here is an example of using stacked dimensions. The vertices of the dimensions were specified from A to D.

Remember that the distance between the stacked dimension lines is determined by the Stack Offset value found in the Dimension Lines category.

DIMENSION ANGLE SIZE

The Dimension Angle Size tool is just one of many tools available for doing angular dimensions. The direction of measurement is set to **counterclockwise**, so the order in which you select the start and endpoints of the angle is extremely important. This tool is similar to the Dimension Size with Arrows tool in that it will let you continue dimensioning angles incrementally so that one angle is dimensioned from the previous angle.

As always, be sure to read the prompts:
1) Select start of dimension.
2) Define length of extension line.
3) Enter point on axis.
   This would be the vertex of the angle.
4) Select dimension endpoint.

If you want only one angle dimensioned, use the Reset button to end the command.

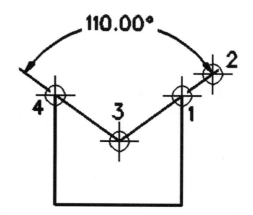

Continue to Place Dimensions

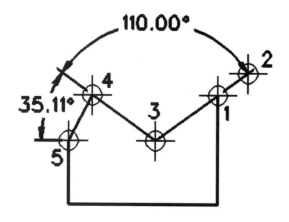

After specifying the endpoint (4), suppose that you wanted to continue dimensioning angles? In that case you would select dimension endpoint (5) as shown here. Notice that this angular dimension uses the same vertex (3). Its first extension line is actually the second extension line of the previous angular dimension (comparable to the Dimension Size with Arrows tool). They are all treated as one dimension element—if you delete one, they all go.

DIMENSION ANGLE LOCATION

Just as the Dimension Angle Size tool acted like the Dimension Size with Arrows tool, the Dimension Angle Location tool acts like the linear Dimension Location (Stacked) tool. With this tool for angular dimensions, the dimensions all refer to the same start of dimension, use the same vertex, and are stacked one on top of the other. Guess what? Again, they are all one element so you can't delete just one!

Here is the same object as shown in the previous example. The prompts are the same. Just the resulting dimensions look different.

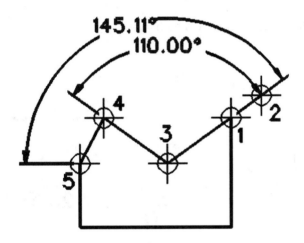

DIMENSION ANGLE BETWEEN LINES

The Dimension Angle Between Lines tool allows you to just select linear elements rather than specify the start of the dimension and the dimension endpoints. It also measures **counterclockwise** from the first line selected to the second line. The prompts are very brief. It asks for the first line, then the second line, then <u>absolutely nothing</u>, which is a bit misleading. What it actually is doing is waiting for a data point specifying where you want the dimension to be placed! So watch the screen, and you'll see that as you move the cursor, the angular dimension will show up in different places.

Here are two different cases of using the Dimension Angle Between Lines tool.

1) Specifies first line.
2) Specifies second line.
3) Data point specifying where the dimension should be placed. In this case, an **interior** angle is dimensioned.

Using the tool a second time:

5) Specifies first line.
6) Specifies second line.
7) Data point specifying where the dimension should be placed. In this case, an **exterior** angle is dimensioned.

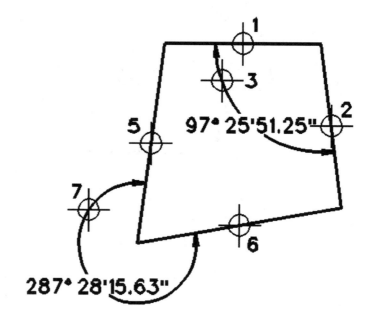

Notice that the Angle Format Display setting (found in the Unit Format category) has been changed to degrees/minutes/seconds for these dimensions.

DIMENSION RADIAL

There are five different Mode options of the Dimension Radial tool, so this one tool can do a lot. It is used for indicating the sizes of radii and diameters. It is also used for placing a Center Mark indicating the center of circles and arcs.

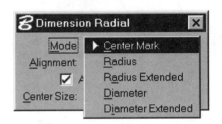

In following standard graphics conventions, circles representing holes will be dimensioned as diameters, and the Radius modes will be used for arcs.

Center Mark

Center Mark was discussed earlier in regard to the Center Size setting in the Placement category. Whether the Center Size has a plus or minus value determines how the Center Mark will look. When placing a radial dimension, MicroStation does not <u>automatically</u> place a center mark. So usually the first step is to use the Center Mark Mode of the Dimension Radial tool to place a center mark. A radius is usually just the + mark. A diameter is done with a negative Center Size allowing for the symmetry lines to extend past the edge of the circle. The following examples will not include a center mark, even though convention would warrant it.

Radius

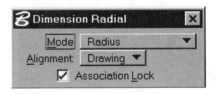

The Radius Mode actually places a leader that is radially directed. First you identify the arc, and then you can dynamically see where the leader will be placed by moving the cursor. You may choose to place the dimension "inside" the arc near its center or "outside" the arc pointing toward the center of the arc. A data point will actually complete the placement of the radius dimension. Notice that the letter R (according to convention) will automatically precede the numerical value. The dimension text Orientation is set to Horizontal, and that results in the leader having a horizontal shoulder.

1) Identifies the element.
2) Selects the "inside" of the arc.

3) Identifies another arc.
4) Selects the "outside" of the arc.

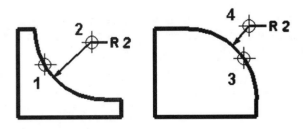

Radius Extended

With the Radius Extended Mode the leader line is actually extended to the center of the arc instead of stopping at the element. Again you identify the element and then select the placement of the dimension either "inside" or "outside" the arc. The line is **extended** to the center of the arc (radius point) in both cases.

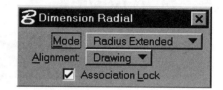

1) Identifies the element.
2) Selects the "inside" of the arc.

3) Identifies another arc.
4) Selects the "outside" of the arc.

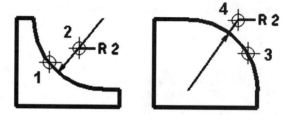

Diameter

The Diameter Mode will place a leader that is radially directed and dimensions the diameter value. The size value will be preceded by the conventional diameter symbol. The procedure is the same as when dimensioning a radius. You identify the element with a data point and then dynamically see how the dimension will look depending on where the cursor is. A second data point either inside or outside the circle will place the diameter dimension.

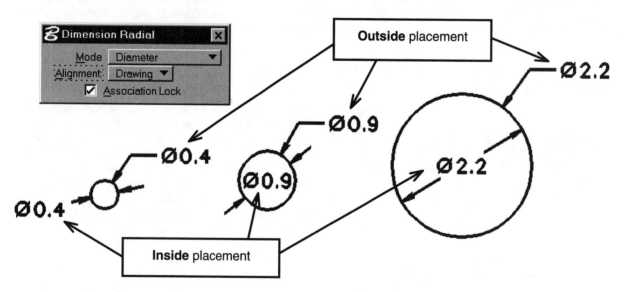

This example shows how the **size** of the circle affects how the dimension looks especially for the inside placement. Whether the placement was inside or outside was determined by the second data point.

Diameter Extended

When placing the dimension outside the circle, the Diameter Extended Mode looks different from the Diameter Mode. The leader line is extended through the center of the circle and two arrowheads point to the object as shown in the illustration. There isn't much difference between the two modes when doing an inside placement.

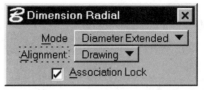

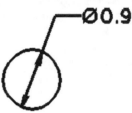

UPDATE DIMENSION

This tool is a lifesaver when you've decided to change the dimension settings but don't want to place all of the dimensions again. It updates the selected dimension to the Active dimension settings. So if, in the middle of a project, you decided that dimension text Orientation should be set to In-Line instead of Horizontal, you would use this tool. First you would make the changes needed to the dimension setting. Then click on the Update Dimension tool and you'll be prompted to Identify the element. Upon selection, the dimension element will now reflect the new dimension settings.

 The Change Dimension to Active Settings dialog box that is evoked by the Update Dimension tool does **not** have the option of using a fence. That means that if you want to update more than one element at a time, you must use an Element Selection tool to select them **before** you activate the Update Dimension tool.

GEOMETRIC TOLERANCE

The Geometric Tolerance tool is a handy way of adding text or notes regarding geometric tolerances. It works with specific fonts that are standard geometric symbols. You have two different Tool options: Place Note or Place Text. In both cases, the Text Editor box opens and by selecting on the buttons in the Geometric Tolerance dialog box, it actually inserts the character into the text editor for you. You don't have to figure out which letter represents the diameter symbol—just click on the symbol and it will type the "n" for you.

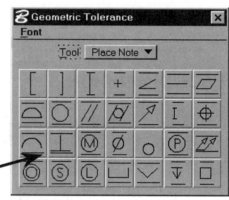

The current font in the illustration at the right is 101 - Feature Control Symbols. It has lines above and below the symbol so that the feature control frame is automatically made around the symbols. The other font available in the Geometric Tolerance FONT pull-down menu is 100 - ANSI Symbols. Its symbols are the same except that they don't have these lines.

These buttons will place the correct character into the text editor box.

Place Note

Activating the Geometric Tolerance tool with its Tool setting of Place Note Tool will bring up both the Text Editor and the Place Note dialog boxes. To actually place the note, you will be asked to specify the start of the leader and then its location. Shown here is a counterbored hole note that was done using the Geometric Tolerance with Place Note Tool.

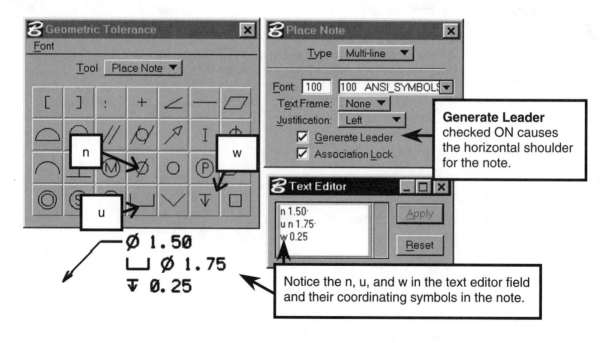

Generate Leader checked ON causes the horizontal shoulder for the note.

Notice the n, u, and w in the text editor field and their coordinating symbols in the note.

QUESTIONS

① Explain the importance of the Text Height setting when doing dimensioning.

Text height dictates extension, text margin, terminator, leader size

② Under what dimension setting category do you find the setting to Show Trailing Zeros?

UNITS FORMAT

③ What dimensioning tool allows you to avoid having to replace dimensions that you've already done but whose settings are wrong and need to be changed?

Update dimension

④ Sketch a center mark that results from the following Center Size settings:
a) Positive Value
b) Negative Value

⑤ You've drawn a detail that is 2 times the real size of the object and you need to dimension it correctly. What do you need to change in the dimension settings? Give the category, the setting name, and the value it should be. *Dimension Settings Units — Scale Factor ÷ 5*

⑥ Name the **three** options for the orientation of dimension text relative to the dimension line. *In-line / above / Horizontal*

⑦ Explain what an **associative** dimension is and why it is useful. *Lock ☑*
— dimensions linked to element so if element is modified then the dimension also changes i.e.

⑧ Use a sketch to show the difference between a Dimension Location (Stacked) and a Dimension Size with Arrows on this object:
a) Location Stacked *— Baseline dimension*
b) Size with Arrows

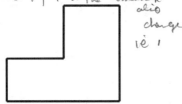

⑨ Name the tool(s) and any specific settings needed to place dimensions A and B illustrated at the right.

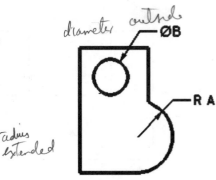

diameter outside ØB
radius extended R A

⑩ For both I and II indicated on this dimension element, give the following information:
a) Name of the setting and under what dimension setting category it is found.
b) Units that the setting is based on.
c) Actual distance, assuming that the setting's value is 0.5, the master units are inches, and the text height is 0.25 master units.

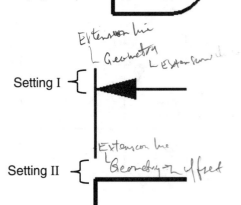

Extension line Geometry Extension
Extension line Geometry — offset

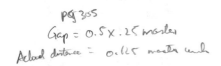

pg 355
Gap = 0.5 × .25 masters
Actual distance = 0.125 master unit

EXERCISE 14-1   Simple Bracket

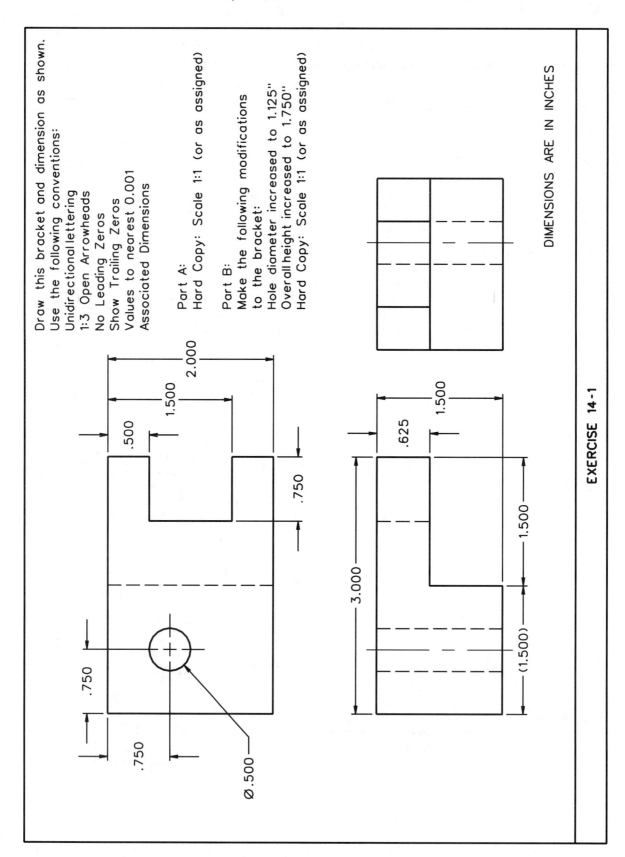

Dimensioning

EXERCISE 14-2   Simple Rod Support

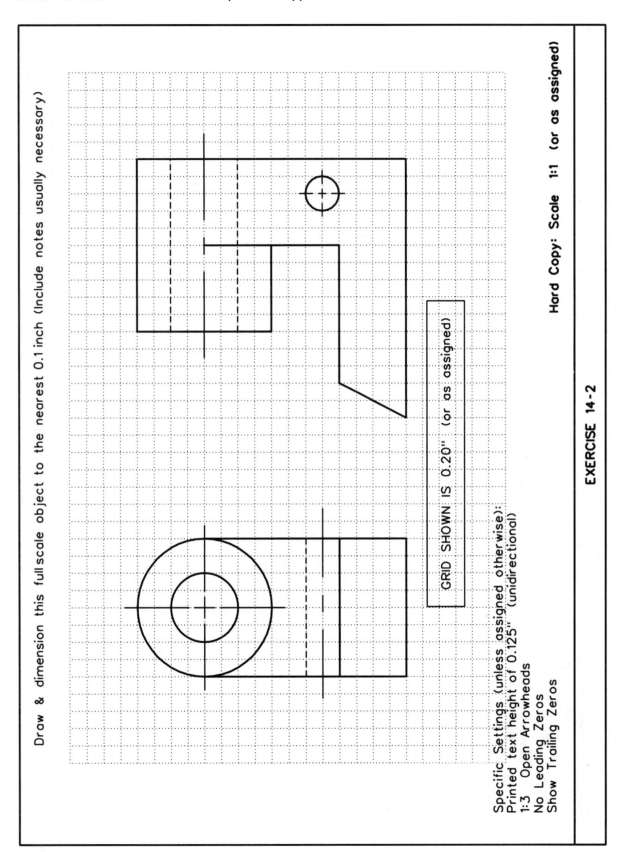

EXERCISE 14-2

EXERCISE 14-3   Pond Renovation

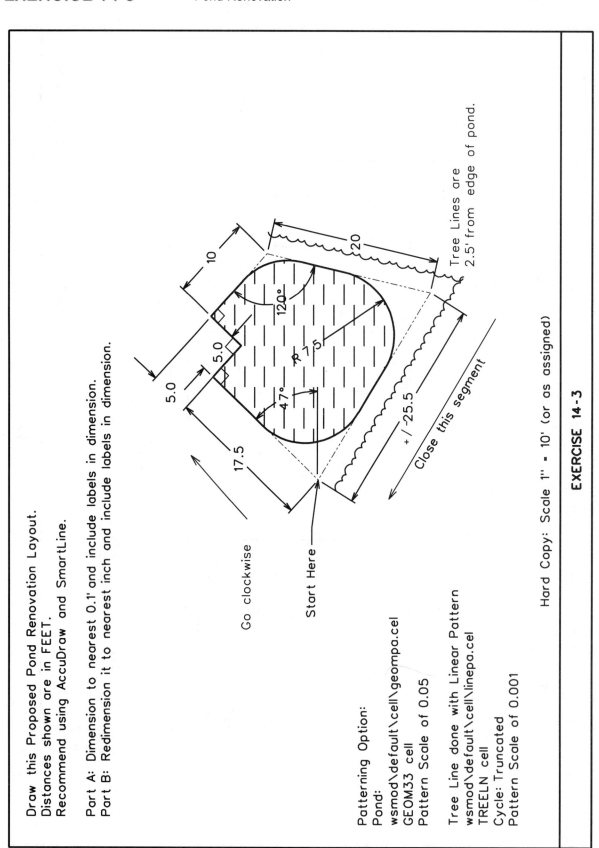

EXERCISE 14-3

Draw this Proposed Pond Renovation Layout. Distances shown are in FEET. Recommend using AccuDraw and SmartLine.

Part A: Dimension to nearest 0.1' and include labels in dimension.
Part B: Redimension it to nearest inch and include labels in dimension.

Patterning Option:
Pond:
wsmod\default\cell\geompa.cel
GEOM33 cell
Pattern Scale of 0.05

Tree Line done with Linear Pattern
wsmod\default\cell\linepa.cel
TREELN cell
Cycle: Truncated
Pattern Scale of 0.001

Hard Copy: Scale 1" = 10' (or as assigned)

EXERCISE 14-4   Rural Intersection

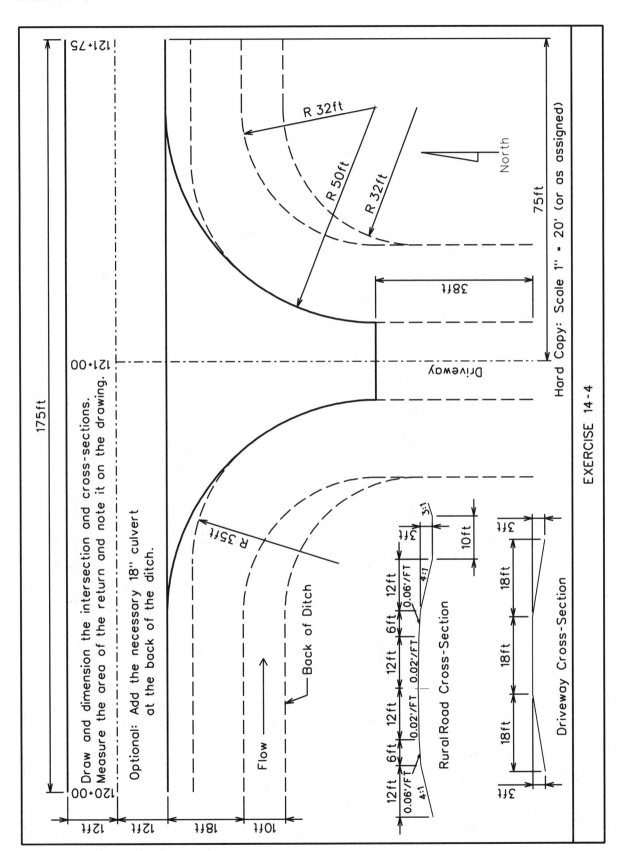

EXERCISE 14-4

EXERCISE 14-5 Symmetric Bracket

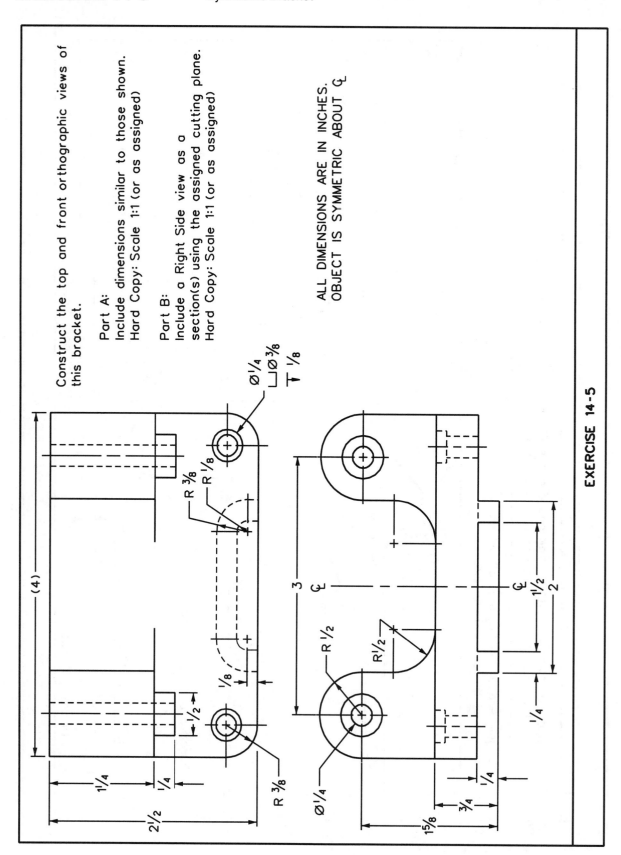

EXERCISE 14-5

Chapter 15: Reference Files

SUBJECTS COVERED:

- Capabilities and Limitations
- Accessing Reference Files Tools
- Attach Reference File
- Reference Files Settings
- Manipulation Tools for Reference Files
- Move Reference File
- Scale Reference File
- Copy Attachment
- Merge Into Master
- Detach Reference File
- Attach URL
- Reload Reference File
- Other Related Tools
- Active Design File Referencing Itself

A reference is a source whose information is available for you to use. For instance, an encyclopedia is a reference that is used in writing a report. The information found in the encyclopedia aids in writing the report. The report itself doesn't affect the encyclopedia's contents—it uses the reference but doesn't alter it. A design file uses the graphical information in its reference files to aid in its construction. The elements (and their properties) of the reference files are available for coordinate information, display, etc., but are not subject to changes or edits. The ability to reference design files and use their information is a major advantage, especially when a group is working on a project. Each person can be working on part of a design individually, but by using reference files, the group can put it all together as one design.

CAPABILITIES AND LIMITATIONS

There are certain things that you can and cannot do with reference files. The capabilities outnumber the limitations.

With reference files you have these **capabilities**:
- Use a tentative point to snap to the elements in the reference files.
- Control the display of levels, symbology, etc., of the reference files independently of the design file.
- Print what is displayed in both the design file and its reference files.
- Copy elements from a reference file into the design file.
- See the updates made to a reference file.
- **Reference the active design file onto itself** (a powerful capability since it allows you to view/print more than one portion of the active design file in a single view window).

Limitations of reference files:
- **You cannot manipulate or modify the actual elements of the reference files.** Even if the file is referencing itself, only the original elements at their original location can be modified. You can't modify the referenced ones.

There are three files that will be used as our example for reference files. First is the **active** design file (named Plumbing) that is going to contain the plumbing for a new bathroom. Currently it doesn't contain any elements at all. The existing items such as the basement wall and water lines are contained in a file called Bathroom that will be used as a reference file (illustrated bottom left). The new fixtures for the bathroom have been done by another person and are contained in a design file called Fixtures (illustrated bottom right). The Fixtures file will also be referenced into the active design file (Plumbing). These files will be used as our example for reference files.

BATHROOM.dgn   **FIXTURES.dgn**

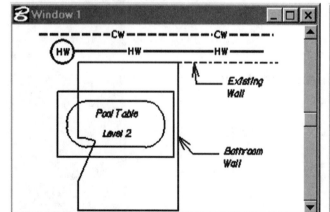 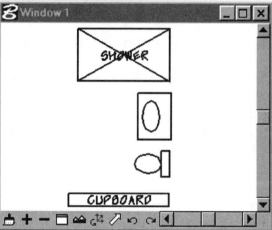

Reference Files 339

ACCESSING REFERENCE FILES TOOLS

As usual, you can get to the tools that deal with reference files in more than one way. There is a tool box available and then also a dialog box with pull-down menus. In this case, the tool box has a lot of tools involved, and often the icons are confusing. The settings available with the pull-down menus are more straightforward—so that's what we'll concentrate on. However, let's show you how to get to the Reference Files tool box, so that if you want to go there later, you can.

Reference Files Tool Box

The Reference Files tool box doesn't show up in the Main tool box. Instead you need to go to the main pull-down menu TOOLS >REFERENCE FILES as illustrated below. A checkmark indicates that the Reference Files tool box is open. It has thirteen tools, so it is easy to spot!

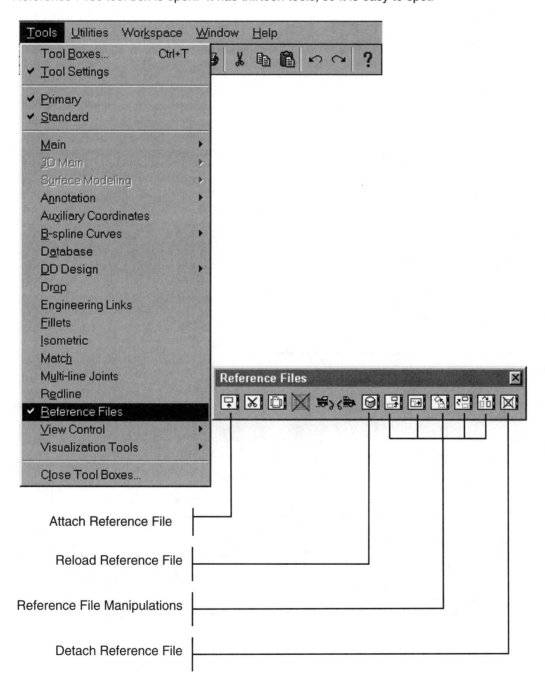

Reference Files Dialog Box

The Reference Files dialog box is often easier to use because its pull-down menu is easy to follow. The Reference Files dialog box is accessed under the main pull-down menu FILE>REFERENCE. The dialog box also includes an area that shows details about the reference files that are attached.

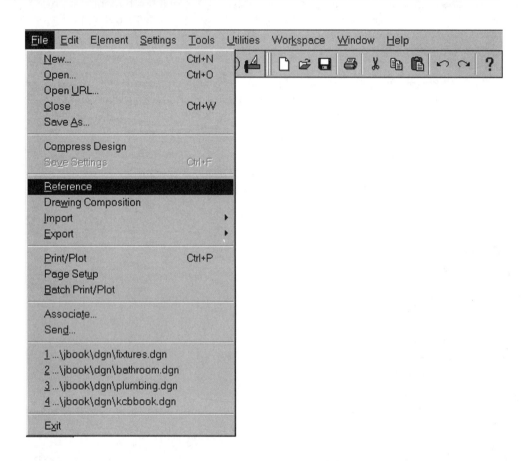

Tools, Settings, Sort, and Display make up the **Reference Files** pull-down menus.

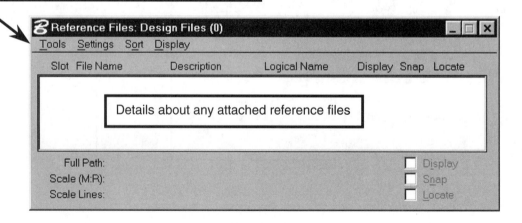

ATTACH REFERENCE FILE

The first thing you must do is to attach the reference file that you want to use. Using the Reference Files pull-down menu, go to TOOLS>ATTACH.

Notice that the other items under Tools can be found in the tool box as icons. Most of them deal with manipulation and control of reference files that are already attached. That is why they are grayed out here—they don't work until you've attached a reference file.

Here are some important points to know about attaching reference files:
- You need to specify one file to attach at a time—but more than one file can be attached; you just need to attach each individually.
- You can't reference 3D files into 2D files.
- Some of the attachment settings (such as Display, Snap, and Locate) are preset, so you may want to modify them <u>after</u> you've made the attachment.

The Preview Reference dialog box is illustrated below. It is similar to the MicroStation Manager box used to open a design file. The file to be attached is highlighted and shows up in the Name field. For example, the bathroom.dgn file is the one that will be attached.

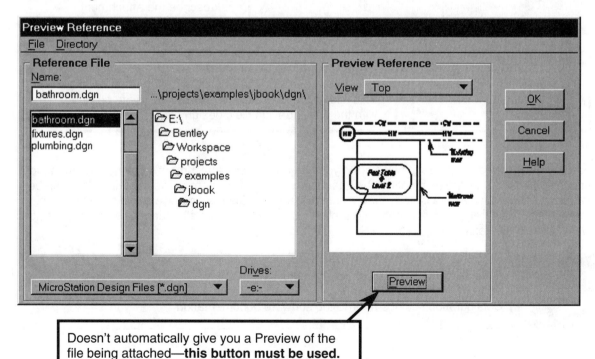

Doesn't automatically give you a Preview of the file being attached—**this button must be used.**

Once you've chosen which reference file to Attach, then you'll have some choices to make in the attachment. This is done in the next dialog box that comes up automatically. It is the Attach Reference File dialog box illustrated below. Once you have the changed the settings in this dialog box, remember to use the OK button to accept them and complete the attachment.

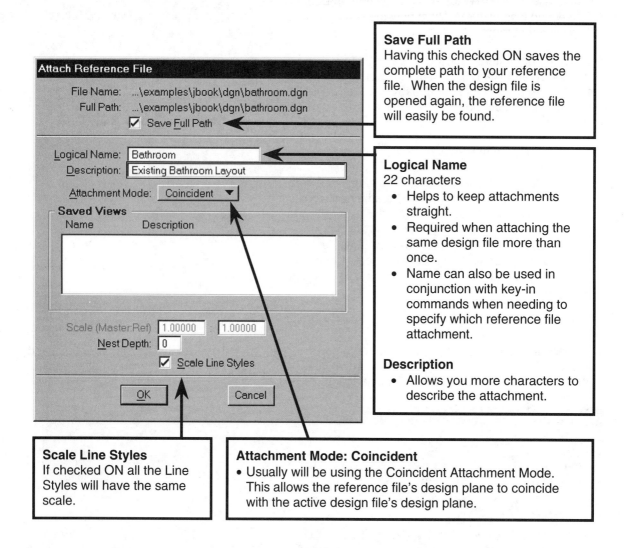

Save Full Path
Having this checked ON saves the complete path to your reference file. When the design file is opened again, the reference file will easily be found.

Logical Name
22 characters
- Helps to keep attachments straight.
- Required when attaching the same design file more than once.
- Name can also be used in conjunction with key-in commands when needing to specify which reference file attachment.

Description
- Allows you more characters to describe the attachment.

Scale Line Styles
If checked ON all the Line Styles will have the same scale.

Attachment Mode: Coincident
- Usually will be using the Coincident Attachment Mode. This allows the reference file's design plane to coincide with the active design file's design plane.

Once you've attached the reference files, then you can adjust each attachment's settings. Even though you can't edit or change the elements in the attached reference file, there are many things that can be controlled. Things such as the Levels, View Attributes, and the Display and Scale can be set for each reference file. Each reference file can be controlled individually, so that all attached reference files don't have to appear the same.

 If the Save Full Path is unchecked (OFF) then the path to look for an attached reference file is specified in the Configuration settings. For more information about Configuration settings refer to the software documentation.

REFERENCE FILES SETTINGS

Here is an example of the active design file Plumbing that has two reference files attached. The title bar of the Reference Files dialog box indicates that as Design Files (2). The first one attached was Bathroom and the second was Fixtures. Right now the settings are the same for both reference files—but we will be modifying some of them. The settings be changed as often as you want—again you're not affecting the actual elements in the reference files, just how they will be displayed, etc.

The highlighted reference file is the selected one. The information and settings shown in the gray area of the dialog box will refer to the selected reference file. Its Display, Snap, and Locate settings can be toggled on/off with the check boxes. Another way to toggle the Display, Snap, and Locate settings of any of the attached reference files (not just the selected one) is to directly click on the X that shows up in the listing.

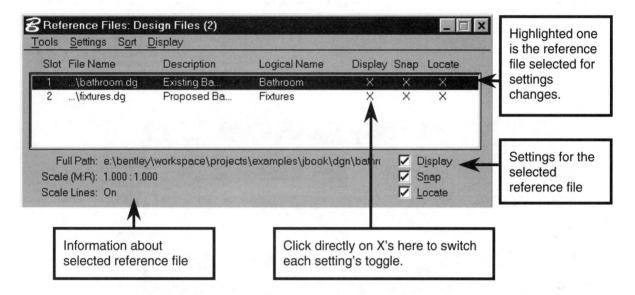

Illustrated below are the active design file (Plumbing) and its two attached reference files (Bathroom and Fixtures).

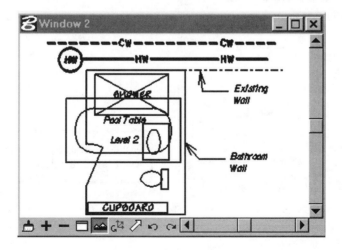

Display

This setting controls whether or not the reference file will even be displayed. Most of the time you'll want to leave it checked ON. In some projects, the same reference files are attached automatically. If you're not using all of them, you may want to have Display unchecked (OFF). Then the reference files won't slow down the response time of the active design file.

Snap

The Snap setting checked ON allows you to snap to and use keypoints of entities in the reference file while you are placing entities in the active design file. This is extremely handy and one of the main reasons to use reference files—so you'll usually want Snap checked ON.

Locate

The Locate setting checked ON allows you to copy elements (and other fence manipulations) from the reference file into the active design file. This isn't always necessary, so you may want to have it unchecked (OFF) to avoid inadvertently getting new elements in the active design file.

Let's adjust one of the reference file's settings in our example. Illustrated below is what occurs when the Display, Snap, and Locate are unchecked (OFF) for the **Fixtures** reference file.

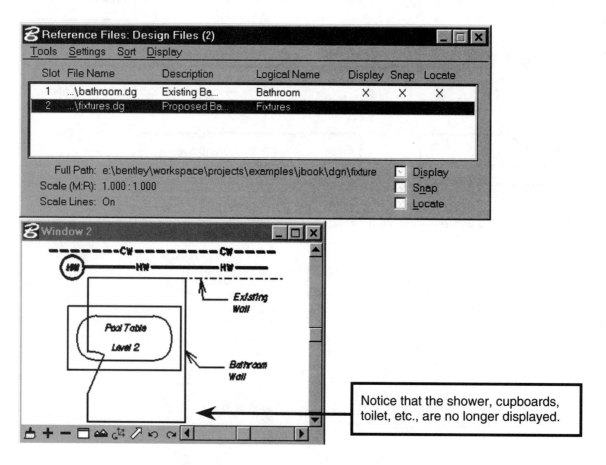

Notice that the shower, cupboards, toilet, etc., are no longer displayed.

Reference Files   345

The existing pool table is on Level 2 of the Bathroom reference file. The display of the pool table isn't really necessary, since it isn't going to be part of the plumbing in the bathroom! However, if you toggle off the Display setting of the Bathroom reference file—you'll also lose the display of the existing water line (Level 1 elements). We need those to help establish our new plumbing. What a dilemma! The solution is that the display of the levels in each attached reference file can be turned on/off in the active design file. This reference file level control is done using the Reference Levels dialog box. It is accessed from the Reference Files pull-down menu SETTINGS>LEVELS.

Reference Levels Dialog Box

The Level display can be controlled independently for each attached reference file. This means that the Bathroom reference file can have Level 2 **off** at the same time that the Fixtures reference file has Level 2 **on**. It also doesn't matter what the active design level settings are—each reference is independent. For even more control, the view windows of the active design file can each have their own Reference Level settings. This means that in Window 2, the Reference Level settings can have Level 2 in the Bathroom reference file **off**, while at same time the Reference Levels setting for Window 1 can have Level 2 in the Bathroom reference file **on**.

Let's look at using the Reference Levels dialog box to turn **off** the display of **Level 2** of the Bathroom reference file in **Window 2** of the active design file.

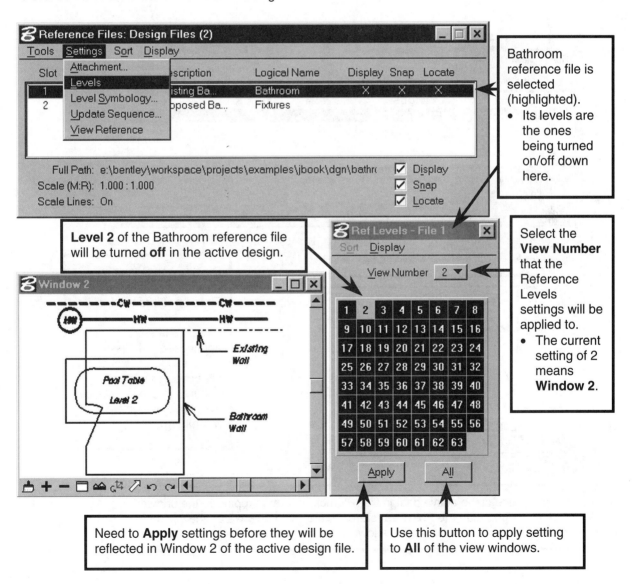

Once the Apply button is clicked on, the elements on Level 2 will no longer be displayed in Window 2 as illustrated here on the right.

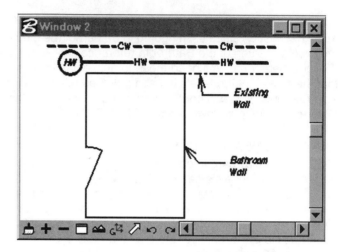

Snapping to a Reference File's Elements

Here is a prime example of the advantage of using reference files. The Fixtures reference file elements can be snapped to in order to place the new water lines in the correct position.

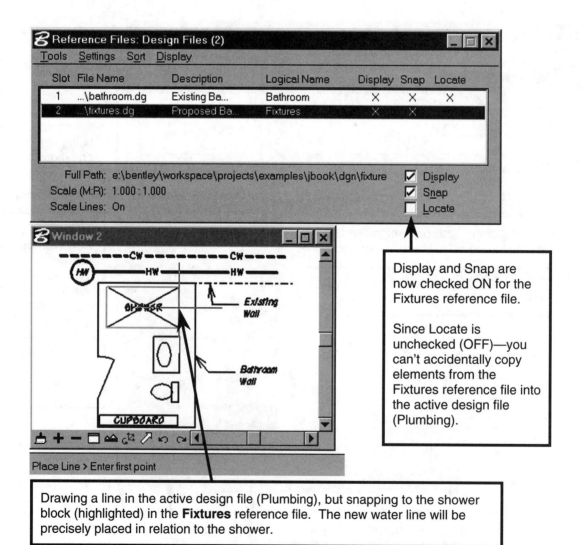

Display and Snap are now checked ON for the Fixtures reference file.

Since Locate is unchecked (OFF)—you can't accidentally copy elements from the Fixtures reference file into the active design file (Plumbing).

Drawing a line in the active design file (Plumbing), but snapping to the shower block (highlighted) in the **Fixtures** reference file. The new water line will be precisely placed in relation to the shower.

Reference Files 347

MANIPULATION TOOLS FOR REFERENCE FILES

Now that there are reference files attached, there are more tools available under the Reference Files pull-down menu TOOLS as shown on the right. They are similar to the regular manipulation tools such as Move, Scale, and Mirror—but the **elements in the reference file are not actually changed.** The manipulation tools for reference files treat the whole reference file as a single element.

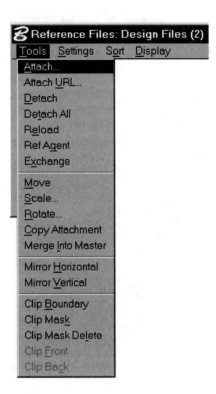

When you use the pull-down menu, the reference file that will be manipulated will be the one selected (highlighted) in the Reference Files dialog box.

If you're using the tools in the Reference Files tool box then you'll be prompted to Identify (or key in) Ref File. You can identify the reference file that will be manipulated in two ways:
- Graphically picking on an element in the reference file.
- Keying in the Logical Name specified for the reference file in the Key-in field.

MOVE REFERENCE FILE

In this case, we'll use the Reference File tool box's Move tool in order to do a **precise** move. You can also use the Reference dialog box pull-down menu TOOLS>MOVE (from the list shown above). The Fixtures reference file will be selected for the Move by snapping to the toilet's ellipse as shown below on the left. It will be moved to the right and it ends up looking like what is displayed on the right.

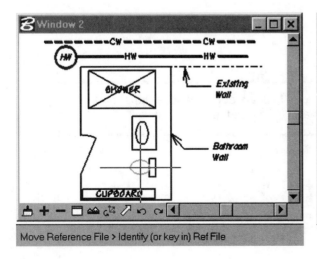 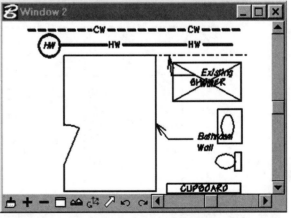

SCALE REFERENCE FILE

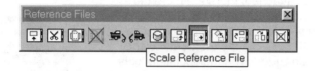

This is a manipulation that will also be used frequently, especially in the case of a design file referencing itself. The ability to Scale a reference file allows you to set the relationship between the **M**aster units in the active design file and the **R**eference file's master units.

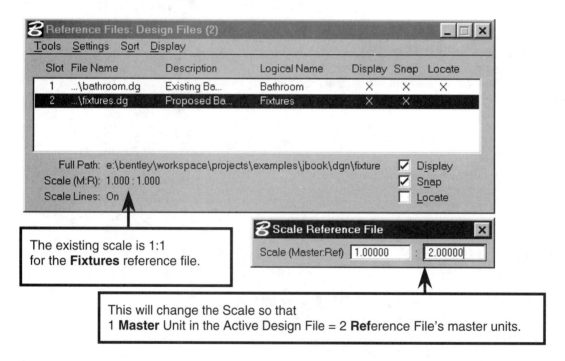

The existing scale is 1:1 for the **Fixtures** reference file.

This will change the Scale so that
1 **Master** Unit in the Active Design File = 2 **Ref**erence File's master units.

Once you've set the new scale ratio, you'll need to specify the point to scale about. Again this is like the typical Scale Element tool. The result of scaling the Fixtures reference file (with the Scale Reference File settings as shown above) is illustrated below.

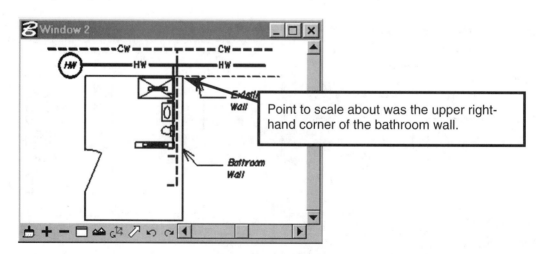

Point to scale about was the upper right-hand corner of the bathroom wall.

COPY ATTACHMENT

Copy Attachment is not found in the Reference Files tool box; rather you can access it by using the Reference Files pull-down menu TOOLS>COPY ATTACHMENT as illustrated on the right. Don't be misled by this option—this does not copy elements from the attached reference file into the active design file. It actually takes the selected reference file and makes another attachment of it. This could prove handy if you've adjusted an attachment just right (correct levels and such) and find yourself needing another attachment, but you don't want to have to readjust it again. The Copy Attachment would work great whereas just using Attach again would require to make readjustments.

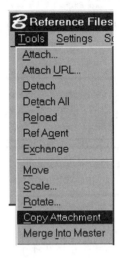

You get a choice of how many copies you want as illustrated on the right. Shown below is the result of making one copy attachment of the reference file Fixtures.

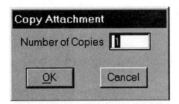

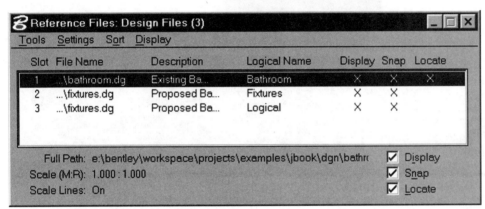

MERGE INTO MASTER

Merge Into Master also needs to be accessed by using the Reference Files pull-down menu TOOLS>MERGE INTO MASTER as shown on the right. Now this tool actually does copy elements from the attached reference file into the active design file. However, there are some important things to keep in mind:
- **All** of the elements in the attached reference file are copied (even those that you don't see).
- It **automatically** detaches the reference file copied from.

Because this tool can introduce a large number of elements into the active design file in a single stroke, an Alert (illustrated on the right) pops up to make sure you really want to go through with the merge.

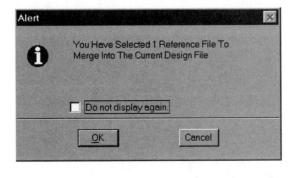

DETACH REFERENCE FILE

When a reference file is no longer required or needed, you can Detach it.

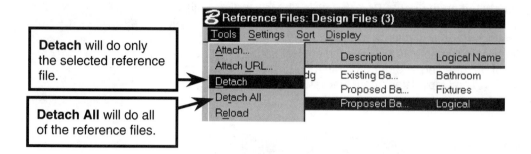

Detach will do only the selected reference file.

Detach All will do all of the reference files.

You will be given a chance to change your mind. An Alert will show up checking to be sure that you know what you're doing.

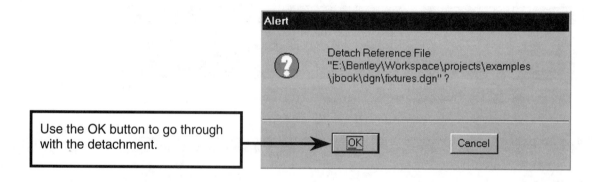

Use the OK button to go through with the detachment.

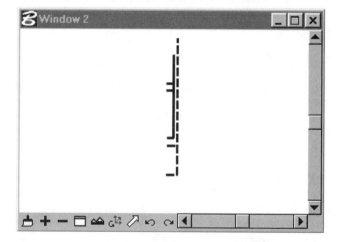

If **Detach All** were used, then all that we would be left seeing are the elements in the Active design file. Shown at the right are the new elements in the Plumbing design file.

ATTACH URL

It is also possible to attach a reference file by specifying its URL (Universal Resource Locator). This would allow you to use the Web to share design file resources. By using the Reference File pull-down menu TOOLS>ATTACH URL (as illustrated on the right), you get the Select Remote Design File to Attach dialog box as shown below. The implementation of Attach URL will be left for advanced users to experiment with.

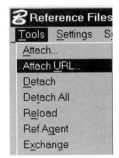

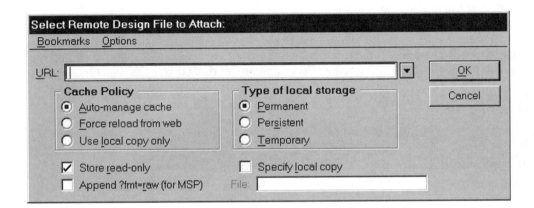

RELOAD REFERENCE FILE

Also available is a tool that will Reload the selected attached reference file. This doesn't "undo" a Detach. When you Reload a reference file, it means that any new elements or modifications that may have been made to the file being referenced will be shown in your attachment of it. Reloading a reference file is often necessary on a project where you are referencing other people's design files and they are actively working in those design files.

OTHER RELATED TOOLS

Some of the regular tools discussed previously need to be revisited to see how they function concerning reference files.

Fit View

The Fit View tool in the view controls located at the bottom of the view window allows you to control the display of the view window. It gives you the Files options as shown here.

- **All:** Fits both the Active design file and any attached Reference (or Raster) files. **Preferred Option**
- **Active:** Fits only the Active design file and will ignore the limits of the attached Reference files.
- **Reference:** Fits only the Reference files attached and will ignore the limits of the Active design file.
- **Raster:** There is also an option of attaching Raster files (which aren't .dgn files) and this option will fit only those attached files.

Print

Printing is a case of "what you see is what you get." The display of attached reference files will be printed right along with the active design file—you don't have to do anything special.

Copy

You can copy elements from a reference file into your own active design file. The reference file must have the setting **Locate checked ON** in order to allow this. The elements will come in with the attributes that they had in the reference file—they will not reflect the current active design file's Active Attributes.

Steps to copy from an attached reference file:
1. Select (or fence) the elements you want to copy.
2. Select the regular Copy tool from the Main tool box.
3. Define the origin of the original elements.
4. Enter 2nd point to define the origin of the copies (it can be the same place as the first point).

The illustration below shows that the Bathroom reference file has been attached again, so we can copy elements from it into our active design file (Plumbing). The attachment has Locate checked ON.

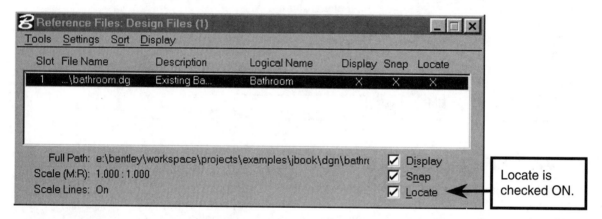

Now using the regular Copy Element tool (found in the Main tool box) with Use Fence checked ON, we can copy the existing waterlines from the Bathroom reference file into our Plumbing design file.

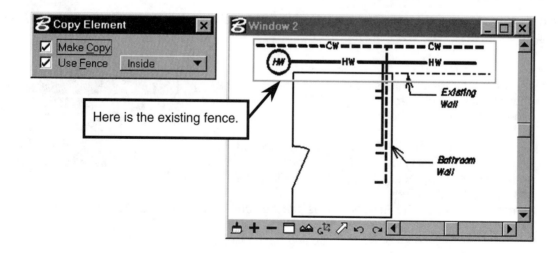

By detaching the Bathroom reference file, you can see that the elements have been copied into the active design file. Now our Plumbing design file contains the elements representing both the new and the existing waterlines. Using the Copy Element tool was better than using the Merge Into Master option because only the existing waterlines were copied. With Merge Into Master, all of the elements in the Bathroom design file (including the Pool Table) would be put into the Plumbing design file.

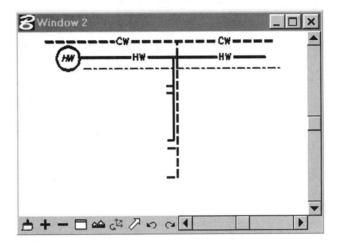

 There is a handy key-in that will save the elements selected by the fence to a NEW design file. It is called fence file and its shortcut key-in is ⌨FF=*filename.* You need to **read the prompts** because you must data point to Accept the fence to file copy. It works really slick.

ACTIVE DESIGN FILE REFERENCING ITSELF

As mentioned before, MicroStation allows the active design file to reference itself. This allows for various areas of a design file to be laid out together in one view. By using this concept, you can include a close-up detail in the same view without having to create elements that are "not to scale." Since it is all in one view, it can easily be printed out on a single hard copy.

Shown below are two separate windows displaying the contents of the active design file called Columns. Window 2 has zoomed in to show the reinforcement in the columns. By attaching the file itself as a reference file, we can get the detail with the layout.

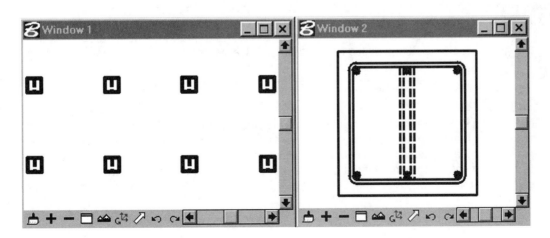

Illustrated below is the resulting view in Window 1 of the active design file (Columns.dgn). Window 1 displays the original elements in the left portion of the window and the attached reference file (of itself) in the right portion of the window. Please realize that the reference file attachment underwent many modifications in order for this Window 1 display to appear with **both** the column layout and the column detail. The attachment was scaled and moved (other advanced reference file tools such as clipping and masking may also be needed in these cases) until it was satisfactory. Dimensions and labeling were then added. **This is not necessarily a simple process, but it is effective.**

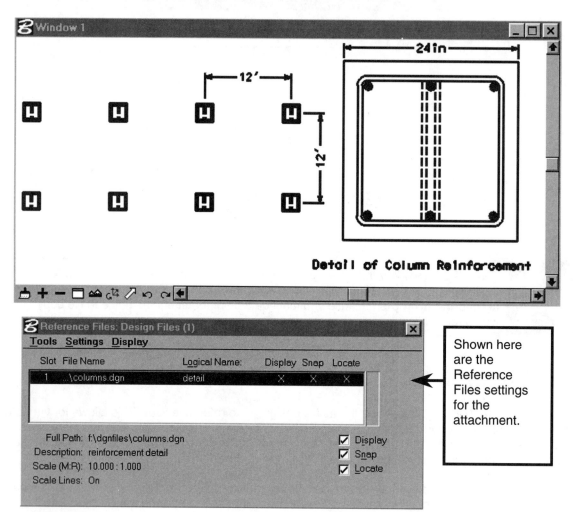

Shown here are the Reference Files settings for the attachment.

If you plan on dimensioning to a reference file, go to the Placement category of the Dimension Settings dialog box and be sure Reference File Units is checked ON.

Drawing Composition

Drawing Composition is another method of getting various views of a design file into one special view called a Sheet View. It actually uses the capability of a design file to reference itself but has special tools that streamline the process. Drawing Composition can be used with two-dimensional files, but it is indispensable when it comes to generating a single layout of numerous standard views from a three-dimensional model. It's a bit too complicated—for now. Drawing Composition is discussed in more detail in Chapter 21.

QUESTIONS

handwritten: a) File → Reference

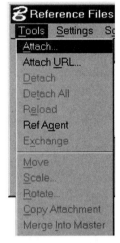

① Under which main pull-down menu do you find the —
   a) Reference File **dialog** box.
   b) Reference File **tool** box.
   handwritten: Tools → reference File

② Under which of the three pull-down menus (Tools, Settings, and Display) shown at right would you find the reference file level display controls?
   handwritten: Tools

③ Why would you want to have the Save Full Path setting checked ON before attaching a reference file? *handwritten: When design file is opened again file is easily found*

④ What does it mean to Detach a reference file? *handwritten: Deletes file*

⑤ Explain the difference between using the regular Copy Element tool and using the Merge Into Master option to get elements from a reference file into the active design file. *handwritten: Copy - make another copy of reference file. Merge copies elements of attached file into design file*

The following questions deal with the Reference Files dialog box shown below.

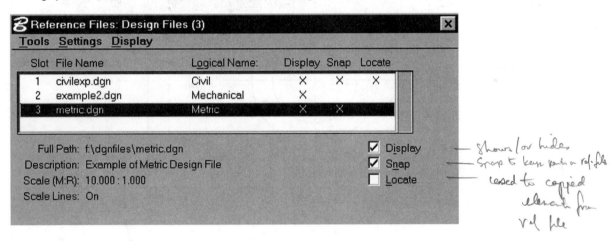

⑥ How many reference files are attached? *handwritten: 3*

⑦ What is the Logical Name of the selected reference file? *handwritten: Metric*

⑧ What does the Scale (M:R): 10.000:1.000 mean?
   handwritten: MU Ref File MU

⑨ Indicate the Logical Name(s) of the reference file(s) whose elements can be copied into the active design file. *handwritten: Civil, M*

⑩ Indicate the Logical Name(s) of the reference file(s) that will allow the active design file to recognize the keypoints of its elements. *handwritten: Civil*

356　Chapter 15

EXERCISE 15-1   Part A: Elliptical Brace

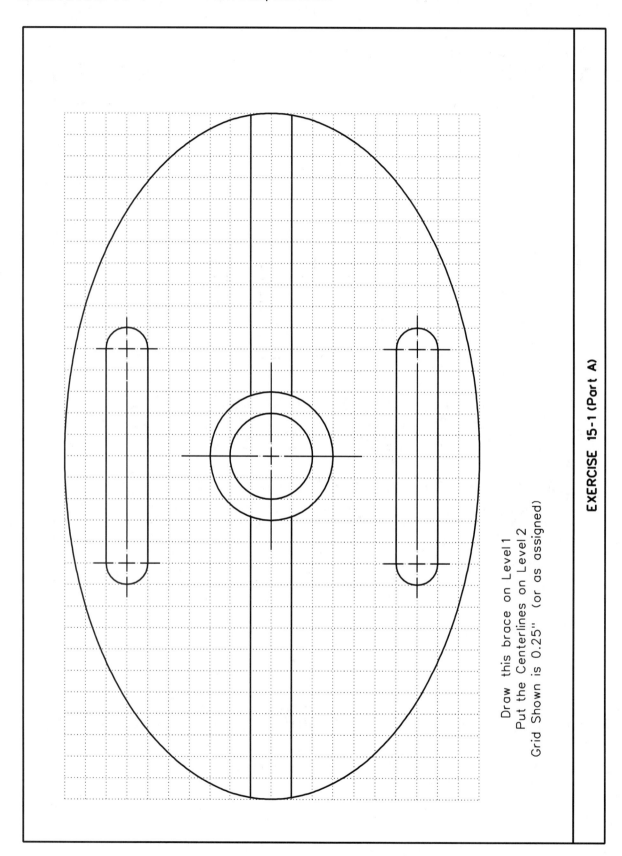

EXERCISE 15-1 (Part A)

Draw this brace on Level 1
Put the Centerlines on Level 2
Grid Shown is 0.25" (or as assigned)

EXERCISE 15-1 (Continued)   Part B: Poles and Bolts with Bracing

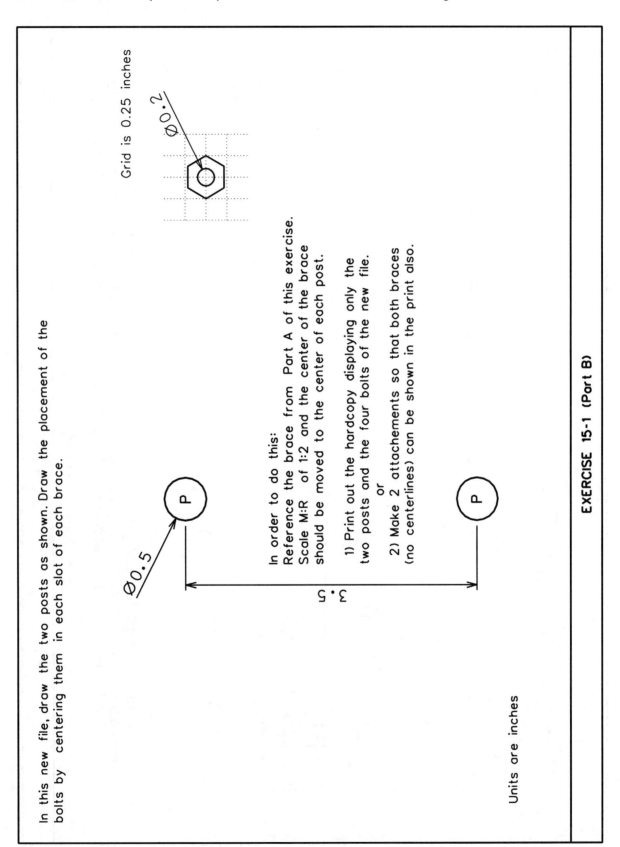

EXERCISE 15-1 (Part B)

EXERCISE 15-2 Classroom Desks Added to Existing Plan

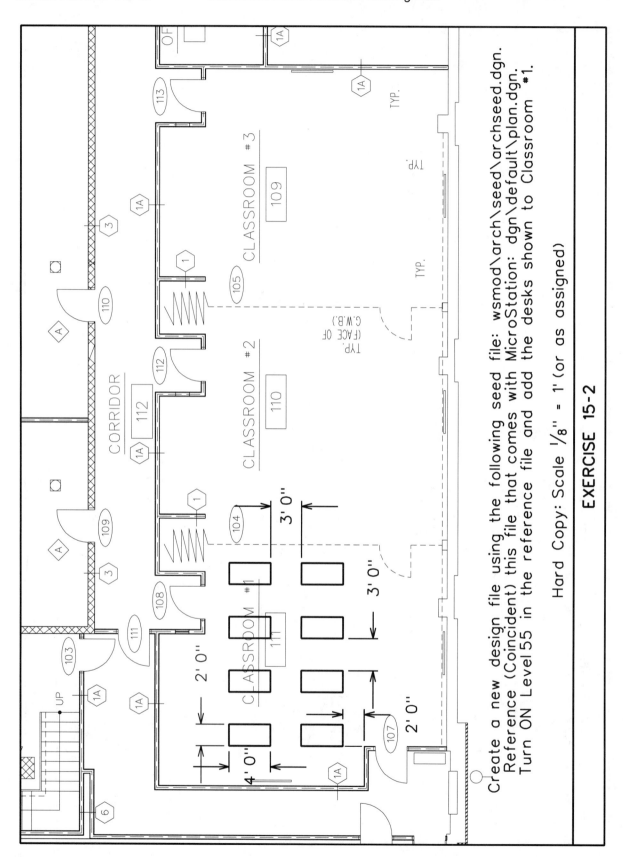

EXERCISE 15-2

Create a new design file using the following seed file: wsmod\arch\seed\archseed.dgn. Reference (Coincident) this file that comes with MicroStation: dgn\default\plan.dgn. Turn ON Level 55 in the reference file and add the desks shown to Classroom #1.

Hard Copy: Scale 1/8" = 1' (or as assigned)

EXERCISE 15-3 Golf Course—New 15th Hole Green

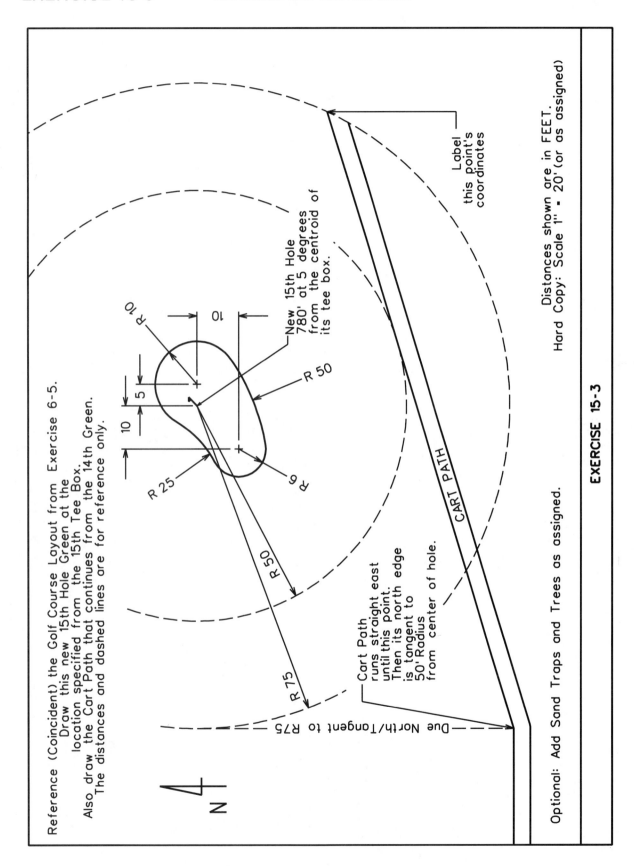

Chapter 16: Special Considerations of 3D Space

SUBJECTS COVERED:

- Positive Right-Hand Coordinate System
- 3D Coordinate System
- 3D Seed Files
- 3D View Controls
- 3D View Control Tool Box
- Zoom In/Out
- Change View Perspective
- Active vs. Display Depth
- Set Display Depth
- Set Active Depth
- Show Display Depth
- Show Active Depth
- View Rotation
- Change View Rotation
- Rotate View
- Set View Display Mode

Up until now, we've been working in two dimensions. The coordinate system has been an X-Y plane and Z has been assumed as zero. Now moving into three dimensions (3D), Z will not necessarily be zero. You need to be familiar with a 3D coordinate system and its orientation, so we'll review that. That way you will have a better understanding of the screen and drawing coordinates, which now become extremely important. Three dimensions are tougher to keep track of, so you must be prepared to pay close attention to details. Since we're no longer working in a single plane, there are additional view controls involved in a 3D space. Different seed files are available for 3D files, including workspaces especially for three-dimensional work. Lots of new stuff, and the intent is to give you a general feel for 3D capabilities—not to make you an expert.

POSITIVE RIGHT-HAND COORDINATE SYSTEM

In the positive right-hand coordinate system, when using your right hand (sorry lefties—this works only with the right hand), you can visualize your thumb pointing in the positive X-axis. Your index finger then would point in the positive Y-axis, and your middle finger then bends into the positive Z-axis.

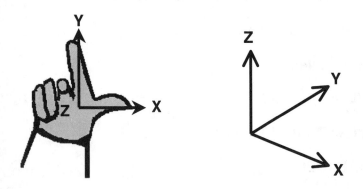

You can use your right hand to help determine the positive axis from the other two positive axes. Point your fingers in the direction of the positive X-axis. Now curl them in the direction of the positive Y-axis (it's like pushing the X-axis into the Y-axis)—your thumb will be pointing in the direction of the positive Z-axis. This would also indicate a positive rotation about the Z-axis.

This also works with the other combinations, but there is a certain order that you must follow:
- X curled into Y will give you Z.
- Y curled into Z will give you X.
- Z curled into X will give you Y.

3D COORDINATE SYSTEM

Some seed files such as seed3d.dgn will result in a 3D design file that has the four view windows already open and in the orientation shown. Some seed files also have a 3D cube that helps with visualizing the 3D coordinate system used. Notice that the four view windows follow the engineering graphics conventions of front, top, and right side views with a pictorial. In the illustration shown below the windows are named Top, Front, Right, and Isometric after the standard conventions. The Isometric view shows a left-front isometric pictorial.

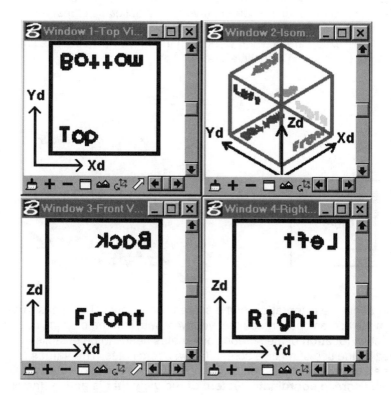

Drawing Coordinates

We've been working in the X-Y drawing plane or Top view for our 2D drawings. Now with a Z direction, we have to consider the X, Y, **and Z** coordinates of the drawing. The drawing coordinates are **not** dependent on the view that you are working in. They are the global coordinate system. The illustration above indicates the orientation of the 3D drawing coordinates. The letter **d** stands for the **drawing coordinates**. **Zd** would indicate the **Z**-axis for the **d**rawing coordinate system.

- For **absolute** rectangular 3D drawing coordinates you can still use the key-in of
  XY=*xcoordinate,ycoordinate,zcoordinate*

- For **relative** rectangular 3D drawing coordinates you can also still use the key-in of
  DL=△*xcoordinate*,△*ycoordinate*,△*zcoordinate*

Screen Coordinates

Each view also has its own local coordinates. If you are working in one particular view, you can enter coordinates and orient elements according to the **view's** individual coordinates. These are oriented with +X to the right, +Y to the top, and +Z coming out of the screen at you. These are often referred to as the **screen** coordinates (but may be called view coordinates too). In the Top view, the screen coordinate system and the drawing coordinate system are coincident. The illustration below on the left shows the screen coordinates for the standard Top, Front, and Right Views. The lowercase letters stand for the specific screen coordinate system. **Yfs** stands for the **Y**-axis of the **f**ront view's **s**creen coordinates. A ● indicates the positive Z-axis coming out of the screen at you. The oblique pictorial on the right gives you a three-dimensional look at the drawing and screen coordinates' positive axes in relation to each other and to the standard views. Be extra careful when working in the Front view because its positive Zfs axis is in the opposite direction of the drawing's positive Y-axis.

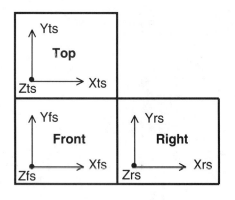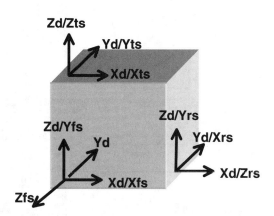

Now this really talks only about the **orientation** of the screen coordinates. There has to be some way for the software to transform the screen coordinates that you work with into the drawing coordinates that it stores. Therefore, the actual **location** of the screen coordinates needs to be established. For screen coordinates, it is handled by the use of relative screen (view) coordinates. **The values keyed in are relative to a known existing point** (established by snapping to an existing element or placing a tentative point) but are based on the **screen coordinate system of the view that is active** (the one being worked in).

The key-in for **relative view** coordinates is
⌨DX=△screenXcoordinate,△screenYcoordinate,△screenZcoordinate

Example of Relative Screen (View) Coordinates

A 3D line has one endpoint entered as ⌨XY=8,22,7. While working in the **Right view**, you specify the other endpoint of the line as ⌨DX=5,10,15. What are the actual **drawing** coordinates of the second endpoint?

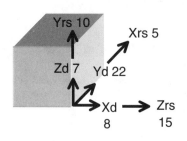

This is how the two positive coordinate systems are oriented together in the Right view. The values established by the key-ins are shown. By adding them together (**paying attention to their sign**), you can determine the second endpoint and report it according to the global drawing coordinates.

Xd= 8+15 = 23
Yd= 22+5 = 27
Zd= 7+10 = 17

Drawing coordinates for the second endpoint are (23,27,17).

3D SEED FILES

There are numerous 3D seed files that are available for use. Using the Select button in the Seed File section of the Create Design File dialog box can access the different seed files. The choices in the Select Seed File dialog box are illustrated below. Notice they are found in the directory of Bentley\Workspace\system\seed.

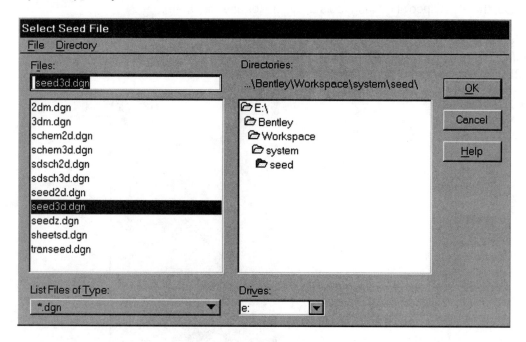

- **seed3d.dgn** This is the one that has the 3D cube for visualization. If you use this file, you can delete the elements that form the 3D cube and its labels. The cube is on Level 1. However, the labels of the cube can be suppressed since they are Construction types. Toggle off the Construction setting in the View Attributes dialog box for each view window. This is the seed file that will be used for this text.

- **3dm.dgn** This seed file is for metric units. It also has an existing 3D cube like the one found in seed3d.dgn.

Remember that the Appendix contains Seed File Information such as View Attributes and Active Setting.

3D VIEW CONTROLS

There will be situations when you may not necessarily be creating the 3D design file. However, you may be expected to view 3D designs that were created by third-party software that uses MicroStation to do its graphics. An example of this would be a digital terrain model (DTM), a 3D surface made up of triangles that represents an area of ground. In any of these cases, whether you create the 3D design file or have one given to you, there are special viewing tools available for use with them.

Accessing the 3D View Control Tool Box

The view controls for 3D design files are found under the main pull-down menu TOOLS>VIEW CONTROL>3D. This is illustrated below.

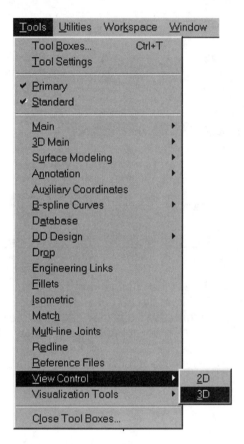

3D VIEW CONTROL TOOL BOX

Here are the tools available to help you view any 3D files. They deal with zooming in and out as well as selecting different view rotations. You can also specify that only certain portions of the 3D design be displayed. A tool for setting up a perspective view of your object is found here. There are virtual camera settings available, but they should be left to advanced users. Render allows you to color and shade the object for a more realistic 3D look. It has a tool in this tool box, but the subject of rendering will be discussed in Chapter 18 covering visualization tools.

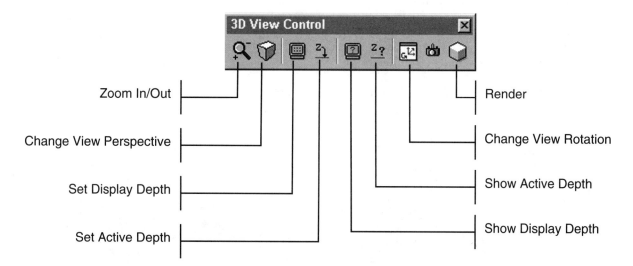

For our examples, we'll just use the 3D cube that shows up with the **seed3d.dgn** seed file. It's easy for you to get to, and the labels on the cube will help you keep things straight. The cube is color-coded too, which will help if you follow along while actually having the design file open. Most of these tools require that you use more than one view to accomplish your objective. This is because one orthographic view handles only two of the dimensions (such as height and width in a front view), and the other dimension is found in an adjacent orthographic view (such as depth in a right side view). The Isometric view is great for observing the total 3D package.

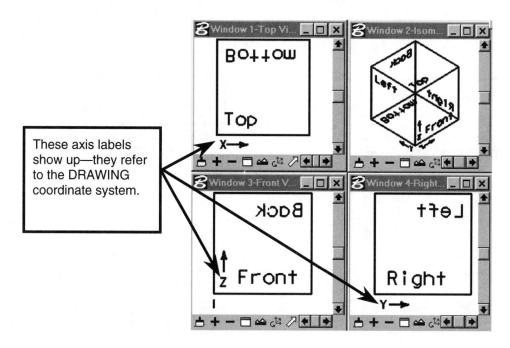

ZOOM IN/OUT

This 3D View Control's Zoom In/Out tool doesn't work like the regular view controls do (although you can still use those Zoom In and Zoom Out tools and may prefer to!). This Zoom In/Out tool requires the following: You select the view you want; specify the volume you want to work with, and finally define what new volume size you want displayed. The first two can be selected in any view. However, the view in which you define the new volume size that you want to display will be the view that is zoomed in or out.

Here is an example of zooming out in the **Front** view. The prompts will help you follow the steps that were taken.

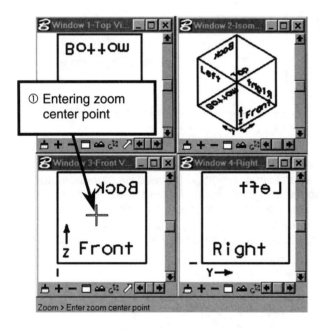

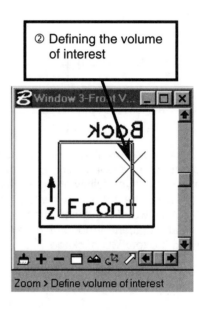

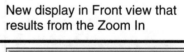

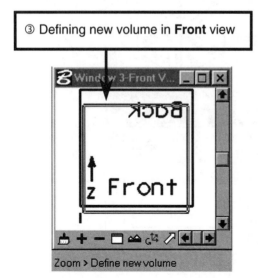

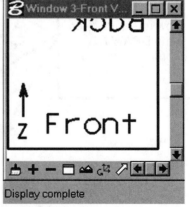

CHANGE VIEW PERSPECTIVE

When a view window displays one of the standard isometric views, it follows the convention of having parallel surfaces remaining parallel. The parts of the object that are farther away from the viewer's eye don't seem smaller (even though they should). In a more realistic display, such as a perspective, the object actually appears to get smaller the farther away it is. The Change View Perspective tool allows you to change a view into a true perspective rather than a standard pictorial convention. It is not recommended for standard engineering drawings. However, it is great for architectural purposes.

Shown here is using the Change View Perspective tool used on the standard Isometric view. First you'll be asked to select the view. The data point that you use to select a view will also specify the point that you want to move your eye **from**. You are prompted to Define new perspective angle. If the Dynamic Display setting is checked ON, then as you move the cursor, you see a dynamic view of your changes. The data point will specify the point that you want to move your eye **to**. One resulting perspective view is shown at the right.

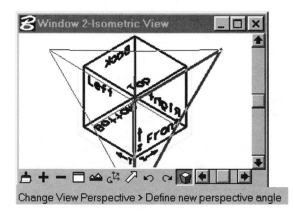

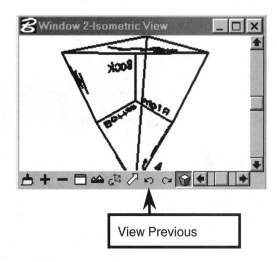

View Previous

 In truth, this tool is not recommended for beginners. If you accidentally use the tool and want to get out of it, use the View Previous tool in the individual view controls. If it is too late for that, then to eliminate the perspective you need to use the Change View Perspective tool. The from point (1st data point that is used to select the view) should be near the **edge** of the view window, and then the to point (2nd data point) should be near the **center** of the view window. This should eliminate the perspective.

How this 3D View Control tool reacts is affected by the view's Active Depth setting. Other tools are also affected by the Active Depth setting. Therefore, we better look at what the Active Depth is and how it affects a view's display.

ACTIVE VS. DISPLAY DEPTH

There are four tools dealing with depth. Two refer to Display and two refer to Active. In each view, we know the orientation of its screen coordinate system: +X is to the right (horizontal), +Y is up (vertical), and +Z is out of the screen. The screen shows us its X-Y plane but we don't have a set position of **where on the Z-axis that X-Y plane actually is**. This is what is meant by the "depth" for a screen. An Active Depth can be set for a view, so that if you draw in that view, the position of its X-Y plane along its Z-axis is established. You now know the depth of the view plane that you are drawing on—it has a specific Z screen coordinate. The screen coordinate values can be transformed back into the drawing coordinate plane. So Active Depth refers to the screen's Z coordinate plane that is being drawn on.

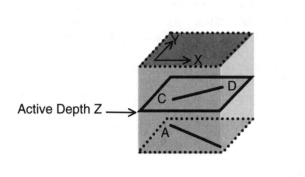
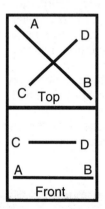

Shown above is Line C–D being drawn in the Top view with the Active Depth as shown. In the Top view, you can't tell which elevation Line A–B and Line C–D are at. Only by looking at the **Front** view will you see that they are each in a different "depth plane."

The term of the Display Depth actually refers to the distance between the front and back clipping planes. The view will display only what lies between its front and back display clipping planes (or within the Display Depth). You actually can control the location of the front and back clipping planes for each view. These two planes have two different screen Z values. You aren't necessarily drawing on either of these clipping planes—you draw on the Active Depth plane.

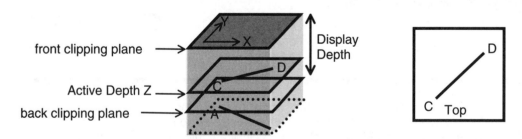

If the front and back clipping planes were defined as shown, then Line A–B wouldn't show up in the Top view. It hasn't been deleted, but it isn't located within the Display Depth so it won't be displayed.

Don't get confused by all of this. Remind yourself that each view can have its own coordinate system (called the screen coordinates), but in the end it all has to go back to the drawing coordinate system. Therefore, each screen's coordinate system has to be defined specifically enough so that it can be internally transformed back to drawing coordinates. If you are drawing in the Top view, you have to be specific about what elevation you're drawing at—high or low or in between. That's what the Active Depth is for. However, sometimes you don't want to see everything you have in a 3D file, and that's where the Display Depth comes in (if it isn't within the Display Depth—it isn't displayed).

SET DISPLAY DEPTH

 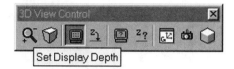

The Set Display Depth tool actually lets you specify the location of the front and back clipping planes. You will need to define the location of clipping planes for a specific view, **not in the view itself but in one of its adjacent views**. The adjacent view will actually show the clipping planes on edge, so it is easy to specify the depth between them. In setting up the front and back clipping planes for a Top view, you would need to go to either the Front view or the Right view where the clipping planes appear on edge. Likewise, to set the Display Depth for a Front view, you would need to define the clipping planes in the Right or Top view. The term "front" clipping plane does **not** mean the Front view—it means the first clipping plane. The same goes for the back clipping plane—it means the second one (**not** the Back view).

For Example: Here the Display Depth will be set for the Top view. It will be set so that the top portion of the cube will not be displayed.

① **Select the view for the Display Depth.** The **Top** view was selected by doing a data point in it.

② **Define the front clipping plane.** In this case, the clipping plane for the Top view will appear as an edge in the Front view.

Data point in the Front view at the elevation wanted for the front clipping plane.

The front clipping plane is being defined in the middle of the cube, between its top and bottom plane.

It is easy to see the front clipping plane in the Isometric view. It will appear dynamically.

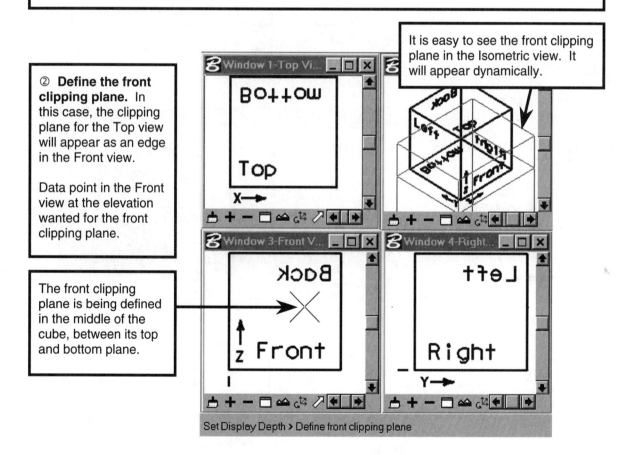

Example (Continued):

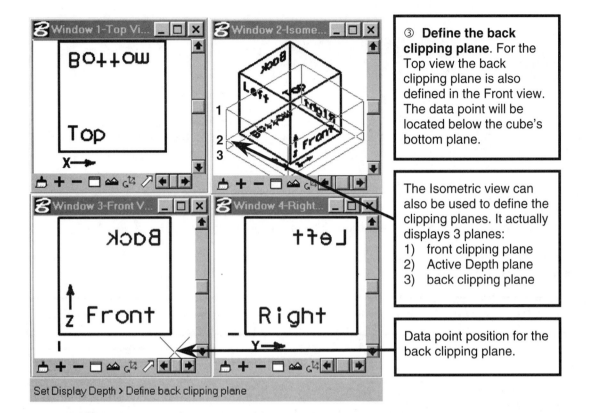

③ **Define the back clipping plane**. For the Top view the back clipping plane is also defined in the Front view. The data point will be located below the cube's bottom plane.

The Isometric view can also be used to define the clipping planes. It actually displays 3 planes:
1) front clipping plane
2) Active Depth plane
3) back clipping plane

Data point position for the back clipping plane.

Once the back clipping plane has been defined, then the Top view will no longer display the top plane of the cube—see the display at the right.

The cube isn't changed or modified. We're just controlling the display of a specific portion.

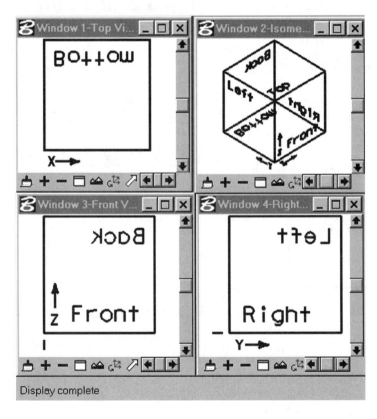

Fit View

Now suppose you don't want to worry about front and back clipping planes for all of the views. Instead you just want to see everything. No problem—use the regular Fit View tool in the view's control bar. For 3D design files it appears with two check boxes. By checking ON the Expand Clipping Planes setting, the clipping planes will be defined **automatically** to show all of the 3D graphics.

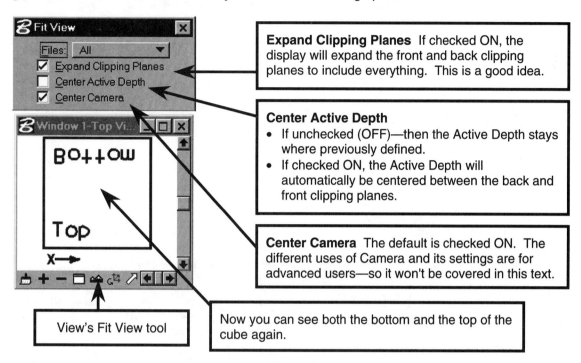

Expand Clipping Planes If checked ON, the display will expand the front and back clipping planes to include everything. This is a good idea.

Center Active Depth
- If unchecked (OFF)—then the Active Depth stays where previously defined.
- If checked ON, the Active Depth will automatically be centered between the back and front clipping planes.

Center Camera The default is checked ON. The different uses of Camera and its settings are for advanced users—so it won't be covered in this text.

View's Fit View tool

Now you can see both the bottom and the top of the cube again.

SET ACTIVE DEPTH

The Set Active Depth tool lets you define the position of the view's X-Y plane on the Z-axis. The Active Depth needs to be within the Display Depth. That way you aren't drawing on a plane you can't see. Again, you set the Active Depth for each individual view. The first thing you need to do is specify which view you want to set the Active Depth for. Then you'll be prompted to enter the Active Depth point. In this case, the Isometric view is the easiest one to work in to enter the Active Depth point. You can actually snap to an existing element if you want to. You can also use the key-in of ⌨AZ= to set the Active Depth.

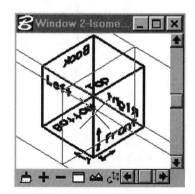

Shown here is the Set Active Depth for the **Front** view. The Isometric view shows the Active Depth plane (parallel to Front view) that can be established. It dynamically moves with the cursor.

By snapping to a point on the cube's back plane, the Active Depth for the Front view has been set. When drawing in the Front view, you'll actually be drawing on the back plane of the cube.

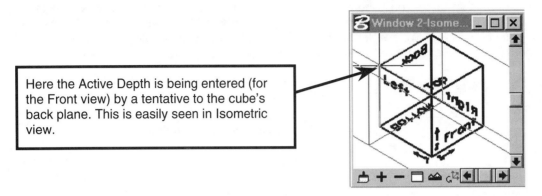

Here the Active Depth is being entered (for the Front view) by a tentative to the cube's back plane. This is easily seen in Isometric view.

Now when a line is placed graphically in the Front view, it shows up in the other views at the Active Depth.

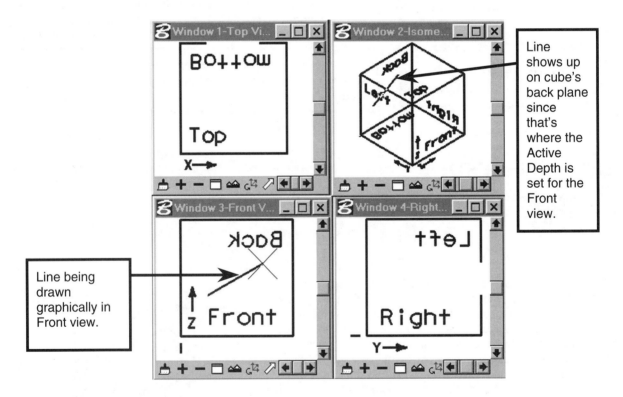

Line shows up on cube's back plane since that's where the Active Depth is set for the Front view.

Line being drawn graphically in Front view.

Remember the following:
- Each view has its own Active Depth that can be set individually.
- Snapping or key-ins will override the Active Depth setting.
- Active Depth (and clipping planes) can also be redefined anytime you like.
- Fit View (with Center Active Depth checked ON) will automatically reset the Active Depth.

SHOW DISPLAY DEPTH

The Show Display Depth tool looks a lot like the one that sets the Display Depth, but it has a question mark in the center of the computer screen indicating a search for information. The tool requires you to select the view that you want to know the Display Depth for. It doesn't give you a graphic display indication of where the Display Depth is located. Instead it will actually just indicate the coordinates in the status bar.

These are the **drawing** coordinates of the front and back clipping planes for View 1:

| Show Display Depth > Select view | View 1: Display Depth=-1.9715,2.7829 |

SHOW ACTIVE DEPTH

This tool is Show Active Depth. It will return the drawing coordinates of the Active Depth of the specified view to the status bar. Like the Show Display Depth tool, it has a question mark indicating a search for information. Again, there is no graphic display of where the Active Depth is—just a display in the status bar indicating the drawing coordinates of the Active Depth of the view selected.

| Show Active Depth > Select view | View 1: Active Depth=0.0000 |

This is the Active Depth for View 1. Notice that its value lies between the front and back clipping planes.

Both the Show Display Depth and the Show Active Depth tools just return information to the status bar. This information can be valuable in knowing where you're drawing and what you're looking at.

VIEW ROTATION

The ability to rotate a view display is invaluable. What good is having a three-dimensional object if you can't take advantage of looking at it from any direction you want to? There are two different methods for rotating the view: You can 1) Change View Rotation or 2) Rotate View. I know they sound like the same thing and in most respects they are. However, each tool also has its very own unique settings that allow you different control. One thing that the view rotation tools do have in common is that they do NOT rotate the actual 3D elements. Instead they behave like moving yourself around a sculpture. The sculpture stays put but looks different depending on where you are and your line of sight. No matter which tool you choose, you'll find that view rotation is easy to do and is extremely useful.

CHANGE VIEW ROTATION

 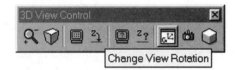

This tool is found in the 3D View Control tool box. When you use the Change View Rotation tool, you'll actually get a View Rotation dialog box that allows you many settings. It gives you a cube visually representing your 3D object. This cube will dynamically adjust according to your changes. It acts as a preview. You can see effects on the view rotation before you actually apply it to the view. Note that it shows only a cube, you don't actually see your object in the preview. The changes that you want can actually be applied to the 3D space axis that is displayed on the right side of the View Rotation dialog box. By clicking on the arrows, the rotation is controlled in Step increments of degrees of rotation. As you click, the cube on the left will change. This is a great feature, because previewing the cube is much easier than trying to come up with actual space coordinates.

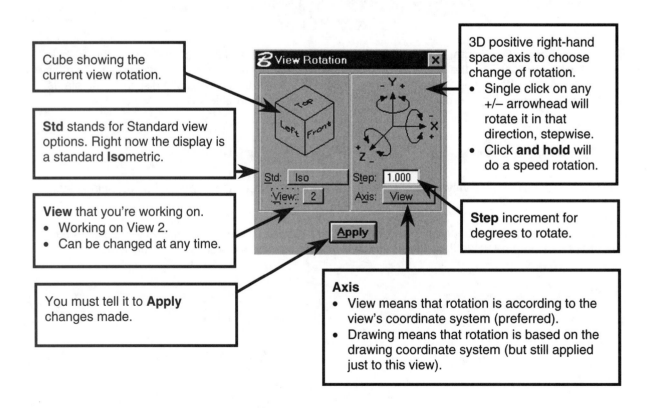

ROTATE VIEW

The Rotate View tool shows up on each view's window view control tool bar along with Update View, Zoom In and Zoom Out, etc. This Rotate View tool will not bring up the dialog box with the dynamic visual cube. Instead, it will have a simple Rotate View dialog box with a single option button. Using this tool requires that you still specify the view that you want to work with. Even though the tool is found with each view window, it doesn't know what view it belongs to! Read the status bar—it will prompt you to select the view (to rotate).

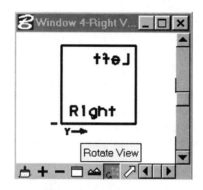

It seems as if of a lot of Method options are available with the Rotate View tool. Six of them deal with the "glass-box" concept of standard orthographic views (shown here in third-angle projection). Two are the standard isometric pictorials.

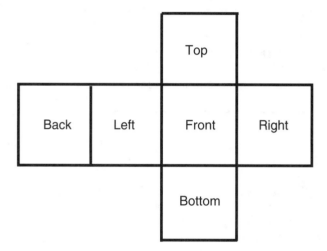

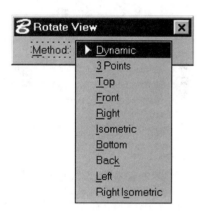

For these, all you need to do is first choose the Method of rotation and then select the view that you want to be affected. Shown here is the result of using the Right Isometric Method and selecting the Window 4 view.

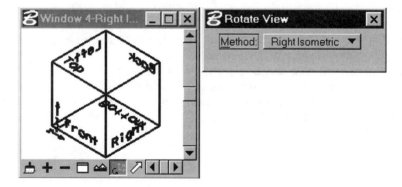

 Sometimes after rotating the view, the object doesn't all show up in the display. Just use the Fit View tool with Expand Clipping Planes checked ON, and you'll see it all.

Dynamic Method

For this Rotate View Method, you're prompted to select the view right away. This is because the Dynamic Method actually lets you see a dynamic 3D display in the view. You need to select the view first so that it knows how to show the cube. As you move the cursor, you'll see the view change dynamically. When it looks the way you want, then data point, and that will set the view rotation. If Dynamic Display is checked ON then you actually see the 3D object rotate in the view as illustrated below. If it is unchecked (OFF) then you'll only get to see a 3D display cube rather than the object. The front plane of the 3D display cube will be darker so that is easily distinguishable. This method isn't as accurate as the axis available in the Change View Rotation dialog box, but it works fine for a quick rotation that doesn't fall into one of the standard view options for Method.

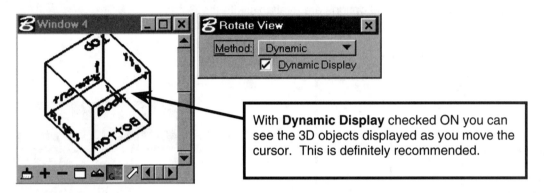

With **Dynamic Display** checked ON you can see the 3D objects displayed as you move the cursor. This is definitely recommended.

3 Points Method

This method requires you to specify a first point (which acts like an origin point). Then you specify an X-axis direction and then a Y-axis direction. These can be selected in any view, but the view that will be rotated is the one that the tool was selected on. This can get tricky but you can experiment with it on your own.

With any of these Rotate View Methods you need to remember the following:

- You are rotating only how the object is displayed. The 3D elements are not being moved.
- The screen (view) coordinate system stays as +X to the right of the screen, +Y to the top of the screen, and +Z coming out of the screen at you. So even though you aren't moving the elements, you are changing the relationship between that view's screen coordinate system and the drawing coordinate system.
- When printing, you can still print only one of the views at a time. If you want to print a sheet that shows the three standard orthographic views in standard conventional layout, you'll need to use Drawing Composition (covered later on in Chapter 21). You could also use reference files and reference the design file to itself.

SET VIEW DISPLAY MODE

If your view window is large enough, the Change View Display Mode tool will show up at the very end of the window's view control tool bar. It looks similar to the Render tool but has a lightning bolt going through it. Clicking this tool will bring up the Set View Display Mode dialog box. Its default setting is a Wireframe Display Mode as illustrated at the right. This is what we've been seeing. **However, by changing the Display Mode setting you can change the display in each view independently.** What's even better is that even when you use another view control (such as rotate view), the display will still appear according to the Display Mode setting. Illustrated below is View 2 shown with a Hidden Line Display Mode. The display of hidden edges can be toggled on or off.

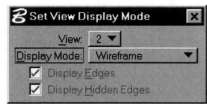

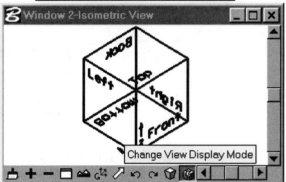

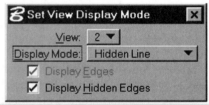

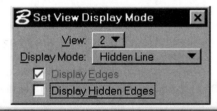

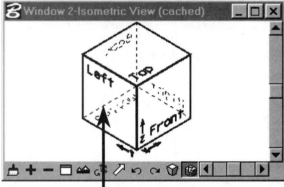

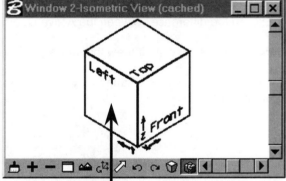

The Display Hidden Edges setting is checked ON so that the hidden edges are displayed as dashed lines.

The Display Hidden Edges setting is unchecked (OFF) so that the hidden edges aren't displayed at all. The cube appears as solid.

You don't necessarily get a plot of exactly what you see displayed in a view. Only with the Display Mode set to Wireframe does a plot of the view actually appear the same as what is displayed. In the other display modes, you will get a plot of a shaded rendered image, which may (or may not) be what is displayed.

QUESTIONS

① Sketch a three-orthographic-view layout as shown and include the following:
   a) The Xd-Yd-Zd positive drawing axis orientation in ALL three views.
   b) The Xfs-Yfs-Zfs positive screen axis for and in the Front View.
   c) The Xls-Yls-Zls for the positive screen axis for and in the Left View.

Then also sketch an isometric cube as shown, and label those 3 X-Y-Z positive axes on the isometric cube.

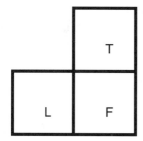
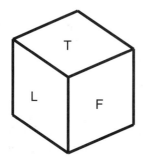

② If you are working in the **Right** view of an object and you start a line with the key-in of ⌨XY=3,20,0 and then key-in the endpoint of the line as ⌨DX=5,10,15, what will the drawing coordinates of that endpoint be? Include a sketch with the axis and values used to determine the endpoint.

③ If you are working in the **Front** view of an object and you start a line with the key-in of ⌨XY=5,32,17 and then key-in the other endpoint of the line as ⌨DX=30,-12,15, what will the drawing coordinates of that endpoint be? Include a sketch with the axis and values used to determine the endpoint.

④ If you are working in the **Front** view of an object and you start a line with the key-in of ⌨XY=5,32,17 and then key-in the other endpoint of the line as ⌨DL=30,-12,15, what will the drawing coordinates of that endpoint be? Include a sketch with the axis and values used to determine the endpoint.

⑤ Explain the difference between the Active Depth and the Display Depth. Use a sketch if needed.

⑥ What is the setting in Fit View that helps you see all of the 3D graphics easily?

⑦ When you're using the Change View Rotation tool in the 3D View Control tool box, what is the benefit of this area?

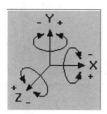

⑧ What is a visual cue to the 3D View Control tools that will show you information but don't actually change any settings?

⑨ Discuss the advantage of having the Dynamic Display option that is available in some of the view control tools checked ON.

Chapter 17:
Auxiliary Coordinate Systems and AccuDraw in 3D

SUBJECTS COVERED:

- ACS Triad
- Uses of ACS
- ACS Key-Ins
- ACS Tools and Utilities
- ACS Plane Lock and Snap
- Define By Element
- Define By Points
- Saving an ACS
- Define By View
- AccuDraw in 3D
- AccuDraw and ACS

Auxiliary Coordinate Systems (called ACS for short) are one more way of controlling the coordinates used in your design file. In two dimensions, we considered drawing coordinates and entering values based on relative and absolute coordinates. In three dimensions, the concept of screen coordinates was discussed. Auxiliary Coordinate Systems come in when you want more control over the orientation and location of the coordinate system than these other options give you. ACS lets you specify the origin's position and the positive X-Y-Z axis orientation in both 2D and 3D files. Numerical values for coordinates can then be entered according to the active ACS rather than the drawing coordinate system. Once again, the software will translate those coordinates back into the regular drawing coordinate system, but that goes on behind the scene. ACS are not set in stone. You can redefine them at any time. Switching from drawing, screen, and ACS coordinates is allowed and encouraged—whichever one suits your needs at the time. There are three types of ACS coordinate systems: Rectangular, Cylindrical, and Spherical. We'll stick with the Rectangular Type, but remember that the others are there if you ever need them.

 ACS tools were available in the software long before AccuDraw. AccuDraw was introduced (in part) to make it easier to draw at different locations and orientations without having to use an ACS.

ACS TRIAD

To refresh your memory, a visual cue of the coordinate system is available. The ACS Triad setting in the View Attributes dialog box controls the display of the ACS triad in the view windows. This triad shows you where the positive X-Y-Z axis is located. Whenever you are working with ACS, you will want to have this setting checked ON for all views.

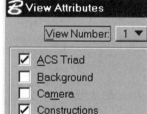

USES OF ACS

In two dimensions, an ACS is used in situations that would require extensive use of relative coordinates. Using an ACS, you can move the origin. The X-axis and Y-axis can remain orthogonal or can be rotated at an angle.

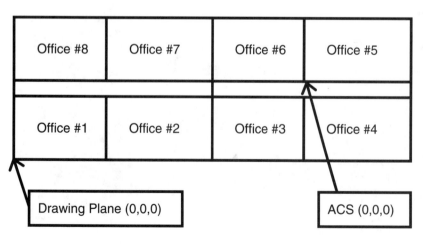

For a very simple example, consider laying out the furniture in these offices. For Office #1 the coordinates are easily based on the drawing plane. For all the other offices, the coordinates would have to be figured as relative or with coordinates of the drawing plane. This is not extremely efficient. Instead, defining the origin for the ACS at the corner of the office you're working on would allow you to avoid relative dimensions. You can work with coordinates based on the office's lower left corner. Such is the case for the ACS shown, which would be handy for working with Office #5.

There are more uses for an ACS in three dimensions. For example, an ACS is handy when you need to draw on an inclined or oblique plane. Not all the action goes on in nice orthographic planes that can use drawing and screen coordinates. Shown below is an ACS that has been defined to match the inclined plane. The text was then placed. Notice how it lies on the inclined plane. The ACS triad shows the ACS origin location and the orientation of the ACS X-Y-Z axis.

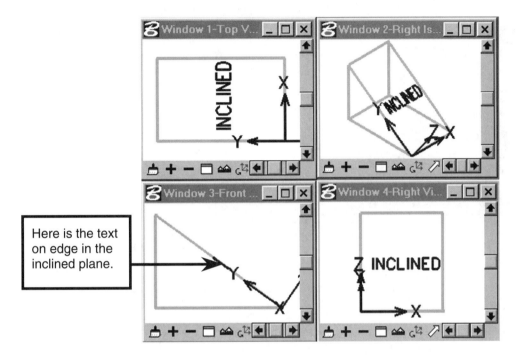

Here is the text on edge in the inclined plane.

ACS KEY-INS

There are two key-ins that are based on the Auxiliary Coordinate System:

⌨AX= will use the **absolute** coordinates of the ACS.

⌨AD= will use the **relative** coordinates of the ACS.

Shown below is an example of a line drawn on the inclined plane. Endpoint A was specified with absolute auxiliary coordinates of ⌨AX=3,3,0. Endpoint B was specified with relative auxiliary coordinates ⌨AD=2,4,0.

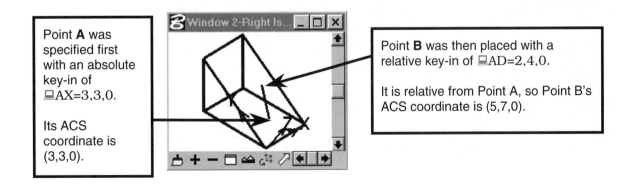

Point **A** was specified first with an absolute key-in of ⌨AX=3,3,0.

Its ACS coordinate is (3,3,0).

Point **B** was then placed with a relative key-in of ⌨AD=2,4,0.

It is relative from Point A, so Point B's ACS coordinate is (5,7,0).

ACS TOOLS AND UTIILITIES

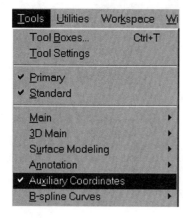

There is an ACS tool box, but it isn't part of the Main tool box. You must bring it up by using the main pull-down menu TOOLS>AUXILIARY COORDINATES as shown on the left. The tools found either define the ACS or modify an existing ACS. We'll be using the first two tools: Define ACS (by Element) and Define ACS (by Points).

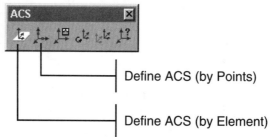

Define ACS (by Points)

Define ACS (by Element)

In this case, there is more information to be found in the Auxiliary Coordinate Systems dialog box. This is accessed under the main pull-down menu UTILITIES>AUXILIARY COORDINATES and will appear as shown. The tools that were in the tool box can be found here in the Auxiliary Coordinate Systems pull-down menu TOOLS. This menu is often easier to read than trying to decipher the icons of the tool box.

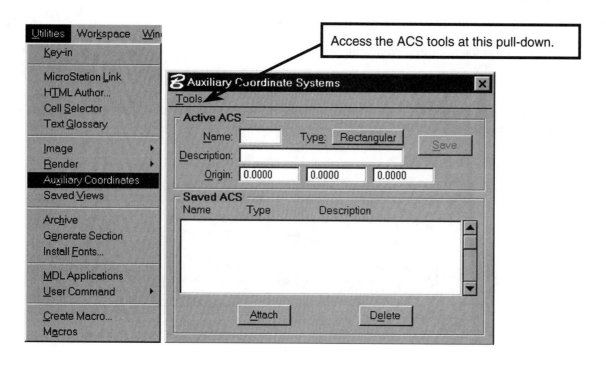

Access the ACS tools at this pull-down.

ACS PLANE LOCK AND SNAP

When using the tools from the tool box to define the ACS, you will see two check boxes: ACS Plane Lock and ACS Plane Snap. These settings are available when working in a 3D file and are grayed out for 2D files. The ACS Plane Lock and the ACS Plane Snap settings affect the placement of data points and the ability to snap with respect to the ACS. The settings are Locks and can be accessed with the other Locks settings (through the Locks pop-up menu, full Locks dialog box, etc.). So even if they don't show up in the Auxiliary Coordinate Systems dialog box, they are still important to the use and definition of an ACS. As shown below, the settings are unchecked (OFF). **It is recommended that you leave them toggled off for now. If they are checked ON, then their constraints can make it more difficult to define a new ACS.**

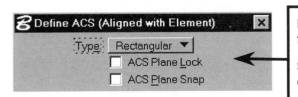

Lock: If checked ON, then data points are forced onto the ACS X-Y plane.

Snap: If checked ON, then snaps will be constrained to the ACS X-Y plane.

For Example: Let's look at the case where both the ACS Plane Lock and the ACS Plane Snap are checked ON. An ACS has been defined to match the side slope of the road and the ACS triad shows its origin and orientation.

A line is being placed. Shown here is the tentative point in the Top View that is being used to specify the line's first endpoint. It is snapped to the bottom of the ditch.

Once the tentative point is accepted, the first endpoint of the line shows up—underneath the ditch bottom! The line itself is constrained to the ACS plane as illustrated below.

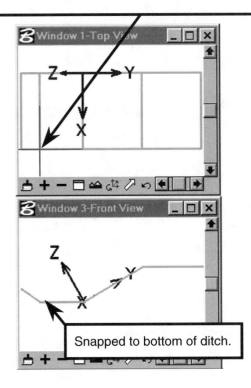

Snapped to bottom of ditch.

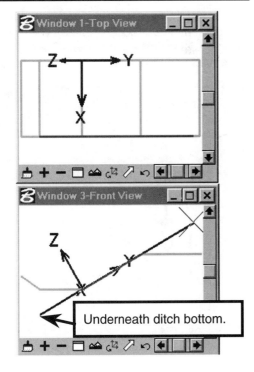

Underneath ditch bottom.

DEFINE BY ELEMENT

  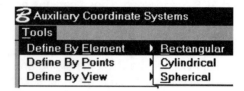

This method of defining the Auxiliary Coordinate System uses an existing element to set the origin and orientation of the ACS. The data point used to identify the element determines the origin, so you'll want to tentative snap so that the Auxiliary Coordinate System's origin (0,0,0) is precise rather than arbitrary.

For Example: Illustrated below is a 2D line that will be used as the element for defining a new ACS.

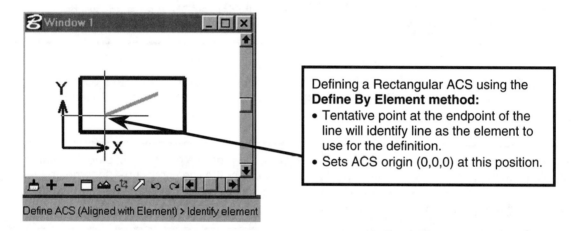

Once the tentative is accepted, then you'll need another data point to accept selection of the element. Now the ACS is defined, and the ACS triad will help you see the definition of the ACS. Because this is a 2D element and file, there isn't a Z-axis shown.

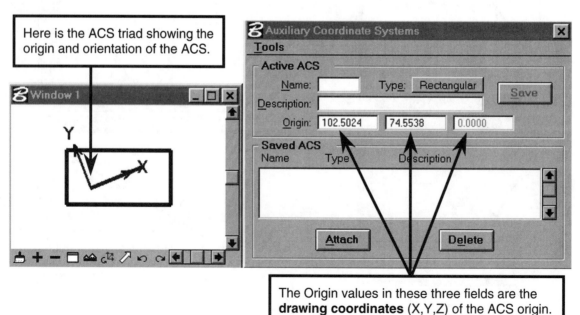

DEFINE BY POINTS

 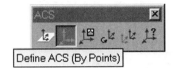 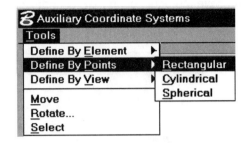

This is also a useful method for defining an Auxiliary Coordinate System. The Define By Points method gives you more control because you specify the actual origin point (by key-in or graphically) first and then the positive X-axis and the positive Y-axis. It is important to read the prompts so that you make the correct selection. Let's look at the 3D example and see how the ACS was defined to match the inclined plane.

There are three steps to defining an ACS by points:
1. Define (0,0,0) origin of ACS.
2. Define Positive X-axis of ACS.
3. Define Positive Y-axis of ACS.

Illustrated below from left to right are these steps used to define an ACS on the inclined plane. The status bar indicates what is being identified with a tentative snap.

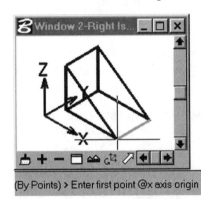 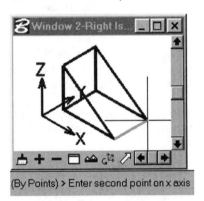 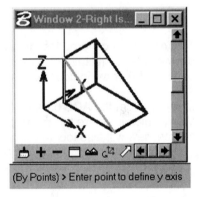

Here is the result of the Rectangular Type ACS defined on the inclined plane using the Define By Points method.

SAVING AN ACS

You can save the origin of an ACS so that it can be attached later. This is found in the Auxiliary Coordinate Systems dialog box. The Save button is available only after you have named the ACS.

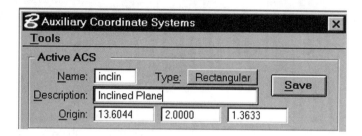

DEFINE BY VIEW

  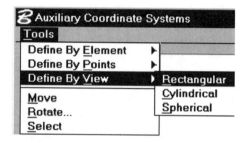

This method is covered last because it can be used to set the ACS plane back to being the same as the drawing plane. It isn't necessary to do this, but it does get the visual cue of the ACS back to the actual origin of the drawing coordinates. This can be reassuring to a beginner.

In order to get the ACS back to the drawing plane, the view you want to align with is the standard Top View, which is where the drawing and screen (view) coordinates coincide. The data point that selects the source view will also locate the origin of the ACS.

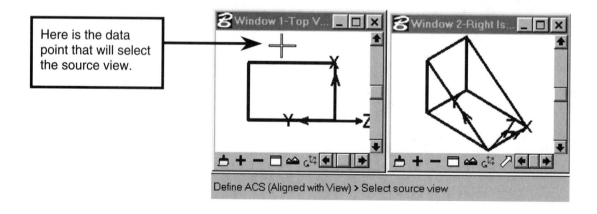

Now the ACS is **oriented** the same as the drawing plane, but its origin is not at the (0,0,0) of the drawing coordinate system. That is why you see the numbers in the Origin fields of the Auxiliary Coordinate Systems dialog box. If you change each of these numbers to 0 then the ACS origin will move to the drawing plane's origin. The ACS triad will be there too!

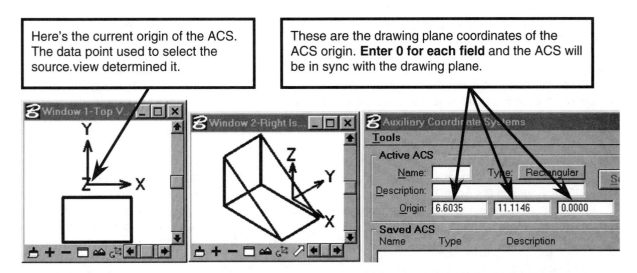

Here's the current origin of the ACS. The data point used to select the source.view determined it.

These are the drawing plane coordinates of the ACS origin. **Enter 0 for each field** and the ACS will be in sync with the drawing plane.

Once the Origin fields have been given the value of 0, an Update View is necessary to see the ACS triad at its new location. Shown below is the ACS that has been redefined to match the drawing coordinate system.

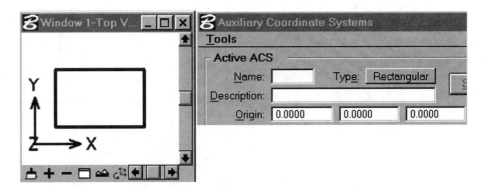

Remember, even if an ACS is defined, you can still use the drawing or screen coordinate systems. You don't have to set the ACS to be in sync with the drawing plane coordinates—only if you're more comfortable seeing the ACS triad there.

ACCUDRAW IN 3D

Using an ACS is one way to adjust the coordinate system to your needs. AccuDraw is also suited for precision input at relative coordinates, and we saw its capabilities in two dimensions in Chapter 11. AccuDraw is even more impressive in 3D. AccuDraw has its own coordinate system, and when AccuDraw is active, data points are constrained on the AccuDraw X-Y plane (referred to as the AccuDraw drawing plane). The power of AccuDraw in 3D lies in the fact that its drawing plane can quickly and easily be reoriented and redefined to whatever suits your needs. The AccuDraw coordinate system can be oriented according to the View (screen) coordinate system. With this setting, the view that the cursor is in determines the AccuDraw coordinate system. But even more useful is that the AccuDraw drawing plane can be set to align with a specific **standard view** coordinate—no matter what view you are working in! This is indeed a great advantage when working in 3D. For instance, you can be working in the Right Isometric View but be drawing in a plane oriented to the Front View's coordinate system.

Here are several important concepts to remember regarding AccuDraw:
- The AccuDraw compass gives you a visual cue as to its drawing plane's origin and orientation.
- The values in the AccuDraw fields are based on the AccuDraw drawing plane and its origin.
- Focus must be in the AccuDraw dialog box in order for the keyboard shortcuts to work. Shortcuts are not case-sensitive. They are noted as <*keystroke*>, and you don't type in the < or the >.
- The <*spacebar*> will toggle you between a polar and a rectangular compass.

Coordinate System Rotation: View Option

AccuDraw can be set so that its coordinate system is based on the View that you are working in. The keyboard shortcut for this is <*V*>. You can also see this Coordinate System Rotation setting for the AccuDraw drawing plane in the AccuDraw Settings dialog box.

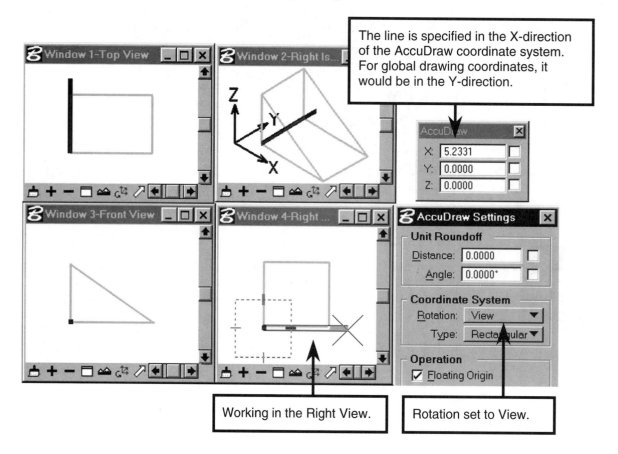

The line is specified in the X-direction of the AccuDraw coordinate system. For global drawing coordinates, it would be in the Y-direction.

Working in the Right View.

Rotation set to View.

Auxiliary Coordinate Systems and AccuDraw in 3D 391

Now if the cursor is moved to the Right Isometric View, the AccuDraw drawing plane and its coordinate system will change so that it is oriented according to this view. This gives it an unusual orientation as illustrated below.

Now cursor is in the Right Isometric View. The AccuDraw compass is orthogonal to the screen, but look at how the line appears in the other standard views—a unique orientation!

Coordinate System Rotation: Top, Front, and Side Options

Instead of having AccuDraw readjust to the cursor movement, you can use the shortcut keys to quickly change its drawing plane to match the standard view axes. This allows you to work in the Right Isometric View where it is easy to see the whole 3D layout but, with a single keystroke, change the coordinate system to being on the Top <*T*>, Front <*F*> or Side <*S*> planes. Shown below are the different rotations for the AccuDraw coordinate system while working in the Right Isometric View.

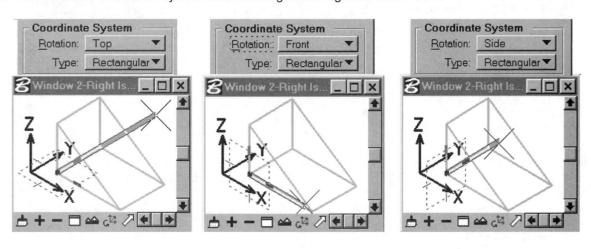

ACCUDRAW AND ACS

AccuDraw also has shortcuts to save and recall Auxiliary Coordinate Systems (only Rectangular Type). This allows you to use AccuDraw and its shortcuts to get the AccuDraw coordinate system in the orientation required and then save it as an ACS.

Rotate ACS

The AccuDraw shortcut <*R*><*A*> stands for Rotate ACS, and it lets you define the AccuDraw coordinate system in the same method used in Define By Points for a Rectangular Type of ACS. You specify the origin, then the X-axis, and then the Y-axis. Illustrated below is the last step of using AccuDraw and Rotate ACS to define the ACS so that it lies on the slanted plane.

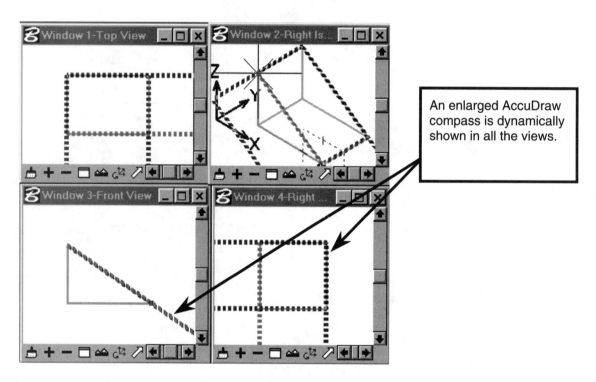

An enlarged AccuDraw compass is dynamically shown in all the views.

After the ACS has been defined with the AccuDraw shortcut, then the Rotation setting (found in the Coordinate System section of the AccuDraw Settings dialog box) is automatically set to Auxiliary. As soon as you do an Update View, the ACS triad will indicate the newly defined ACS as illustrated on the right.

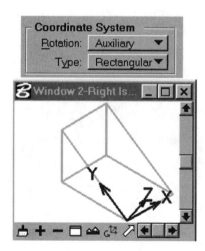

Write to ACS

The AccuDraw shortcut <***W***><***A***> brings up this Write to ACS dialog box, which lets you name the ACS just defined. This allows you to save the ACS for use in later drawing sessions.

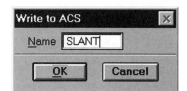

You can check to be sure that the ACS has been saved by using the main pull-down menu UTILITIES>AUXILIARY COORDINATES to get to the Auxiliary Coordinate Systems dialog box as shown at right.

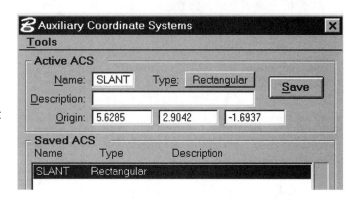

Get ACS

The AccuDraw shortcut of <***G***><***A***> just recalls a saved ACS, but you have to know its name. If you can't remember, then you'll need to use the Auxiliary Coordinate Systems dialog box (just discussed above). The two settings of Origin and Location deal with what happens to the current AccuDraw drawing plane origin and rotation when the saved ACS is recalled. If the setting is checked ON, then the AccuDraw drawing plane will be set to match the Auxiliary Coordinate System. If unchecked (OFF), then the AccuDraw drawing plane won't change.

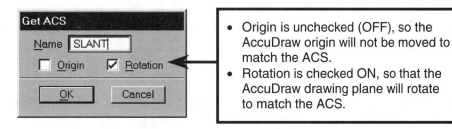

- Origin is unchecked (OFF), so the AccuDraw origin will not be moved to match the ACS.
- Rotation is checked ON, so that the AccuDraw drawing plane will rotate to match the ACS.

Remember, you can always use any of the AccuDraw shortcuts to change the AccuDraw drawing plane at any time. Illustrated below (from left to right) is the use of AccuDraw to place a SmartLine in 3D while always working in the Right Isometric View. Pretty slick, isn't it? If you haven't been using AccuDraw in 2D, this should convince you to try it in 3D.

Start with the ACS.   Using <***F***> for Front rotation.   Using<***T***> for Top rotation.

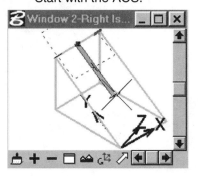 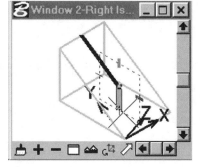 

QUESTIONS

What is an Auxiliary Coordinate System? What is its abbreviation?

Name the **three** different methods discussed for defining an Auxiliary Coordinate System.

Name the key-in that uses the absolute Auxiliary Coordinate System.

Name the key-in that uses the relative Auxiliary Coordinate System.

How do you define the origin of the ACS based on an element?

How many points are involved in a By Points (Rectangular Type) ACS definition? Name and describe them in order.

In the Auxiliary Coordinate Systems dialog box there are three fields that give the values for the Origin of the ACS. What coordinate system do these numbers refer to?

What is this called? How do you get it to be displayed in a view?

Name the AccuDraw shortcuts that would result in the following compass orientation while you're working in the Right Isometric View:

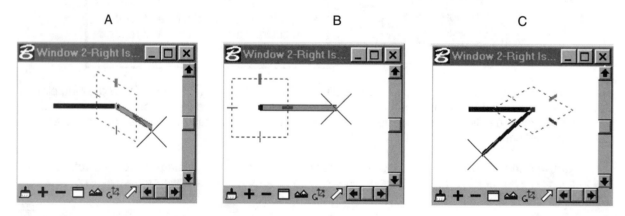

What does the AccuDraw shortcut <*G*><*A*> do? Explain some specific advantages and disadvantages of using this shortcut.

Chapter 18:
Rendering and Visualization

SUBJECTS COVERED:

- Render
- Saving Rendered Images
- Visualization Tool Box
- Define Light
- Place Light Source
- Edit Light Source
- Materials
- Apply Material
- Transparent Materials
- Rendering View Attributes
- Rendering Setup
- Animation
- Example of FlyThrough Animation
- Movies

Rendering is a process that takes your three-dimensional model and, by using lighting, materials, and colors, gives it a realistic appearance of the actual physical object that it represents. There are various methods of rendering, from the very simple and quick to the more complicated and time-consuming methods. Rendering techniques involve applying various materials and patterns to your model and controlling lights and their placement in order to achieve a realistic rendering. Advanced rendering tools have the ability to calculate and show more accurate reflections and shadows, which result in a photo-realistic view of the model. Rather than looking at a single stationary picture, there are also tools and techniques that allow you to make your model "come to life" with animation. Animation involves making a movie of your model in action. There are various animation techniques such as KeyFrame, FlyThroughs, and Parametric Motion Control. Animations can help you to evaluate a model as well as jazz up a presentation, but it takes a lot of hardware to do so.

Just because you have a huge amount of rendering and animation tools at your disposal doesn't mean their use is easily mastered. In order to achieve good results with many of these tools and techniques, you need to be an experienced user. With this in mind, we'll look at the basics of rendering, including how to use lighting and materials to obtain a more realistic rendering. Because of its complexity, animation will be discussed only briefly. The capabilities of the rendering and other visualization techniques are enormous and the possibilities staggering (especially for beginners), but their use isn't beyond your reach. Let's start by looking at the basics of rendering.

For the following discussion, we'll go back and visit our 3D cube from the discussion of 3D space in Chapter 16. It was obtained by using the seed3d.dgn seed file and has labels on its different faces.

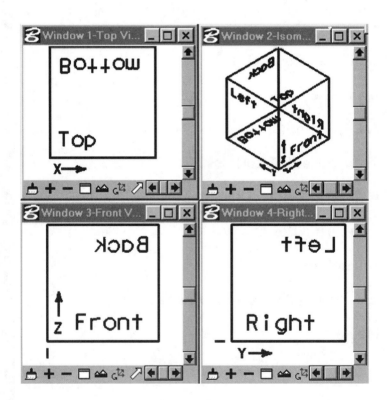

RENDER

There is a Render tool that is found in the 3D View Control tool box (discussed in Chapter 16). This Render tool actually initiates the rendering process that allows for an even more realistic three-dimensional display of the object. How the rendering will appear is determined by a whole lot of rendering tools and settings that are utilized separately. These are great for the advanced user, but for right now, we'll use the default settings for things such as stroke, antialiasing, and fog color (you can see why they're for advanced users—even the names are cryptic). The Render dialog box, which includes three option buttons, is shown below.

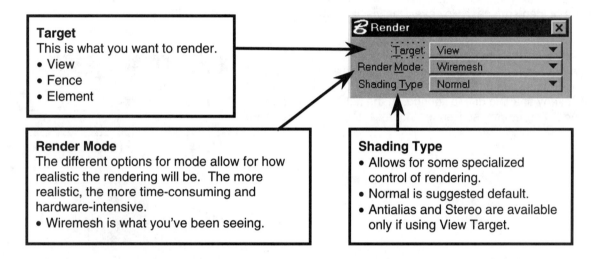

Target
This is what you want to render.
- View
- Fence
- Element

Render Mode
The different options for mode allow for how realistic the rendering will be. The more realistic, the more time-consuming and hardware-intensive.
- Wiremesh is what you've been seeing.

Shading Type
- Allows for some specialized control of rendering.
- Normal is suggested default.
- Antialias and Stereo are available only if using View Target.

Target

Depending on what Target you have chosen to render, you'll either have to select a View, or accept a fence (which needs to exist), or identify the element to render. By now this should be pretty self-explanatory. Usually the Target of choice is **View**. The status bar will show a little line that will rotate like the hands on a clock to indicate it's working. The status bar will also say when the display is complete. If you do a Fit View or other changes, the rendering will be gone, and again you'll just see a Wiremesh display.

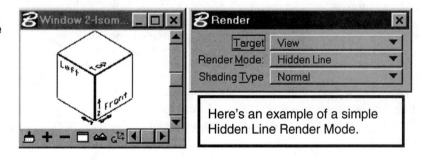

Here's an example of a simple Hidden Line Render Mode.

 The larger the view is, the longer it will take to render! Everybody loves to maximize a view and then render it—just be prepared to wait (especially if the computer system you're using is hardware-challenged). If the rendering is taking too long, use the Reset button (right mouse button) to stop the rendering process.

Render Modes

Shown at the right are the basic choices for the Render Mode setting. They are listed in order of fastest (simplest) to slowest (but most realistic).

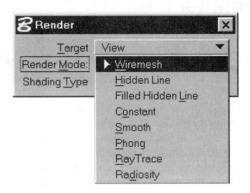

Wiremesh

This is what we have been seeing. The object being displayed is thought of as just wires for lines and intersections of planes. It is the fastest rendering process, but often it is hard to actually know which surfaces are visible and which are hidden.

Hidden Line

This is the example that was shown previously. The visible surfaces will actually cover up the edges, etc., of the hidden surfaces. However, the visible surface will not be filled in with the element's color.

Filled Hidden Line

With this rendering mode, the surface is filled with the element color (in this case gray). However, the fill is uniform. It doesn't reflect any shading due to lighting. At the right is an illustration of the 3D cube with the Render Mode set to Filled Hidden Line.

Constant

With the Constant Render Mode, the surface can be displayed as one or more (as in the case in a curved surface) polygons. Each polygon will be filled with a constant color. The color can be the element's color, but more advanced users will apply a material to the surface (we'll discuss assigning materials later on in this chapter). The main plus for the Constant Mode is that it will take into account the lighting defined for the view. Right now, we're leaving the lighting alone, but if you place a spot light to shine on the object, that surface will be lighter than the others.

Smooth

The Smooth Mode is comparable to Constant—it takes into account material and lighting. The difference is that smooth shading will do curved surfaces more realistically. It will smoothly blend the shading of adjacent polygons rather than just using a constant color for a single polygon.

Phong

This is a time-consuming step up from the Smooth Mode of rendering because it also does shadows. Phong is fairly realistic and gives good results. However, be prepared to spend some time waiting for the rendering to be done.

Ray Trace

The Ray Trace Mode for rendering is really advanced. Ray tracing requires a high level of computation that involves tracing a ray of light from the viewer's eye (in this case, the camera or where you're looking at the object from) and calculating what happens to that ray in terms of reflection and refraction. With this Mode, you get great photo-realistic results since metals will reflect and glass can be transparent. Even the shadows are also more realistic. Renderings with ray tracing take a lot of time—even more than with Phong—but it's worth it.

Radiosity

Radiosity is not a rendering method, but it does affect how a view is rendered. With radiosity, the color of one object can be reflected on another object. If a rendering is done without radiosity, light sources just illuminate an object. The object's color has no effect on the color of the other objects or surfaces nearby. A rendering done with radiosity is much more photo-realistic since it models the true characteristics of light. Lighting is extremely important when working with radiosity, and it is important to note that radiosity is not recommended for use with ambient and flashbulb lighting. The use of a radiosity solution in a rendering mode involves lots of computations and is not for a hardware-challenged system.

Shading Types

The Normal Shading Type is always available and should be used when you are doing just a quick rendering to aid your visualization of an object. The other two options for Shading Type are Antialias and Stereo—these take much more time to render.

Antialias

This is available only if Target is set to View. It reduces the jagged edges that can occur when rendering. Again, it gives a better rendering but takes longer to do. Use the Antialias Shading Type when you're trying to impress someone—such as with presentations and animations.

Stereo

Get out your 3D glasses. Yep—this renders the view so that you can see it in stereo. Great fun if you actually have the glasses—if you don't, it's a waste of the extra time involved.

SAVING RENDERED IMAGES

Even if the view on your screen is rendered nicely, you can't take your computer screen around to show everyone your realistic model. So other than impressing your friend sitting next to you, what do you do with the rendered view of your model? This is where the image utilities come in handy. The actual rendered image can be saved in various electronic formats that can then be used in a variety of ways. Electronic images can be used in presentations, in documents, and on the Web. Getting a hard copy print of the rendered image requires saving the rendered images and doing a little bit of extra work. Printing a rendered image requires a printer/plotter that can handle it. For the most part, beginners don't print a rendered image directly out of MicroStation. Instead, the image is saved in a format that can then be used in other software that has the capabilities of printing it out. At least this is the easiest method if you're just starting out. If you have your heart set on printing rendered images directly out of MicroStation, then start reading about plotter drivers and know the hardware you're using.

It is important to know the other software's requirements in order to save the rendered image in a usable format. Not all electronic formats are created equal. Sometimes it's a case of trial and error to see what format will work in your situation.

Accessing the Save Image Dialog Box

Going to the main pull-down menu and using UTILITIES>IMAGE>SAVE will bring up the Save Image dialog box. The View isn't rendered on your screen. Instead the image is sent to a file. Many different electronic formats are available to choose from, and they are shown on the next page.

We won't get into the other options of the Image Utilities (except that the Movies option will allow you to play animations created with MicroStation). The only other one that beginners may want to work with in conjunction with rendering is Capture—which actually captures what is displayed on your computer screen. Therefore you can actually render the image and then capture the view (or other parts of the screen) as an image file. The quality obtained using this method will depend on the display resolution of your monitor.

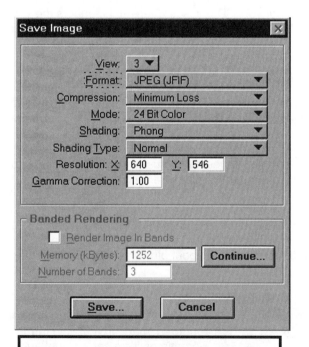

Here is an illustration of the Save Image dialog box.

> **View:** You can do only one view at a time, make sure that it is set to the right one.

> **Format:** Lots to choose from—it will be covered later.

> **Compression:** Usually left as Minimum Loss for best results.

> **Mode:** You can choose between 24 Bit Color and Grey Scale.

> **Shading** is same as Rendering Modes.
> **Shading Type:** Same as before.

> **Resolution and Gamma Correction** are best left alone.
> **Banded Rendering** may be available for some formats—still you should leave it at the default settings.

You must tell it to **Save**—it will have a dialog box pop up to let you choose path and file.

Format Options

Here are the Format options available for saving your rendered images. There are lots to choose from, as you can see for yourself.

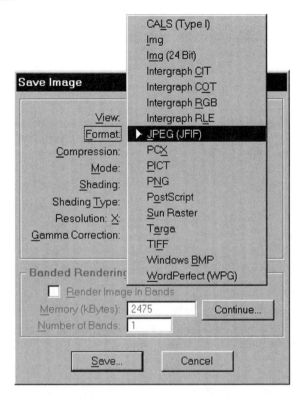

If you click on the Save button, the Save Image As dialog box shows up as illustrated below. Out is the default directory for the files to be saved in. So this kcb3d2.jpg rendered image file will be saved in the projects\examples\jbook\out directory.

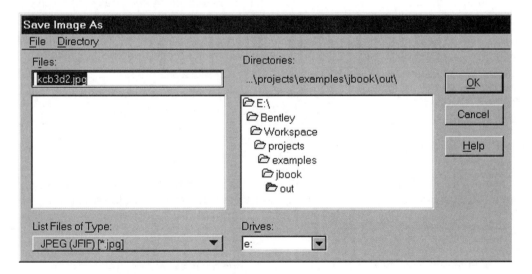

So which one do you use? Well, you're better suited to decide that. Here is a quick rundown on some of them, but again you may need to experiment to see what works the best for your own purposes.

Format	Application
JPEG	For Web use
TIFF (Uncompressed)	General purpose—good quality
Windows BMP	Bit map for Windows compatibility

 Microstation has an HTML authoring utility that will create an html file that can be used on the Web. This utility will create images (.jpg format) of your design file (3D or 2D) automatically and set up the html file to display those images. Pretty cool stuff—find it under UTILITIES>HTML AUTHORING.

EXAMPLE DESIGN FILES

Experimenting with rendering on models is better than the simple little 3D cube used previously. Here's a list of some design files that come with MicroStation that you can use to experiment with rendering. These will be available only if a "Complete" installation was done for MicroStation, which installs all of the Example Projects found in the Workspace Options. These files are found in the **Bentley\Workspace\projects** directory with the path as shown. A brief description of each 3D design file is included. It is only a partial list—but enough to get you started.

File	Description
...\examples\civil\dgn\dtm.dgn	This is a digital terrain model. Renders fairly quickly.
...\examples\arch\dgn\pool.dgn	This is a very fancy model that is very time-consuming to render, but the final product is really impressive. Rendering a small view is suggested.
...\examples\generic\dgn\bearexpl.dgn	This is often used as an example. Not only can you experiment with rendering, but also it is useful for setting display depth.
...\examples\generic\brake.dgn	This file comes up with three views of a brake. It is a good example of lighting sources.
...\examples\visualization\dgn\plangear.dgn	This file is also good for experimenting with display depth and view rotation as well as rendering.
...\tutorials\intro\dgn\bracket.dgn	**This file will be used for the examples in this text that discuss the use of materials and lighting to achieve a more realistic rendering.**

Now that we've discussed the basics of using the Render tool, we need to look at other items that affect the rendering and, in turn, the visualization of the model. Things such as lighting, materials, and camera settings all affect how the model will be rendered. These all work together to help you visualize what the real object that is being modeled can look like.

RENDERING TOOLS

The Render tool can also be found in the Rendering Tools tool box. You can open this tool box under the main pull-down menu of TOOLS>VISUALIZATION TOOLS>RENDERING TOOLS. It is way near the bottom of the options listed. In this tool box there are other tools that affect how a view will render by controlling lighting and material. Many of the specifics of these tools and their related settings are for the advanced user, so they won't be discussed in detail. We will be taking a closer look at working with lights and applying different materials to your model. The View Size tool makes it easier to have consistent view sizes, which is nice when you're doing animations, but it is fairly self-explanatory. Tools such as Define Camera, Photomatch, and Query Radiosity won't be discussed due to their complexity. It is important to remember that most of these visualization tools require a 3D design file—you'll get warnings if you try to use them in a 2D file.

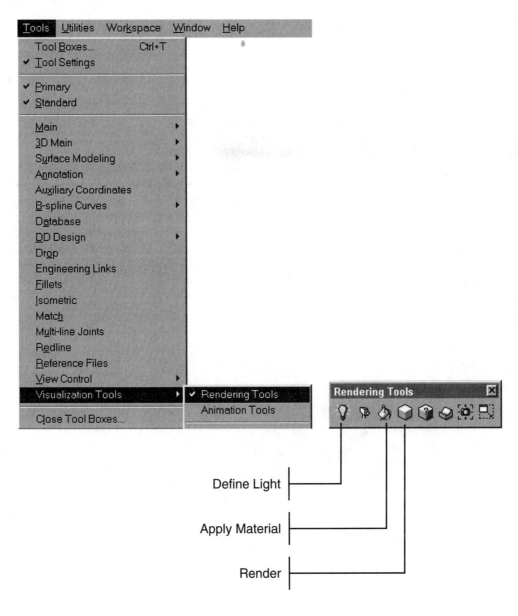

DEFINE LIGHT

Using the Define Light tool will give you access to the Define Light dialog box. Here there are two different modes possible: You can modify the existing lighting, or you can choose to create a new light. Depending on which Mode option is selected, this dialog box will have different settings and buttons. There are two different classifications of types of lights: Global and Source. Global Lighting is made up of Ambient and Flashbulb. Source Lighting includes Distant, Point, Spot, and Area.

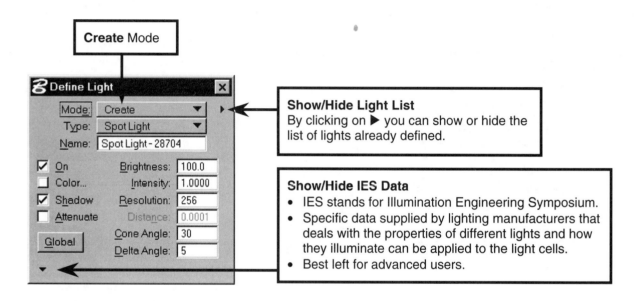

Create Mode

Show/Hide Light List
By clicking on ▶ you can show or hide the list of lights already defined.

Show/Hide IES Data
- IES stands for Illumination Engineering Symposium.
- Specific data supplied by lighting manufacturers that deals with the properties of different lights and how they illuminate can be applied to the light cells.
- Best left for advanced users.

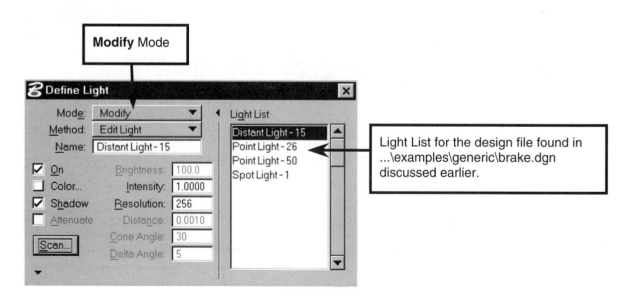

Modify Mode

Light List for the design file found in ...\examples\generic\brake.dgn discussed earlier.

Global Lighting

The settings for Global Lighting can be accessed by using the Global button that appears in the Define Light (Create Mode) dialog box. It can also be obtained from the main pull-down menu by SETTINGS>RENDERING>GLOBAL LIGHTING as illustrated below.

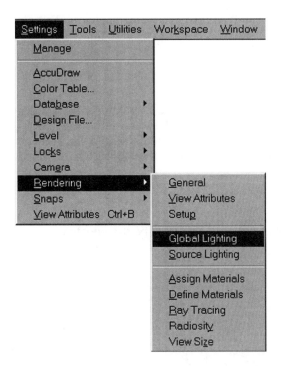

Notice that the SETTINGS>RENDERING pull-down menu contains a large number of options. As mentioned previously, some of these are the same options that will be available using the tools from the Rendering tool box.

Here are a "few" settings to consider for Global Lighting. **It is extremely important to have some type of lighting. Otherwise your rendering will be dark (or opaque). Global Lighting is a good place to start.** You can choose any combination of Global Lighting—you aren't limited to just one type of Global Lighting.

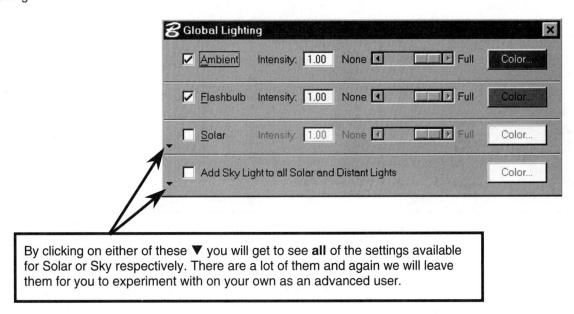

By clicking on either of these ▼ you will get to see **all** of the settings available for Solar or Sky respectively. There are a lot of them and again we will leave them for you to experiment with on your own as an advanced user.

Ambient

This light will shine equally on all surfaces and doesn't cause any shadows. Its Intensity can go from 0 (None) to 1 (Full).

Flashbulb

If you don't have any idea what type of lighting to use, start with the Flashbulb Global Lighting. This type acts like having the flash on for a camera. It gives a point light source from the camera (or eye-point) that is looking at the view, but since it is just a flash, there aren't any shadows. Its Intensity can also go from 0 (None) to 1 (Full).

Solar and Sky

This type of Global Light is really neat because it actually acts like the sun. If you're modeling buildings, you can be extremely specific as to its location—right down to major cities. This is used to do solar studies by looking at different times and the position of the sun and its effect on the building and its surroundings. With the Sky settings, you can define how cloudy it is as well as the air quality. Solar and Sky lighting is definitely beyond the scope of this text.

 You can also control the Color of the light. A white light would be the brightest.

 When you're using Radiosity, it highly recommends that Ambient and Flashbulb Global Lighting **not** be used.

Illustrated below are three different renderings of the bracket model (tutorials\intro\dgn\bracket.dgn). All three were done using the Phong Render Mode and Normal Shading Type but with different background and light settings.

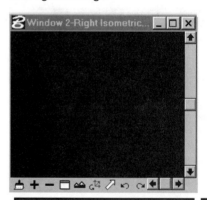
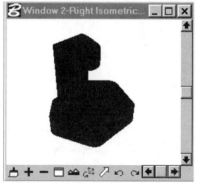

Whoa—you call this a rendered view? Yes, it is. This is how a rendered view can look with
1) Black background
2) **No Global Lighting**

The background has been changed to White (setting in main pull-down menu WORKSPACE>PREFERENCES in the View Windows category). Now this is what the rendered view looks like. **Still no Global Lighting.**

Now this is a bit more like it! The Global Lighting has been turned on so we can at least see the object.
- Ambient—an Intensity of 1
- Flashbulb—an Intensity of 1

There won't be too many illustrations of rendering results since gray scale doesn't do them justice. Rendering is more impressive displayed in color on your computer.

Source Lighting

Going back to the Define Light (Create Mode) dialog box, we can see the options for different types of Source Lighting. You can also reach them through the main pull-down menu SETTINGS>RENDERING>SOURCE LIGHTING.

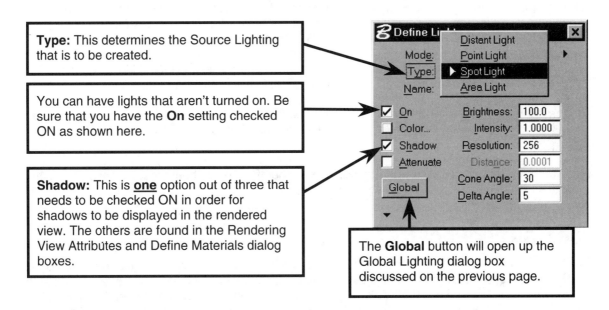

Type: This determines the Source Lighting that is to be created.

You can have lights that aren't turned on. Be sure that you have the **On** setting checked ON as shown here.

Shadow: This is **one** option out of three that needs to be checked ON in order for shadows to be displayed in the rendered view. The others are found in the Rendering View Attributes and Define Materials dialog boxes.

The **Global** button will open up the Global Lighting dialog box discussed on the previous page.

Here's a brief description of each type of source light. MicroStation Help has a more detailed discussion found under the topic of Light Source Types. Placement of a source light requires a good sense of 3D space and a bit of trial and error.

Distant Light

This is like sunlight and requires a direction, but not a distance since all surfaces will be illuminated equally. It can cast shadows.

Point Light

The Point Light source acts like a light bulb. It is placed at a specific position but a direction is not required. It will not cast shadows.

Spot Light

This light is like its namesake—a spot light. It shines a beam of light (conical shape) and requires a position and a direction. The size and shape of the Spot Light's cone can also be controlled. Placement of a Spot Light is easier for beginners because you can actually see the cone of light represented graphically—it looks like a flashlight! It is often brighter than a Point light with the same settings since the light is more focused. This light can cast shadows.

Area Light

This Source Lighting Type allows you to use an existing element's area as the shape of the light. It is a fairly new option, but an Area Light can be useful when there is really a light (such as a floor lamp) in the model.

Shadowing

Shadows can help your model look more realistic. First of all, the rendering method must be capable of calculating shadows (Phong and Ray Trace can). In addition to that, there are three settings that can affect the ability to have shadows cast in your rendering.

1) **Define Light** dialog box with the Shadow setting checked ON (this has already been discussed). Please note that it is the same setting as that found in the Source Light dialog box (accessed by the main pull-down menu SETTINGS>RENDERING>SOURCE LIGHTING).

2) **Rendering View Attributes** dialog box with the Shadows setting checked ON. This dialog box is accessed by the main pull-down menu SETTINGS>RENDERING>VIEW ATTRIBUTES.

3) **Define Materials** dialog box with the Cast Shadows setting checked ON <u>for the material being used</u>. The dialog box is accessed by the main pull-down menu SETTINGS>RENDERING>DEFINE MATERIALS.

We won't venture too deeply into the world of shadowing other than to point out these settings. You can go back and experiment with shadows on your own.

PLACE LIGHT SOURCE

The Define Light tool brings up the Define Light dialog box (Create Mode) that allows you to place a light source. This is illustrated below on the left. Just like Global Lighting, there is more than one way to get to Source Lighting—you can also use the Source Lighting dialog box that is accessed under the main pull-down menu SETTINGS>RENDERING>SOURCE LIGHTING. It is shown below on the right. It has a slider bar for some of the settings and its own Tools pull-down menu. We'll stick with the Define Light dialog box from now on, because it has most of the options and tools that beginners need—all in one spot.

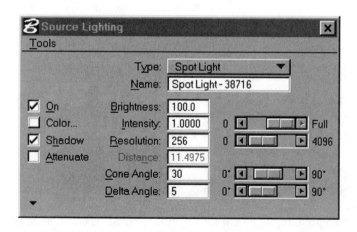

These will be the Source Light settings used for the illustration of the Define Light tool.

Once the Define Light dialog box is open and set to Create Mode (as shown directly above) the status bar will indicate that it's ready to Define light position. Illustrated here is the cursor indicating the position of the Spot Light. Once the position of the Spot Light is defined (by using a data point), you can see the cone of the light dynamically change with the cursor movement, as illustrated below on the right. Another data point will be needed to Define the light target point, which completes the placement of the Spot Light.

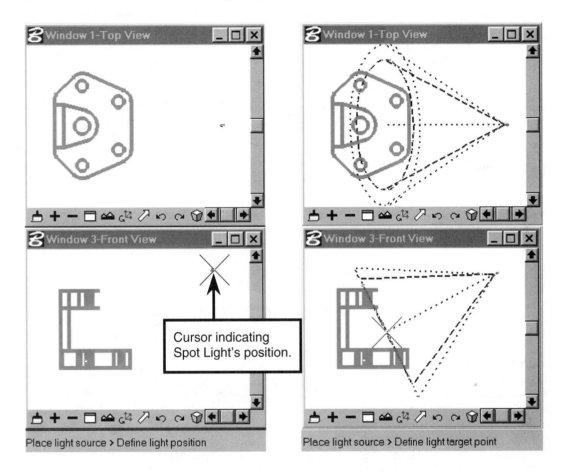

Once the Spot Light has been defined, you can zoom in and actually see a little flashlight cell that indicates the lighting. Other information about the light is written near the flashlight. Shown at the right is the Spot Light that was just defined. All right—so it is upside down. At least the information is there!

All of this information and the flashlight symbol are <u>Construction</u> types. This means that their display can be toggled on/off under SETTINGS>VIEW ATTRIBUTES. Whether or not they are displayed does **not** control whether the lighting is on or off—that is found in the Define Light dialog box.

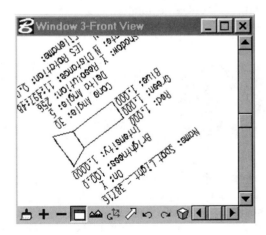

EDIT LIGHT SOURCE

Existing Source Lighting can be modified. The various tools for modifying a light source are available from the Method option button found in the Define Light (Modify Mode) dialog box as illustrated below on the right. Again, we won't go into each and every Method. We will look at how to Scan for existing light sources and then how to use Delete Light. The other methods will be left up to you to try out on your own.

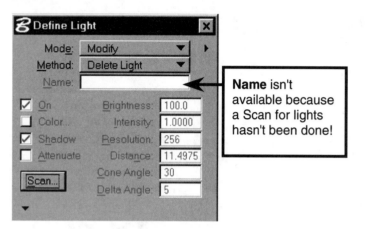
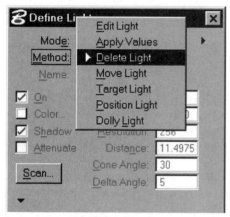

Scan

If you don't know if there are any source lights out there, the Scan button of the Define Light (Modify Mode) dialog box will scan for them. If it finds any, it will prompt you to accept the light found, and, if accepted, then that light will be modified according to the Method chosen. This Name field will also show the light that will be modified. You can also use the Light List to select a specific light from the ones listed. Illustrated at the right is the result of using the Scan button. The Light List is hidden because it wouldn't be a very long list–there's only one light!

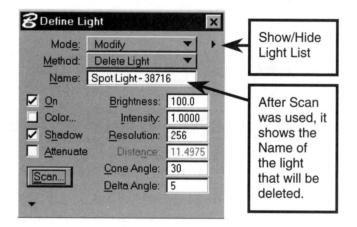

 Even if the Constructions setting in View Attributes is unchecked (OFF) a Scan will still find the light, but you should leave Constructions checked ON while working with lights just to play it safe. You will probably want to toggle off the Construction setting when producing a finished rendered view.

Delete Light

Using the Delete Light Method (Modify Mode), you can select the light to delete by identifying it with a data point on the light cell symbol (if Constructions is checked ON in the View Attributes) and accepting the data point. You may also just double-click on the light in the Light List. An Alert box will appear asking if you're sure you want to delete the light. It is that simple.

That's enough information on lighting to get you started. There is a certain art to being able to light a model to achieve a realistic rendering. Trial and error and a lot of patience are involved, but the results are well worth it.

MATERIALS

The use of different materials assigned to different elements that make up your model is just as important as lighting when it comes to producing a realistic rendering. A concrete T-beam looks more realistic with a rough gray material than it does in plain green. A slab that has been assigned a transparent glass material will render to realistically look like the window that it represents. There is an enormous amount of resources available that deal with materials. Once again, we'll cut it down to the basics. Once you've gained some experience, feel free to go experimenting! Let's start with some material terminology and go from there.

Palette

The different materials that are available are stored in different palette files. These palette files contain all sorts of materials to choose from. It's like a big decorating store where you pick out the different paints, wallpaper, or any other material you may want to use. You can choose materials from any of the palettes available. The palette files have a .pal extension and are found under Workspace\system\materials.

Material Assignment Table

Your design file doesn't need **all** of the different materials available in an entire palette. You don't take the entire store home, just the materials you may want to use! This is where the Material Assignment Table comes in—it contains the materials that you selected from one or more palettes. Each design file has its own Material Assignment Table. It has the same name as the design file but ends with a .mat extension.

Bump and Pattern

Not only do materials consist of colors, but there can also be textures and patterns involved with the material. This is where the terminology of bump and pattern maps comes in. You can mix and match bump and pattern maps to define a more realistic material that can then result in a more realistic rendering.

- **Bump**—this actually simulates the texture of a material. There are high and low spots so that the rendered model can look rough.
- **Pattern**—these are raster images that have a pattern such as a rug or brick pattern.

Now we will just consider applying a material that has already been defined in the palette. If you're really finicky, then there are ways of tweaking an existing material definition or defining your very own material. We won't be doing any of that customizing here.

APPLY MATERIAL

 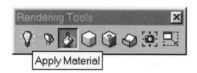

The Apply Material tool that is found in the Visualization tool box brings up the Apply Material dialog box. This dialog box also has two different methods and a lot of different modes that deal with materials. The display on the dialog box's right side shows a preview of the material on a standard display object of your choice. Options and settings found in this dialog box can also be reached through the main pull-down menu under SETTINGS>RENDERING>ASSIGN MATERIALS. We'll stick to using this Apply Material dialog box. Note that the ... on the buttons indicates that clicking on the button will bring up a dialog box.

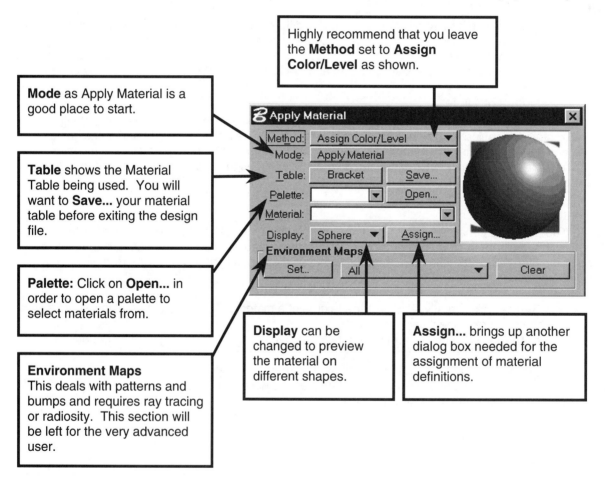

Highly recommend that you leave the **Method** set to **Assign Color/Level** as shown.

Mode as Apply Material is a good place to start.

Table shows the Material Table being used. You will want to **Save...** your material table before exiting the design file.

Palette: Click on **Open...** in order to open a palette to select materials from.

Environment Maps
This deals with patterns and bumps and requires ray tracing or radiosity. This section will be left for the very advanced user.

Display can be changed to preview the material on different shapes.

Assign... brings up another dialog box needed for the assignment of material definitions.

For the most part, saying "apply" material or "assign" material means the same thing. It gives a material definition to an element. The terms "apply" and "assign" may be used interchangeably, but don't get them confused with <u>attaching</u> a material definition—which is **not** recommended. **A material definition is assigned to an element based on the element's Level attribute <u>and</u> the element's Color attribute.** This means that some planning ahead is required when a model is made up of different objects so that different materials can be defined for each of the objects. Those objects that have the same Level <u>and</u> Color attributes will have the same material assigned to them. You'll need to pay attention to what Level and Color the objects are. Let's look at how to open a palette, select a material definition, and then apply it to the same bracket that we've been using.

Open Palette File

By clicking on the Palette Open button in the Apply Material dialog box, the Open Palette File dialog box appears. We'll open the file metal.pal so that a material definition of Aluminum is available for application to our bracket.

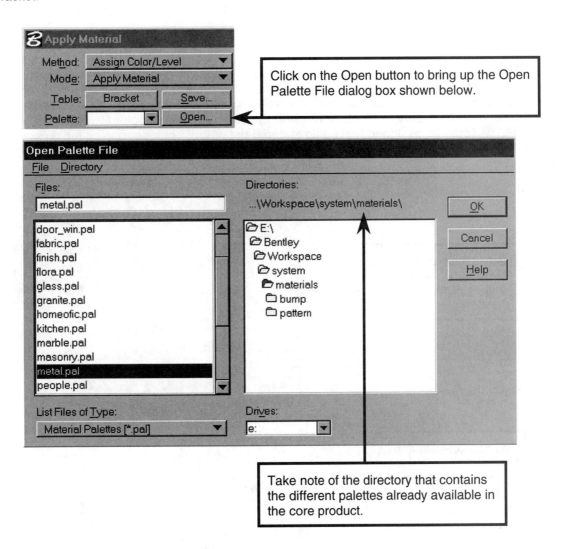

Click on the Open button to bring up the Open Palette File dialog box shown below.

Take note of the directory that contains the different palettes already available in the core product.

Select Material Definition

Once the palette has been opened, we can select a material from it. Shown here are some of the Material definitions that are found in the metal.pal file. You just click on the Material definition that you want. The active Material setting can be previewed on the display object shown in the right side of the Apply Material dialog box.

Assign Material

The prompt in the status bar will ask you to identify the element that the Material should be assigned to. The Level and Color attributes of the element identified (with a data point) will be assigned the Material listed in the dialog box. **Any other elements with the same Level and Color attributes as the element identified will also be assigned that same Material definition!** This seems to be the quickest and easiest method. You can also use the Assign button, but the dialog box that it brings up is a bit overwhelming for beginners and is better suited to complicated models.

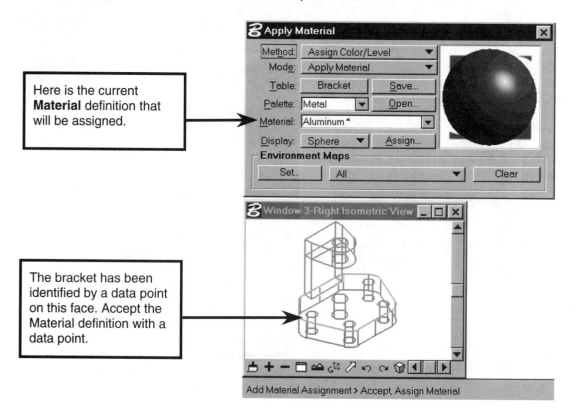

Here is the current **Material** definition that will be assigned.

The bracket has been identified by a data point on this face. Accept the Material definition with a data point.

Now if you render the view, the bracket will look as if it is made from aluminum. This won't be illustrated here; you'll have to try it on your own. Since the Aluminum Material definition is "shiny" you might want to add a few Spot Lights and experiment with the different rendering techniques.

Other Modes Dealing with Materials

Shown here are other Mode options that are found in the Apply Material dialog box. Not only can you use this dialog box to Apply Material but you can also use it to Modify, Remove, or Preview Material. These modes are left for you to explore on your own.

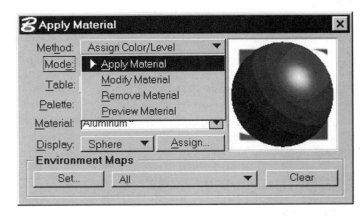

TRANSPARENT MATERIALS

Transparent materials are impressive because you can see through the object. It is well worth taking a quick look at how to get a realistic rendering using transparent materials. The illustration below shows a 3D plane that was placed in front of the bracket by using Place Block.

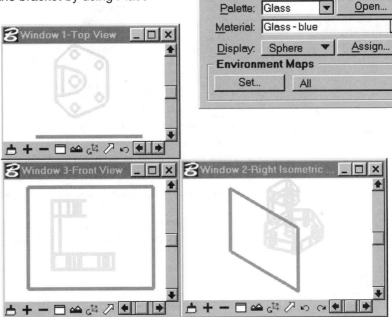

Here are some other important details regarding what was done to obtain the rendered image.
- The plane has a different level attribute than the bracket.
- The palette used was glass.pal which contains material definitions that are especially set up to be transparent.
- The material chosen was Glass-blue, and it was assigned to the plane.

Here's the result of rendering the Right Isometric View using Ray Trace and Normal shading. The corner of the bracket that is hidden by the plane is still visible through the glass. In the color rendering, the plane has a bluish tint that is very realistic.

It is possible that even with assigning a transparent material, the rendering may not show the transparency. Transparency is a setting that can be toggled off. We need to look at the Rendering View Attributes for that option and a few others that are also important.

RENDERING VIEW ATTRIBUTES

Using the main pull-down menu and going under SETTINGS>RENDERING>VIEW ATTRIBUTES will bring up the Rendering View Attributes dialog box that is shown here.

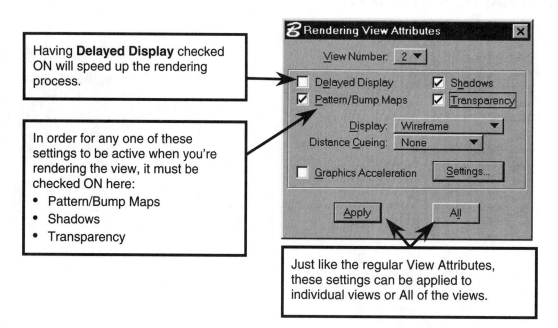

Having **Delayed Display** checked ON will speed up the rendering process.

In order for any one of these settings to be active when you're rendering the view, it must be checked ON here:
- Pattern/Bump Maps
- Shadows
- Transparency

Just like the regular View Attributes, these settings can be applied to individual views or All of the views.

Here's an example of the same rendering, but in this case **Transparency is unchecked (OFF)** in the Render View Attributes for the Right Isometric View. One little setting can make a huge difference.

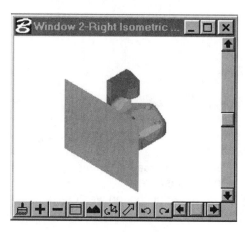

RENDERING SETUP

The Rendering Setup dialog box puts the majority of settings that affect rendering into one dialog box so they are easier to check and change. This gives you the ability to try different combinations of settings and save them until you get the combination you want. The set-up can also be saved and applied to other design files. You can reach the Rendering Setup dialog box illustrated here by using the main pull-down menu under SETTING>RENDERING>SETUP. Clicking on the tabs will bring up the corresponding settings. A majority of these settings found here are not discussed here, but the Rendering Setup dialog box puts them all at the advanced user's fingertips.

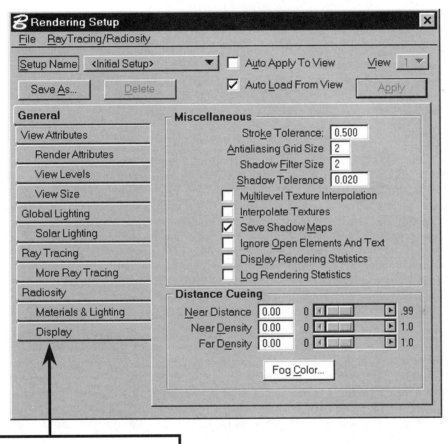

Click on one of the tabs, and the appropriate settings/tools will appear.

ANIMATION

Making an animation involving your 3D model takes even more work than rendering. Animation techniques in CAD borrow heavily from the movie world. You have actors, cameras, scripts, and settings that all work together to make a movie that can be played. Animation works hand in hand with rendering because the frames of the movie are made up of rendered views.

The main Animation tool box shown at the left contains four different tool boxes called Animation Actors, Animation Cameras, Animation Settings, and Animation Preview. You can access this main Animation tool box from the main pull-down menu under TOOLS>VISUALIZATION TOOLS>ANIMATION TOOLS.

Animation is not for the inexperienced user. Let's briefly discuss the three basic types of Animations that can be done: Keyframe, FlyThrough (sometimes referred to as "Paths"), and Parametric Motion Control. Animation tools are accessed by the main pull-down menu UTILITIES>RENDER>ANIMATION. Creating any animation is a very time-consuming process, from the set-up to the final recording, even for a very short movie.

Keyframe

This animation technique is probably the simplest to use. Different "Key" frames are created by changing the position of actors or objects in your model using regular tools. You create a Keyframe from the view, move the actors/objects and then create another Keyframe. You keep repeating this process. Once the Keyframes have been specified, then a script is created. This involves assigning the Keyframes a specific frame number in the animation. For instance, if there is to be a total of 70 frames, then the Keyframes may be assigned to frames 1, 30, 50, and 70. The software (according to the settings you specify) fills in the frames that are between the Keyframes. Once the script has been created, then it can be previewed and edited if necessary. When it's just the way you want it, the script is recorded. This actually renders the view for each frame of the animation and puts it all into a movie file. Be prepared to wait while recording a script since it renders the frames one at a time.

FlyThrough

A FlyThrough animation usually leaves the model in one spot and instead specifies a path for the camera to move along. The frames are established by rendering the view (as seen by the camera) at certain increments along the path. This is done automatically for you by using the FlyThrough Producer. A Path animation is similar to a FlyThrough animation except that actors, objects, and cameras can **all be moving** along their own specified path. This requires defining actors and scripting them, similar to the Keyframe technique. In the next section, an example of producing a Flythrough animation is shown.

Parametric Motion Control

This is the most advanced animation because actual mathematical functions are used to specify the motion. As the name implies the motion is controlled by parameters. For instance, the equations for velocity and acceleration as a function of time can be applied to the actors and objects to specify their position. This type of animation gives very realistic results but is a highly advanced technique.

Remember that these animations can be played using the Movie Player that is accessed by the main pull-down menu UTILITIES>IMAGE>MOVIES.

First let's produce a simple FlyThrough animation. The example won't be a true detailed step-by-step process. Instead we'll consider the general steps of using FlyThrough Producer to set-up the animation and record it. We'll need to look at Movies after that, so that the animation movie can be played.

EXAMPLE OF FLYTHROUGH ANIMATION

For the example, a new 3D design file was created using seed3d as the seed file. In addition to the standard cube that comes with the file, a smiley face "sign" has been added. The sign was created behind the cube and offset to the right of the cube. In order to do a FlyThrough animation, a path to "fly along" is needed. The path needs to be a "joined element" so that there isn't any discontinuity. In order to achieve this, AccuDraw and SmartLine were used to create the light gray path illustrated below.

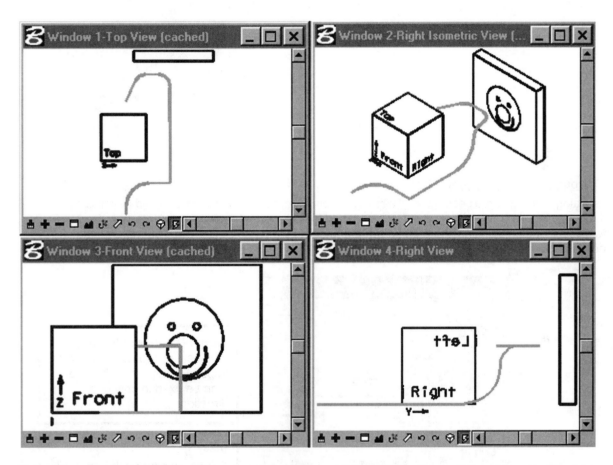

Notice the path starts in front of the cube and goes toward the cube before making a curve to the right. The path then takes a sharp turn and heads toward the front of the sign (moving along the side of the cube). There is a smooth curve that gives a change of elevation (going up in the Z-direction). Before it hits the sign, the path curves to the left and then turns toward the back of the cube.

 The "smiley" face was accomplished by setting the Active Depth for the Front View so that it was at the front of the sign.

Accessing the FlyThrough Producer Dialog Box

Once a path is created and there are some objects to "fly through" the next step is production. Shown at the right is the pull-down path to get to the FlyThrough Producer dialog box. You access it by going to the main pull-down menu UTILITIES>RENDER>FLYTHROUGH.

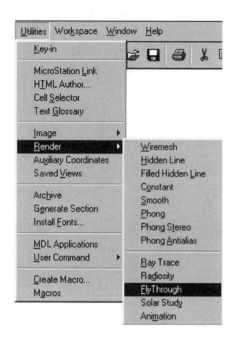

FlyThrough Producer Dialog Box

This dialog box has three sections as illustrated below. Most of these settings should be left as the default, but there are some that have been changed for the example. These are noted as shown. Right now Tools is the only setting in the pull-down menu of the dialog box. There will be more as the production continues.

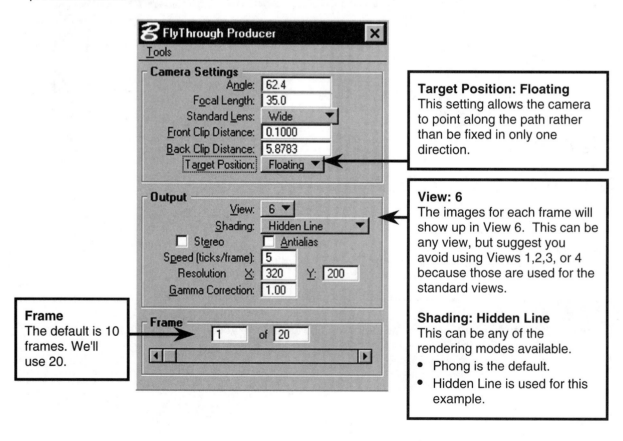

Target Position: Floating
This setting allows the camera to point along the path rather than be fixed in only one direction.

View: 6
The images for each frame will show up in View 6. This can be any view, but suggest you avoid using Views 1,2,3, or 4 because those are used for the standard views.

Shading: Hidden Line
This can be any of the rendering modes available.
- Phong is the default.
- Hidden Line is used for this example.

Frame
The default is 10 frames. We'll use 20.

Define Path

Even though a 3D SmartLine has been made to use as a path, it still needs to be defined. You do this by using the FlyThrough Producer pull-down menu TOOLS>DEFINE PATH as illustrated to the right. You'll be asked to define the starting point of the path and the ending point of the path. This will define the path and set the direction that the camera is going to move in (from the starting point to the ending point).

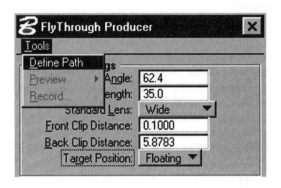

Once the path has been defined, little imaginary cones are displayed. These represent the camera **at each frame**. The more frames specified, the more cones. There are 20 cones automatically appearing on the path (count them if you want!). The point of the cone is the location along the path. The base of the cone represents the direction that the camera is pointing. The cone is kind of like a flashlight beam—if that helps you visualize it. With a Floating Target Position, the camera points down the path as you can tell by the cones. If the Target Position setting was Fixed, the camera would always point at a single target. This would be like turning your head as you drive past a landmark, you've fixed your eyes on a single target rather than looking straight ahead as you drive.

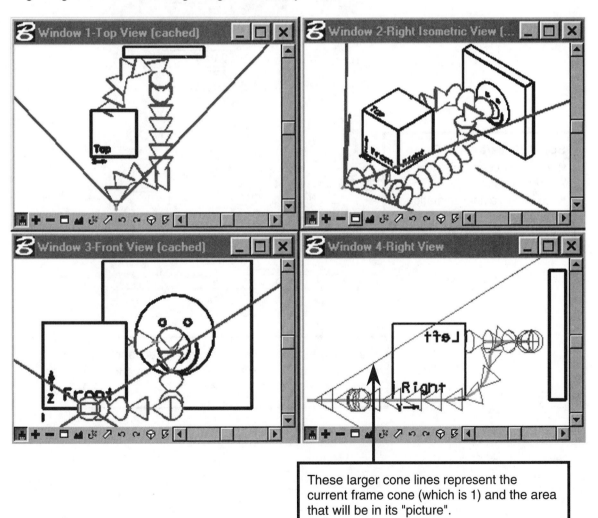

These larger cone lines represent the current frame cone (which is 1) and the area that will be in its "picture".

Preview

Once a path has been defined, then you can preview the animation. There are two choices as illustrated at the right. You can preview the Camera (each cone will highlight). Better yet, you can preview the View—this lets you preview the animation in the View that was selected for output. By selecting the FlyThrough Producer pull-down menu TOOLS>PREVIEW>VIEW and watching in the specified Output View you can actually see what the animation will look like. It will quickly display all of the frames in sequence. If it goes too fast you can slow it down by changing the Speed setting in the FlyThrough Producer dialog box. You can also preview it frame by frame too. This requires the use of the slider bar at the bottom of the FlyThrough Producer dialog box.

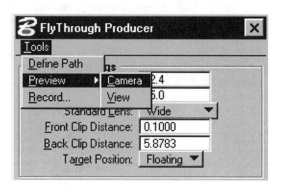

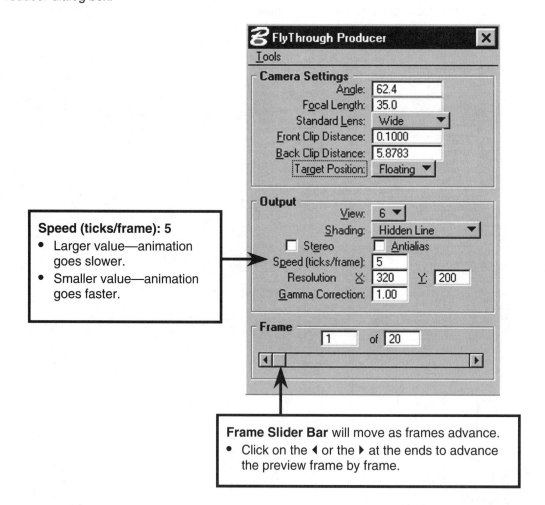

Speed (ticks/frame): 5
- Larger value—animation goes slower.
- Smaller value—animation goes faster.

Frame Slider Bar will move as frames advance.
- Click on the ◄ or the ► at the ends to advance the preview frame by frame.

Here's what the Preview View looks like in Window (View) 6 for the first frame of the animation. The path does appear on this animation so that you have some reference point.

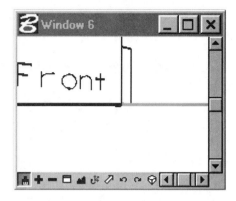

Shown below are some previews from specific frames. The other view windows are shown too so that you can see the specific camera location that is being used for each of the frames previewed.

Frame 5: Camera is pointing to the right so there isn't anything visible except the path.

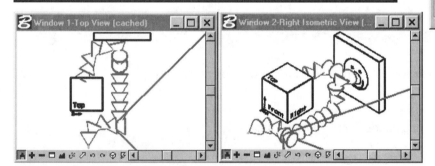

Frame 6: By advancing just one frame, the camera sharply turns back to head towards the sign.

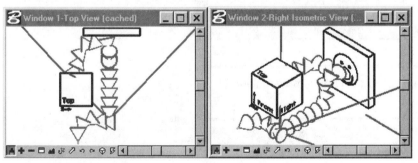
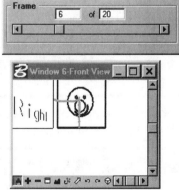

When the camera gets closer to the sign, the path curves upward. This mean that the camera too is angled upward, as you can see in the Right View window.

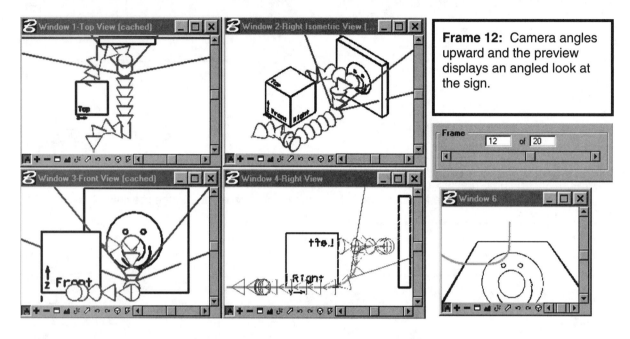

Let's look at two more preview frames of the animation before we look at recording it.

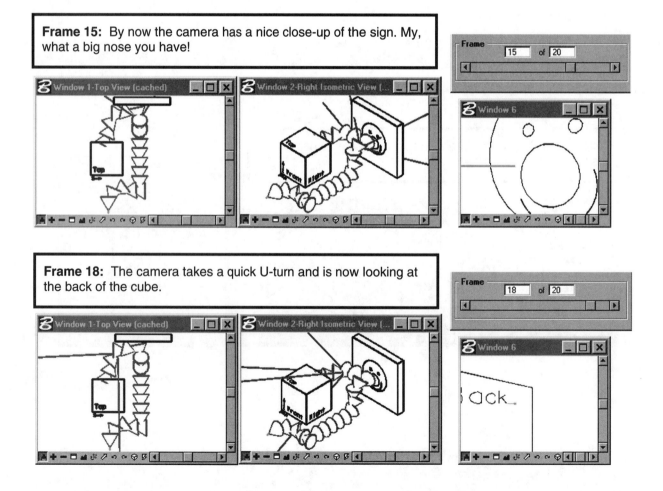

Record

The animation needs to be recorded so that a .fli file (short for flick) can be made. This is the "movie" of the animation. This file will allow you to play your movie over and over again for your friends. To record the movie just go to the FlyThrough Producer pull-down menu TOOLS>RECORD, as illustrated on the right. This will open up the Record Sequence dialog box shown below. Here you get to name the movie. By default, it will put it in the out folder of the directory your design file is in.

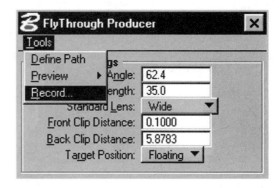

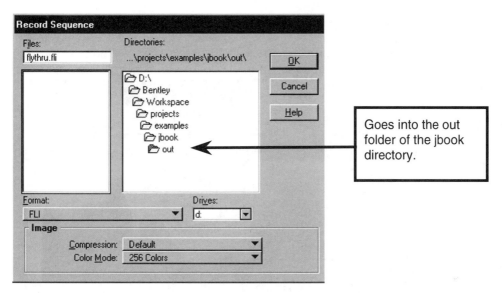

Goes into the out folder of the jbook directory.

By clicking OK in the Record Sequence, the animation will automatically be recorded. This may take some time (especially if you've selected an advanced rendering mode) because it has to render and save each frame's image.

MOVIES

Once the movie flick file has been recorded, you can load the file and play the movie you've just made! MicroStation comes with its own movie player that needs to be used.

Accessing the Movies Dialog Box

This Movies dialog box can be accessed from the main pull-down menu UTILITIES>IMAGE>MOVIES as illustrated on the right.

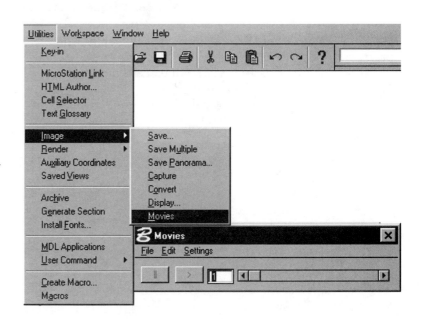

Load Movie

Even though the FlyThrough movie was recorded, it isn't loaded into the movie player automatically. Use the Movies pull-down menu FILE>LOAD as illustrated on the right. Select the .fli file that you want to load. This can also take some time, be patient. The scroll bar in the Movies dialog box moves to the right as it is loading the movie frame by frame.

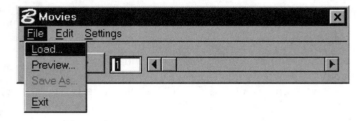

Play Movie

When the movie has been loaded, it will appear at the bottom of the Movies dialog box. Shown below is frame 6 of the flythru.fli movie that was made. A click on the play button (arrow to the right) will play the entire move. Again the slider bar allows you to go through the movie frame by frame.

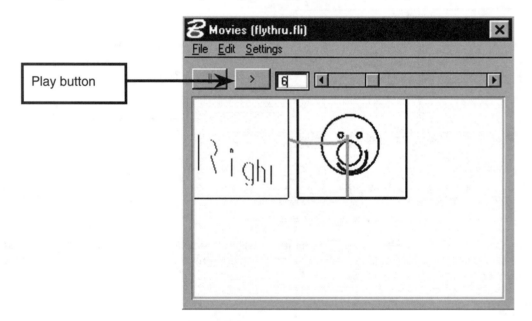

Here are some settings that you can adjust for the movie's playback. These are accessed by using the Movies pull-down menu SETTINGS>PLAYBACK. The Speed setting works the same way as it did in the FlyThrough Producer dialog box (increase the Ticks/Frame to slow down the movie). With the Loop Sequence checked ON as shown, the movie will loop back and keep playing over and over. You can use the button (looks like | |) next to the play button to stop the movie.

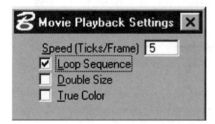

QUESTIONS

① What are some of the benefits of rendering your model? What are some of the drawbacks?

② Which is the faster rendering Mode: Hidden Line or Ray Trace?

③ What are the three different types of Global Lighting? Include a brief description of each.

④ For the display on the right, which gives information about a Source light, explain the following:
   a) What Type of light it is.
   b) Where you go to toggle the display of the light information on and off.
   c) What the Brightness setting is.
   d) How you can determine whether or not the light can cast a shadow.

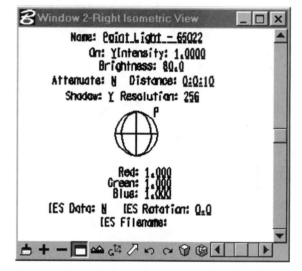

⑤ You render your model and the view is pitch black. What are some of the first things to try or check in an attempt to correct this rendering problem?

⑥ What does it mean to Scan for a light?

⑦ Explain the difference between a file with the extension .pal and one with the extension .mat.

⑧ What dialog box contains the check boxes for Delayed Display, Pattern/Bump Maps, Shadows, and Transparency? How do you access this dialog box?

⑨ Name four different formats available for saving a rendered image.

⑩ What type of solution will allow a rendered view the most photo-realistic appearance—for instance, when a green bracket actually gives a greenish tint to the white wall where it is hanging?

Chapter 19:
3D Modeling with SmartSolids

SUBJECTS COVERED:

- 3D Seed File
- Locating Solids
- 3D Main Tool Box
- 3D Primitives
- Common Settings
- Place Slab
- Place Sphere
- Place Cylinder
- Place Cone
- Place Torus
- Place Wedge
- 3D Modify Tool Box
- Solids to Use for Boolean Operations Example
- Construct Union
- Construct Difference
- Construct Intersection
- Export

Chapter 19

The core MicroStation/J product now comes with SmartSolids used to represent a three-dimensional object. Previous versions had 3D surface capabilities in the core product—true "solid" modeling required using an add-on product called MicroStation Modeler. Now some of the basic solid modeling tools such as Boolean operations, extrusions, and projections can be done fairly easily using SmartSolids. SmartSolids also allows the use of chamfers and fillets to modify the solid model.

3D SEED FILE

The use of SmartSolids does not require any special seed files. You can go ahead and use the seed3d.dgn file that was discussed in Chapter 16. Illustrated below is the Select Seed File dialog box with seed3d.dgn indicated to use as the seed file when creating a new design file. If you don't use a 3D seed file then you won't be able to get to the 3D Main tool box, which has all of your SmartSolids tools in it. It is very important to create your new design file using a 3D seed file.

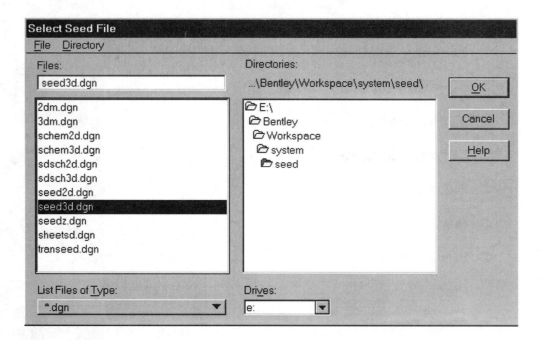

Here are some changes that you may want to make to the new design file.

- Delete the 3D cube that is found in this seed file.
- Change the Active Color to 0—the original Active Color is 2 (green).
- Change the Input Preferences (Input Category) so Highlight Selected Elements is checked ON.
- Change the Input Preferences (Input Category) so that you can locate a solid by using a data point anywhere on it (see the discussion in the next section, Locating Solids).

LOCATING SOLIDS

With the introduction of QuickVisionGL, the views that you are working in may be displaying a solid **rendered** mode. The ability to locate a solid with a data point on a rendered view is now necessary. In the Preferences dialog box (accessed by the main pull-down menu WORKSPACE>PREFERENCES) there is a Locate By Picking Faces setting as illustrated below.

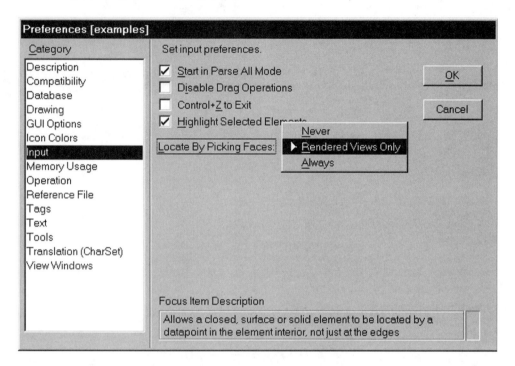

Here are the three options and what they will do.
- **Never**—you can never select the solid by picking on its face; you must pick on an edge or isoline. This is not recommended for beginners.

- **Rendered Views Only**—if the view is rendered, then you may select the solid with a data point anywhere on the solid. If the view is wireframe then the data point must be on the edge or isoline of the solid. This is the default setting.

- **Always**—you can always select the solid by picking on its face. This is not necessarily the best choice because with a wireframe display, sometimes it is hard to tell what face you are actually data pointing on! However, if you use this option you always know that you are free to data point anywhere on the solid and get something!

3D MAIN TOOL BOX

There is a 3D Main tool box. It contains four tool boxes that are used with SmartSolids. They are 3D Construct, 3D Modify, 3D Primitives, and 3D Utility. You can access the 3D Main tool box by the pull-down menu TOOLS>3D MAIN as illustrated at the right. This chapter will discuss the 3D Primitives tool box and the Boolean Operations found in the 3D Modify tool box. Chapter 20 will look at some of the other 3D Modify tool box's tools such as Fillet Edges and Chamfer Edges. Chapter 20 will also cover some of the tools of the 3D Construct tool box.

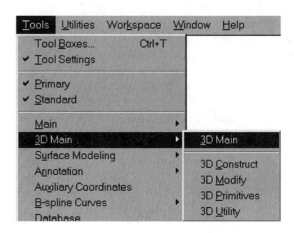

Here is the 3D Main tool box and its tool boxes.

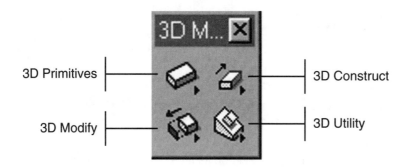

 It is worth repeating that **if you plan on doing extensive solid modeling you may prefer to use the add-on called MicroStation Modeler.** It has more tools and options available for easier solid modeling and modifications of the model. **However, the subjects of primitives and Boolean operations discussed here can be directly applied to those similar tools found in MicroStation Modeler.**

3D PRIMITIVES

There are six basic SmartSolids primitives: Slab, Sphere, Cylinder, Cone, Torus, and Wedge. These are based on mathematical primitives that should be familiar to you. These primitives are the "building blocks" of any model. By using these solid primitives found in the 3D Primitives tool box and some Boolean operations of Union, Difference, and Intersection, you can model a three-dimensional object.

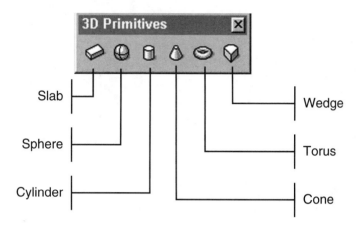

COMMON SETTINGS

Most of the tools for placing the primitives have some common settings. The Type of primitive and the Axis for the primitive are two of these common settings. The first primitive to start with will be a Slab. Shown below is the Place Slab dialog box; we'll use it as an illustration of the Type and Axis settings.

Type

The default Type is Solid as illustrated below. **Leave it that way.** The other choice of Surface will not behave in the same fashion and limits the ability to create a "solid" model. It is highly recommended that beginners leave the Orthogonal setting checked ON as defaulted so that their primitives are not skewed.

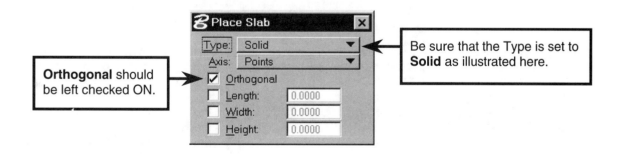

Axis

A primitive has some type of mathematical axis defined for it. This Axis setting can usually be thought of as the "height" of the primitive. The Axis setting determines which axis of the three-dimensional coordinate system will be used to correspond to the primitive's axis. **It is extremely important to keep track of the Axis setting because it controls how the primitive will be oriented.**

The default Axis setting is usually Points—not recommended for beginners. If AccuDraw is running then you'll see the Axis set to Points (AccuDraw) as illustrated at the right. This is still not recommended for beginners because of the difficulty of setting up an oblique axis. However, the use of AccuDraw in 3D does have its advantages. You can still use AccuDraw even if you aren't using the Points (AccuDraw) Axis setting.

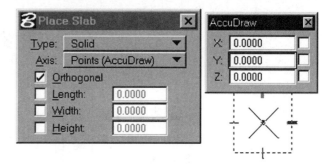

There are a lot of choices for the Axis setting as illustrated on the left. This isn't necessarily good for beginners! Notice that they relate to the discussion of 3D space coordinates and are either Drawing or Screen. The default of Points is the toughest to use, especially for beginners. It uses graphic points to set up an oblique axis. **It is highly suggested that Drawing X, Y, or Z be used (for orthogonal primitives).** When using a screen coordinate, you also get an orthogonal axis, but you need to pay close attention to which view you are working in. Screen X, Y, or Z Axis settings can get extremely confusing.

For Example: Let's look at the Slab with the settings shown. The Axis is Drawing Z, so the Height value will be the measurement in the Z-direction.

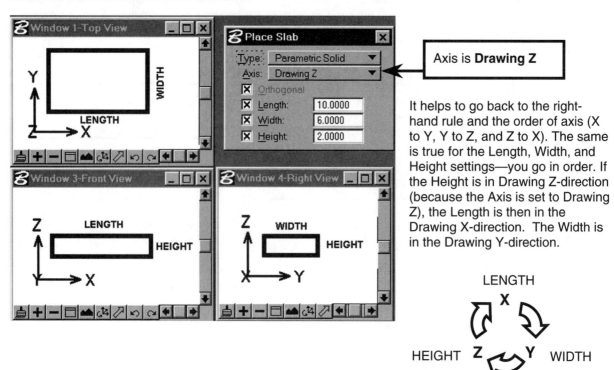

3D Modeling with SmartSolids

 You'll find that having the four view windows displaying the orthographics and a pictorial is the easiest way to keep things straight in a 3D model. It allows you to data point in any of the views but also lets you see how it is being oriented. If you are short on space, then two view windows with one of them set to Right Isometric also works well.

Now let's look at another orientation determined by the Place Slab settings. In this new slab placement as shown at the right, the **Axis is Drawing X.** The Length, Width, and Height settings are the same ones used in the previous slab placement.

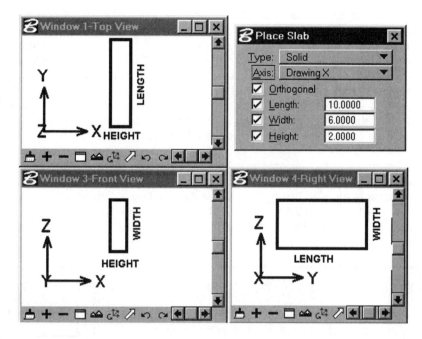

The Height will be in the Drawing X-direction. The Length will be in the Drawing Y-direction and the Width will be in the Drawing Z-direction.

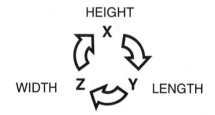

Here are a few additional things that are common to solid primitives and worth noting:

- If you constrain any of the size information (Length, Width, Height or Radius, Angle, Height), then when it prompts you for defining that measurement, it just needs the **direction** since it already has the size. Usually you can just graphically data point in the positive or negative direction.

- The last view that you selected or did a data point in will be the "active" screen when you initiate the placement of the solid. This is extremely important to the orientation when you're using a Screen Axis setting.

- Remember, the key-in DX= is also based on the **screen** coordinate system.

PLACE SLAB

We've already seen this Place Slab tool in use. It places a slab and requires you to enter the starting point and then 1) define the length, 2) define the width, and 3) define the height. They will always be defined **in that order**: Even if the Length setting is in the Drawing Z-direction, it will be still be the first size defined. If the size setting is constrained (checked ON), then the data point to define it is specifying only the **direction** of that constrained size. Do you want it defined in the positive direction or the negative direction from the starting point? The use of negative values for the Length, Width, or Height settings isn't necessary and is not recommended. With a positive value, the direction will be defined on the same side as the cursor. With a negative value, the direction defined will be **opposite** of the cursor's position. This can cause some confusion when you're a beginner.

For Example: Let's look at Place Slab in action. The settings are shown in the bottom right corner of the illustration. The Top and Front Views as well as the **Right** Isometric View illustrate the position and orientation of the slab being placed. The Axis setting of Drawing Z means that the constrained Height setting of +1 will be in the Drawing Z-Direction. The constrained Length setting of +10 will be in the Drawing X-direction. The constrained Width setting of –6 will be in the Drawing Y-direction. Pay close attention when we define the width because of its **negative** value.

❶ The starting point has already been entered.

- Usually will be done with key-in to have specific and precise placement of the slab in the design file.

The **Right** Isometric View will easily display the front, top and **right** side planes (rather than the left side plane).

❷ The prompt reads Define length.

- Shown here is the cursor in the Top View, defining the direction of the length.
- The Length setting is constrained at a value of +10 and will be in the Drawing X-direction. This is indicated by the dashed line.
- Because it is a **positive** value, the direction will be defined on the same side as the cursor. The data point at this cursor position defines the slab's length in the **positive direction from the starting point**.

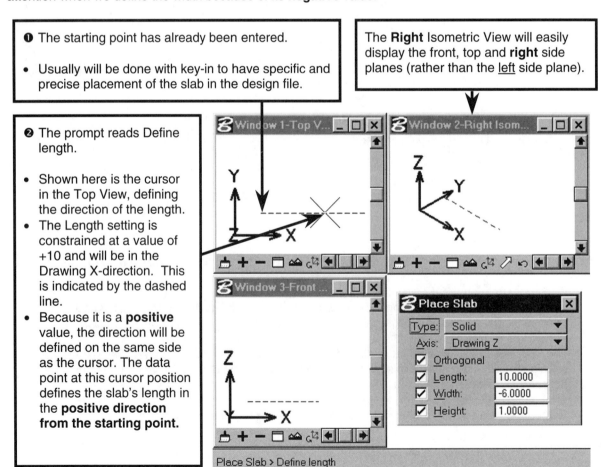

After the length has been defined, the next step will be to define the width. The Width setting is constrained to a **negative** 6.

❸ Now the **width** is being defined using the Top View.

- Since the Width setting is a negative value, the cursor appears on the opposite side of the dashed line.

- A data point with the cursor here will define the width in a **positive** direction from the starting point (as indicated by the dashed lines).

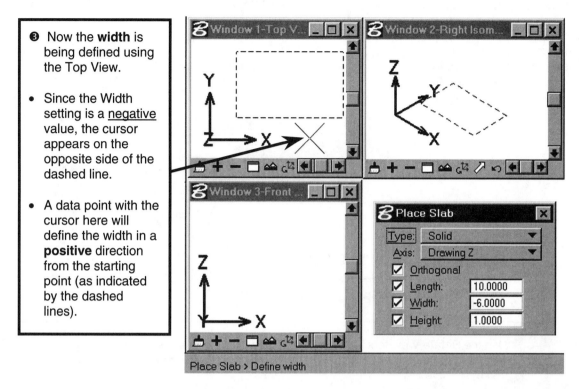

Now all that remains to do is Define height. The Height setting is constrained to a positive 1. Again, it just needs a data point to define its direction. These views' displays are a bit different as you can tell. The Right Isometric View shows the slab the best but you can still see bits of it in the other views (you'll get used to it).

The Top View will show only the 4 corners as points. The Front View shows the height lines as dashed.

❹ Height is being defined in Front view (Right view would show it as well).

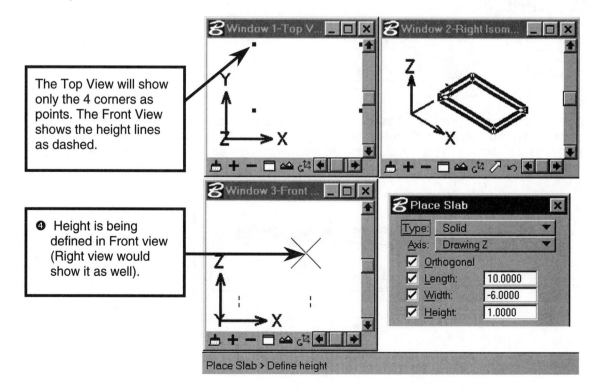

Here's the final result of the Place Slab. Yes, there are ways to edit the slab in case you goofed but those will be covered later on in the book. Right now we're going to assume that you won't make any mistakes (yeah right) or if you do, you'll just delete the slab (use Delete Element) and try again.

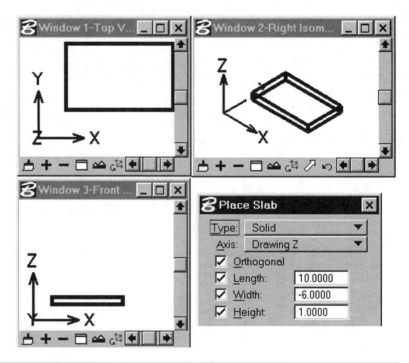

 Use the Rotate View tool in the view's window (it looks like the coordinate axes with an arrow) to change from the default of Isometric to the Right Isometric as shown in these examples.

PLACE SPHERE

The Sphere is the simplest primitive. The Place Sphere tool requires a Radius size and where the center of the sphere should be placed. Not much to it, is there? The Axis setting will determine how the wireframe lines are oriented. This can be important when applying a material to the surface for rendering purposes, but it doesn't affect what we're learning to do now.

Here's what a sphere looks like in wireframe. It doesn't look very spherical. However, it will when it is rendered, so that's what is shown in the lower right-hand corner in the Right Isometric View. This was rendered using a constant shading—cool, huh? This also allows you to see how the Axis setting affects the rendered view.

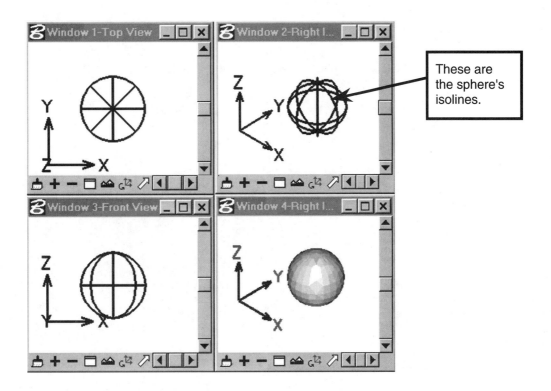

These are the sphere's isolines.

PLACE CYLINDER

The Cylinder primitive is used just as much as a Slab primitive. It requires a Radius (for the circular face) and a Height. The Axis is as indicated—this is a great example of how the Axis setting will quickly change the orientation of the cylinder. By placing a cylinder and then subtracting it, you can form a hole.

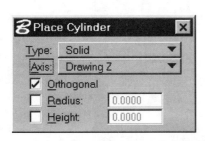 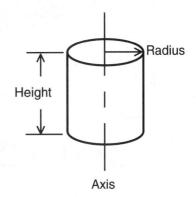

For Example: Here is a cylinder being placed using a constrained value for the Radius and a **Drawing Y Axis**. First the center of the circle of the cylinder needs to be defined. This has already been done, and the circle of the cylinder is in the X-Z Plane. The height will change dynamically as the cursor moves in the Top View. This is because the Height setting was not constrained to a specific value in the Place Cylinder dialog box.

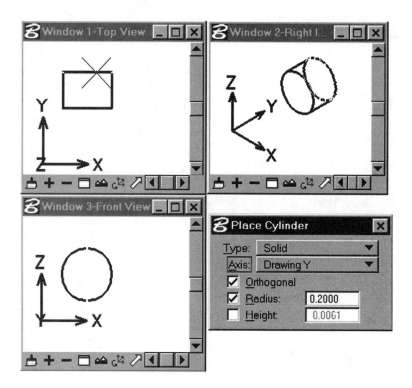

Once the height has been defined, then the Right Isometric View is rendered to give you a better idea of how the cylinder looks.

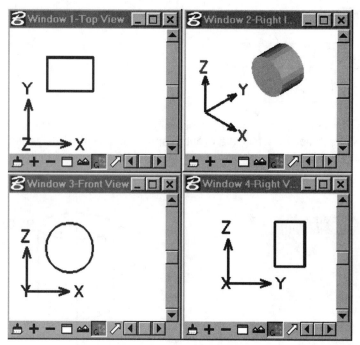

PLACE CONE

A Cone primitive has two radii to be defined: a Top Radius and a Base Radius. If one of the radii is set to 0 then you'll have a pointed cone. A truncated cone results from having non-zero values for both the Top and Base Radius settings. The center of the **Base Radius** will need to be entered. It is similar to the cylinder primitive because the Axis setting determines the orientation of the primitive.

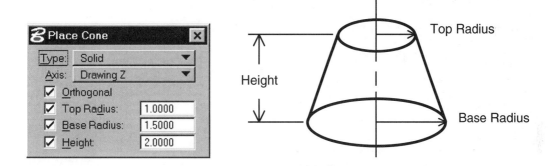

 The Top Radius value doesn't have to be smaller than the Base Radius value. The cone would just be inverted. Just remember that when prompted to Enter Center Point, you're specifying the **Base** Radius center (whether it's the larger or smaller one).

Here is an example of a pointed non-truncated cone. The Axis is set to Drawing Z so the circular base shows up in the X-Y plane.

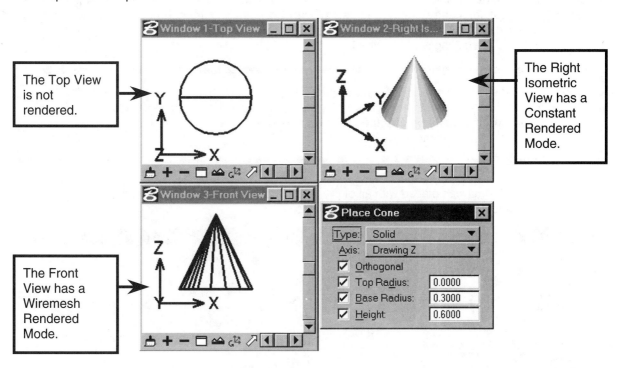

PLACE TORUS

This Torus primitive is not as common as a slab, cylinder, or cone. A torus involves a **circular** cross-section that is then revolved about an axis. Imagine a bagel (or a donut depending on your breakfast preference), and you'll have one form of a torus. An O-ring gasket is a bit more technical example of a torus. The cross-section of the torus doesn't have to be revolved a full 360°, so there is an Angle setting that allows creation of a segment of a torus.

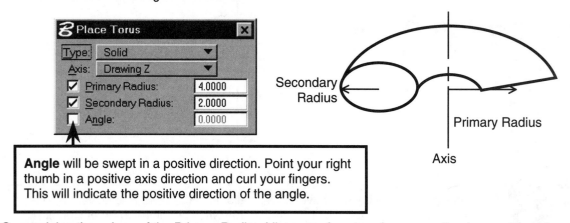

Angle will be swept in a positive direction. Point your right thumb in a positive axis direction and curl your fingers. This will indicate the positive direction of the angle.

Constraining the values of the Primary Radius (distance of center of cross-section from axis) and Secondary Radius (size of circular cross-section) settings allows you to be specific about the size of the torus. The steps to place a torus with both Radius settings constrained are fairly simple. First you will need to specify a start point. The center of the Secondary Radius (center of circular cross-section) will be placed at this start point location. Then the prompt reads Define center point. This refers to the location of the Axis. If the Primary Radius size is constrained, then a circle will dynamically appear indicating the Primary Radius location. (If the Primary Radius size is **not** constrained, then the Define center point step will determine the size of the Primary Radius as well as its location). The last step will be to define the angle (unless the Angle setting is constrained).

 If you try to specify a Primary Radius that is smaller than the Secondary Radius, when you go to place the torus, the constrained value for the Secondary Radius **will automatically be changed to the value of the Primary Radius**. This is so that the torus can maintain its form.

For Example: This torus was placed with all of the settings constrained. The Axis was set to Drawing Z, and a view was rendered so that you can see how it looks.

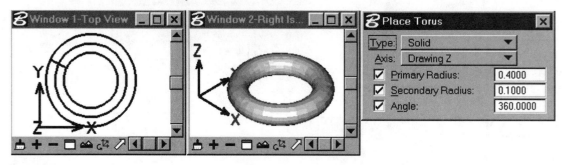

PLACE WEDGE

The Place Wedge primitive is also based on a cross-section that is revolved about an axis. This time it is a rectangular cross-section rather than a circular one used in a torus. Also the axis of revolution is on the inside edge of the cross-section. You can't have empty space inside the wedge. Because the wedge is based on a revolution, its outside boundary is curved. This is sometimes confusing since some people picture a wedge as a triangular cross-section with all straight surfaces—not the case here. Visualize this wedge as a slice of pie (first bagels, now pies…hungry yet?). If the Angle is 360°, then you just get a cylinder shape.

Here's what the settings control. The Radius and the Height actually set up the size of the rectangular cross-section.

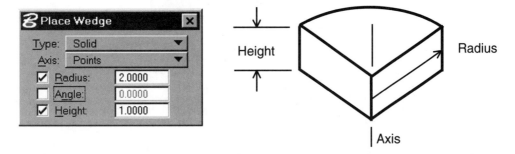

With the Radius and Height values constrained, the sequence for placing a wedge is similar to that of a torus. First you enter the start point. For a wedge, this will be a point on the radius. Next, the prompt will read Define center point. If the Radius setting is constrained, a dynamic circle indicating the size of a full 360° wedge will appear. The circle moves as the cursor indicates the orientation of the center point relative to the starting point. If the Radius is not constrained, then the data point defining the center point will also set up the size of one side of the wedge's rectangular cross-section. The next step is to define the angle (it doesn't matter if the Angle setting is constrained or not). Last but not least, you specify the height. In the case of a constrained Height setting, you'll still need to data point to indicate the direction of the height (similarly to how Place Slab worked).

Here's an example of a wedge placed with the Axis set to Drawing Z and the Angle constrained to 45°.

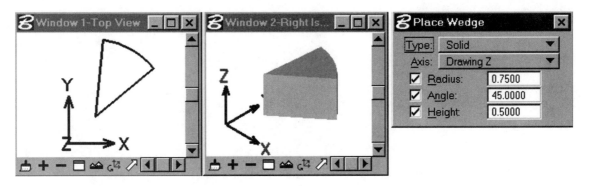

3D MODIFY TOOL BOX

Very few models consist of a single primitive—that would be too easy. Therefore some modifications to the primitives will be necessary. The Boolean operations of Union, Difference, and Intersection allow you to take the primitives that are the "building blocks" of a model and actually get a more complicated model. The three Boolean tools are found in the 3D Modify tool box and will be discussed in this chapter. The other tools of the 3D Modify tool box will be discussed in the next chapter.

These Boolean concepts have been discussed in other two-dimensional tools—those used to pattern or measure areas. The same techniques can be applied to using Boolean operations on parametric solids. Using these tools is similar to 2D. You select the primitives (that have already been placed) with a data point. Then the result will highlight, and you can "see it before you buy it." Another data point will be needed to accept. The Difference operation requires close attention to the order of selection. First you data point on the primitive that is to remain and then you select the primitive to subtract from it.

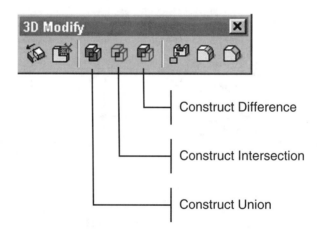

Keep Originals

The dialog box associated to each of the tools will have a single option button. The Keep Originals option deals with what happens to the original primitives that are selected for the Boolean operation. **It is highly recommended that you leave the Keep Originals option unchecked (OFF).** Otherwise you will have both the original primitives and the newly formed solid—two things in one place (not good practice for beginners).

3D Modeling with SmartSolids   445

SOLIDS TO USE FOR BOOLEAN OPERATIONS EXAMPLE

As an example of placing primitives and using Boolean operations to form a more complicated model, we'll create this simple object consisting of a Union of a Slab primitive and a Cylinder primitive. Then we will make a Cone primitive and subtract it from the slab/cylinder using the Boolean Difference operation. Part of the time, AccuDraw will be involved so that you can se how it works in 3D.

Place Slab

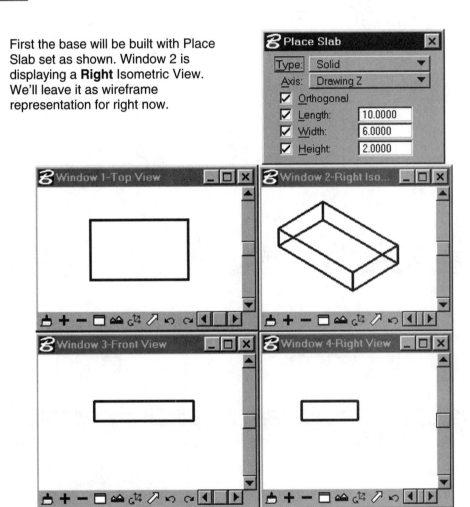

First the base will be built with Place Slab set as shown. Window 2 is displaying a **Right** Isometric View. We'll leave it as wireframe representation for right now.

Place Cylinder

The next step will be to place a Cylinder primitive on the bottom and centered in the slab. AccuDraw will aid in its placement.

 It is important that you use tentative to snap the placement of one primitive to be positioned exactly on another primitive or AccuDraw to achieve a precise placement. Once again DON'T EYEBALL.

When using the Place Cylinder, the Axis is set to Drawing Z and the Radius has been constrained to 3.0000 as illustrated below. AccuDraw will be used to specify the height when needed.

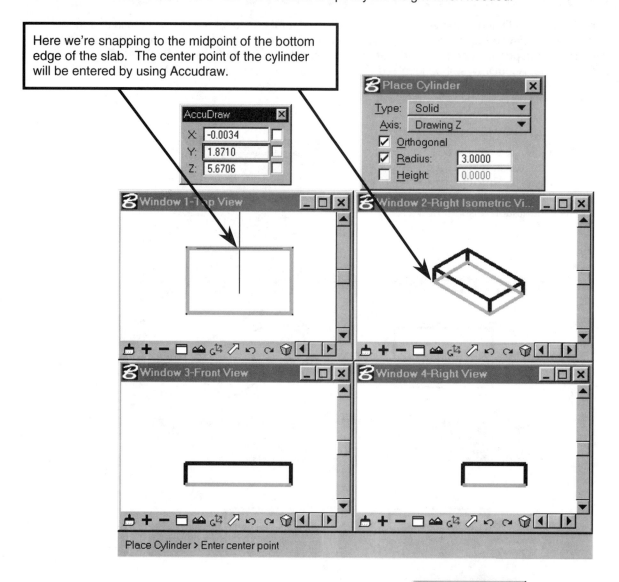

Here we're snapping to the midpoint of the bottom edge of the slab. The center point of the cylinder will be entered by using Accudraw.

Shown on the right is the AccuDraw compass that results from setting the origin at the tentative snap point (remember the <O> for the keyboard shortcut). The value of –3.0000 in the Y: field will put the center of the cylinder in the middle of the bottom of the slab.

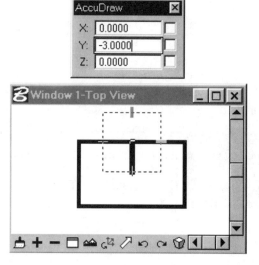

Since the Height setting was not constrained, it will now need to be defined. By moving the cursor into the Right View, the AccuDraw compass reorients itself as illustrated below. Now a value of 4.0000 entered in the Y: field will define the height of the cylinder.

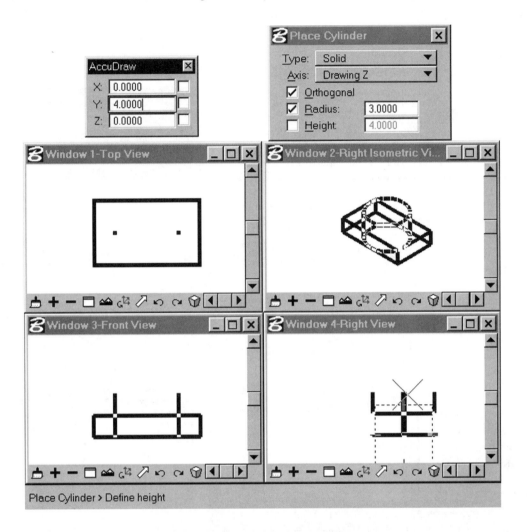

Right now there are two individual primitives as shown on the right. These primitives will need to be unionized together.

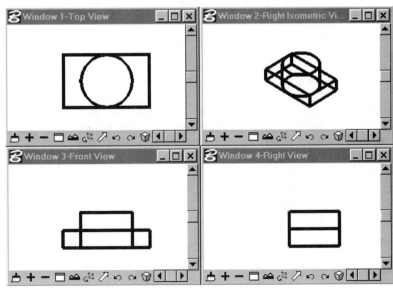

View Display Mode

Working with the Display Mode for the view window set to Wireframe is usually a good idea. However it doesn't necessarily let you easily see what is going on. You may need to use the Change View Display Mode tool at the bottom of the view window (its the one with the lightning bolt) for a different Display Mode.

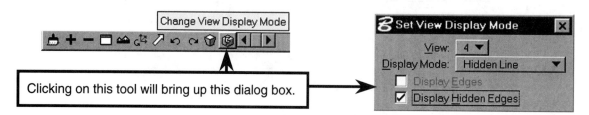

Clicking on this tool will bring up this dialog box.

Shown below are the same Slab and Cylinder primitives displayed in each view with a Hidden Line Display Mode (with Display Hidden Edges checked ON).

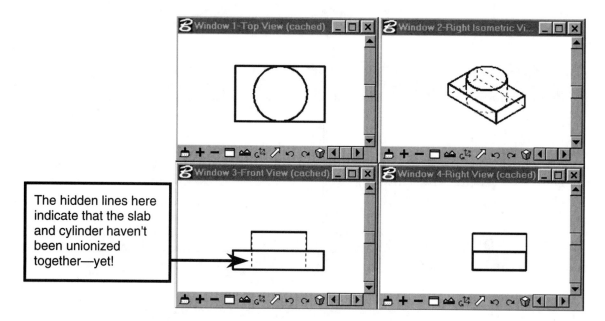

The hidden lines here indicate that the slab and cylinder haven't been unionized together—yet!

The different Display Modes are available because of QuickVisionGL. This needs to be running on Windows 98 or NT for it to work properly. Even then, sometimes the view doesn't display correctly! For instance, you may delete a primitive and have it still appear in the display. Change back to a Wireframe display mode if your view display is not appearing as you think it should. By going back to Wireframe and then back to Hidden Line (if you want) for the Display Mode setting, the display will correct itself.

Also what you see in a view isn't necessarily what you get in a plot! It can be necessary to use Export to get a print of the solid model. We'll look at that later on in this chapter.

CONSTRUCT UNION

Using the Boolean operation of Construct Union will join the slab and cylinder primitive into one solid. We will have Keep Originals unchecked (OFF). You'll be prompted to identify the first solid (use a data point) and then will be prompted to identify the next solid (again use a data point to identify it). After at least two solids have been identified to be unionized, the prompt will ask you to identify the next solid, or data point to finish.

Here is the result of a Union of the Slab primitive and the Cylinder primitive. We'll refer to this as a slab/cylinder.

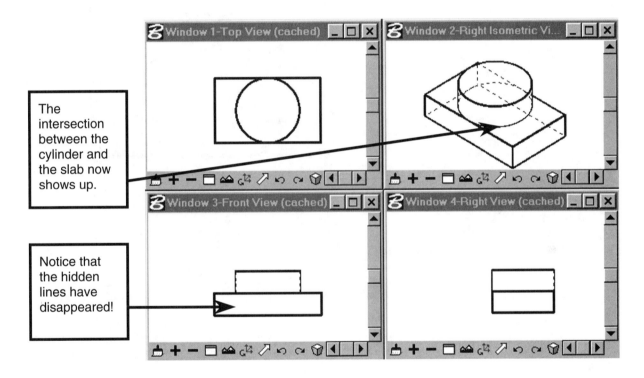

The intersection between the cylinder and the slab now shows up.

Notice that the hidden lines have disappeared!

Place Cone

We will place a Cone primitive. This cone will be subtracted from the slab/cylinder later. The Axis setting will be Drawing Z with the sizes constrained as illustrated below. The placement will be at the top of the slab/cylinder with the Cone primitive extending to the bottom of the slab/cylinder.

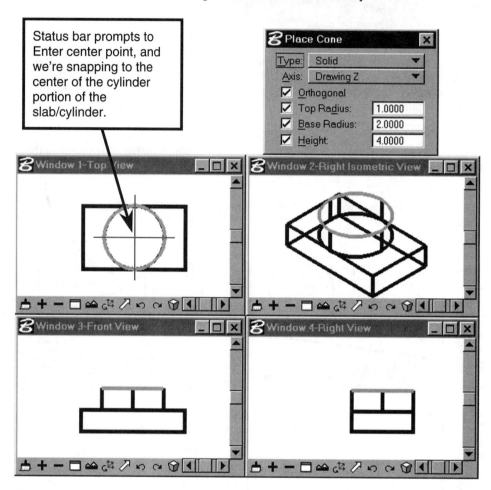

Now the height of the cone will need to be defined. Since the Height value was constrained, it just needs a data point to specify whether it goes above the slab or towards the bottom of the slab as illustrated below.

Once the Cone primitive has been placed, it will need to be subtracted from the other solid (result of slab/wedge union).

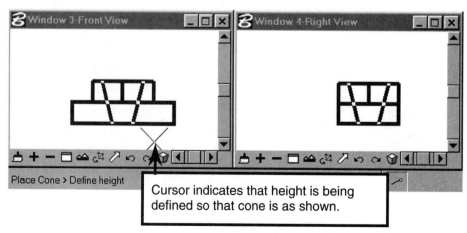

CONSTRUCT DIFFERENCE

The Construct Difference tool requires close attention to the **order** that you select the solids. The first solid will remain and the second solid (and subsequent ones) will be subtracted from it. Think of it as Solid 1 – Solid 2. **Read the prompt**—it will say that you are to identify the solid to subtract **from** first.

Shown at the right is the slab/cylinder that has been selected as the solid to subtract from. It all highlights because it was made into one object with the Union. If only the slab highlights, that would indicate that the model isn't correct. We've switched back to a Wireframe display so that the selection is easier to see.

Now it is prompting for the next solid or surface to subtract (from the highlighted solid is implied!).

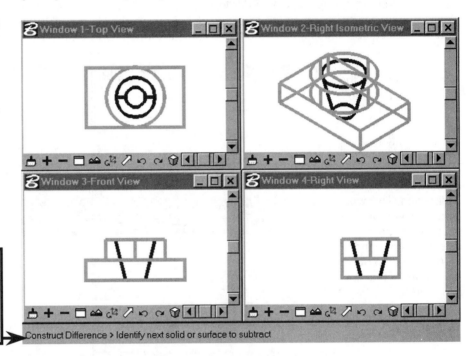

Here is the final result of the Boolean Difference operation with a Hidden Line display mode.

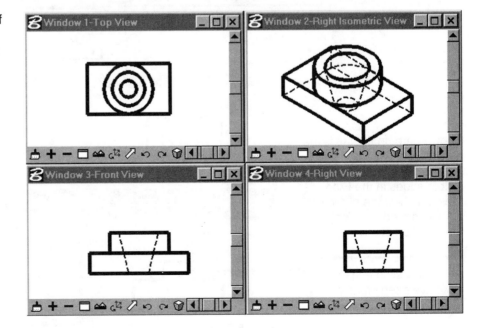

CONSTRUCT INTERSECTION

 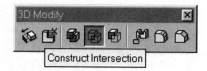

Construct Intersection is the last Boolean operation to discuss. This operation takes two solids and leaves only the solid material that is common to both of them—the intersection of the two solids. This operation is often overlooked when planning how to build a model, but it can be extremely useful. You just need to remember its capability. In order to illustrate this Boolean operation, a new Slab primitive was placed so that it is smaller than the existing solid. By using the Construct Intersection tool, the sides of the slab/cylinder can be cut off so that there is a vertical surface. Only a partial curved portion of the cylinder will be left.

The illustration on the right shows the slab in position and already selected for the intersection. Notice that it is centered on the object in the Y-direction. This was accomplished by snapping to an existing corner and then a key-in to offset the corner of the slab being placed.

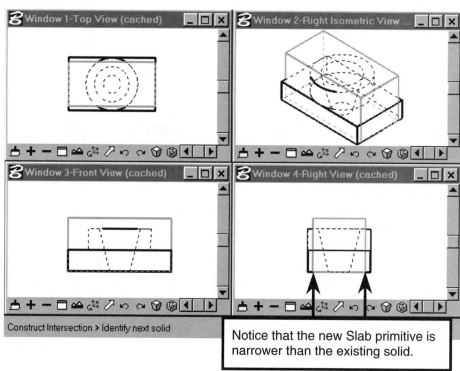

Notice that the new Slab primitive is narrower than the existing solid.

Shown here on the left is the result of the Create Intersection. Note the new visible edges in the Front View and how the other views have changed to reflect the modification of the model.

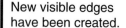

New visible edges have been created.

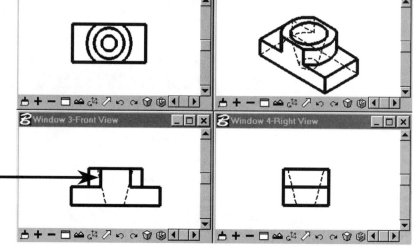

EXPORT

The appearance of the model is important. When a model is shown as a wireframe, it is often hard to visualize which way it is facing. Rendering improves the appearance of the model. It looks more realistic, but printing can be difficult. As stated in a warning earlier, you may need to use Export to get a hard copy of the model at this stage. The use of QuickVisionGL sometimes results in a render print even if the view window displays something else! The use of Export will help to work around that but you are still limited to getting only one view of the model to print at a time.

 Drawing Composition allows you to layout different views of the model **all on one sheet** to print out! Drawing Composition is discussed in Chapter 21 but you may want to jump ahead if you would prefer to use Drawing Composition instead of Export.

Export takes a single view and exports the visible (and hidden, if you want) edges to a file. This file can be a new file or the existing file (on levels that differ from the current model). The edges from the model as seen from that view's orientation become lines, arcs, circles, etc. in the exported file. The other information from the model is not contained with in the exported file. This exported file can then be printed and it can show the visible (and hidden) edges that result from that view of the model. Each time you change the direction from which you view the model, the visible and hidden edges will change. Therefore, a new Export would need to be done for each view. For instance, the Front View of our model will result in different visible edges than the Right Side View of the model.

There are a whole lot of options to be found in the Export Visible Edges dialog box. It has five different sections: General, Output, Hidden, Visible, and Advanced. The following discussion will deal with some of the basic settings that work to get both visible and hidden edges (with different line styles and line weights) exported to a new file. This file then can be used to print the elements that represent the visible and hidden edges. We'll see what can work for beginners and if you want to experiment with the other settings—feel free to explore on your own.

Accessing the Export Visible Edges Dialog Box

The Export Visible Edges dialog box is opened by going to the main pull-down menu as illustrated on the right. It is FILE>EXPORT>VISIBLE EDGES. Some of the other choices here deal with exporting the MicroStation design file to other CAD and modeling packages.

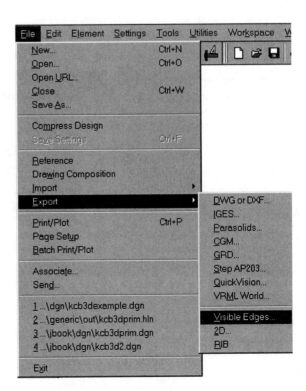

General

The settings found in the General section of the export options deal with which view you want to export and what you want to export. The Export and Preview buttons are at the bottom of dialog box and are not in a specific section. Once you have all the settings in all of the sections set the way you want, then use the Preview button to see a preview of the exported view before actually using the Export button to complete the export operation. Shown below are the recommended settings. You may want to change the View and the Include Hidden Edges settings.

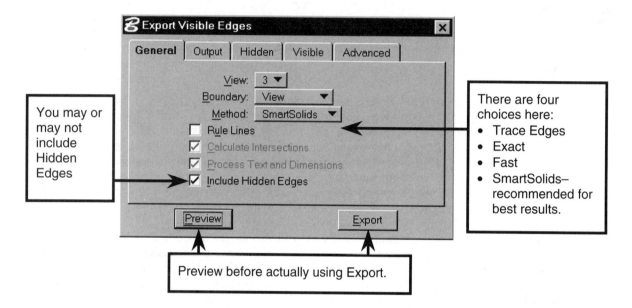

Hidden

This controls the Style and Weight settings of the elements that are exported for the Hidden Edges. In order for the settings in the Hidden Edge Overrides to be in effect they must be checked ON as shown.

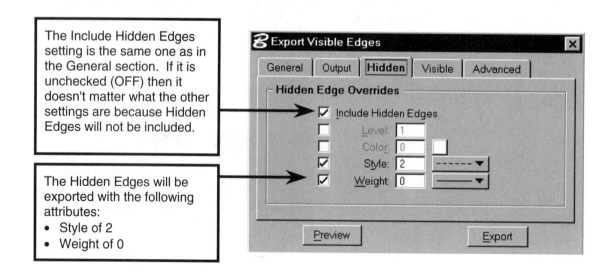

Visible

This section controls the Visible Edge Overrides. Export will automatically include the Visible Edges.

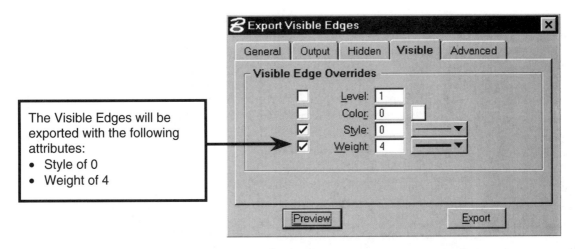

The Visible Edges will be exported with the following attributes:
- Style of 0
- Weight of 4

You may want to use the Level setting so that visible edges are placed on one level and hidden on another. This would allow the option of "turning off" the display of hidden lines even if they've been exported.

Output

This section controls whether the edges are exported to the Active design file or a new file. You can also choose to open that new file immediately upon doing the exporting. If you store it in the Active design file it would be important for the Level attributes to be adjusted. You can choose a 2D or 3D file for the output. A 2D file will give you two-dimensional elements to represent the edges, and a 3D file will result in three-dimensional elements being exported.

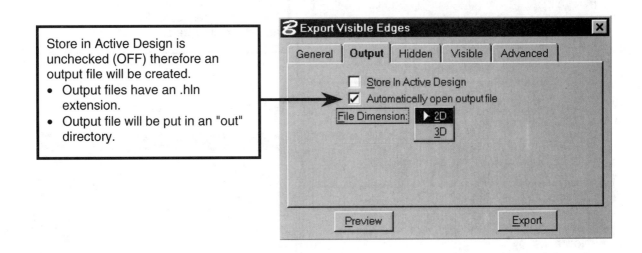

Store in Active Design is unchecked (OFF) therefore an output file will be created.
- Output files have an .hln extension.
- Output file will be put in an "out" directory.

For Example: Let's preview the export of the visible edges from the model that we made with the Boolean operations. The preview will appear in the view specified in the View setting. Shown below is the Right Isometric View that displays the preview—this is because the View setting is set to 2. The Include Hidden Edges is unchecked (OFF) so it doesn't show any hidden lines.

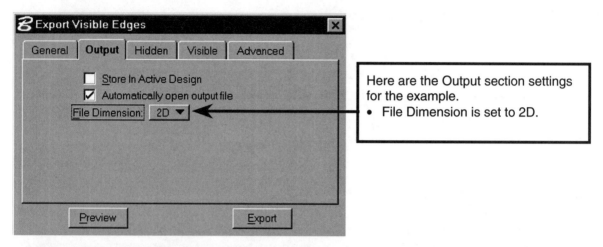

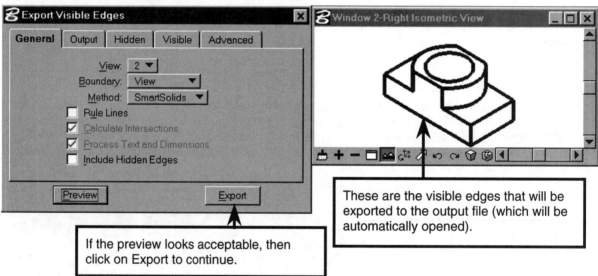

When you actually export, this Save Visible Edges Design File As dialog box appears to let you name the output file.

- File name of rightiso.hln was entered by the user.
- Notice that by default it goes in the **out** directory not a dgn directory.

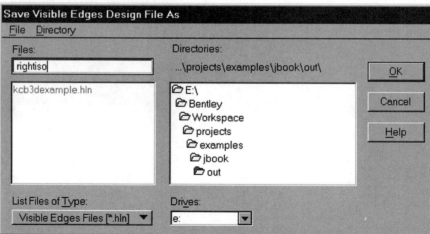

Here's the file that automatically opens (because the setting in the Output section said to.) It is a 2D file and now you can plot the view (still Window 2) or place a fence and plot that.

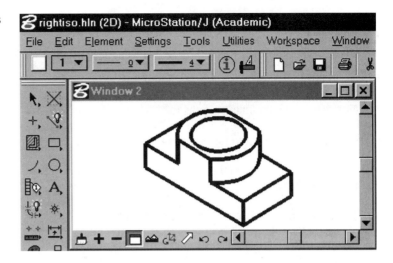

Shown below is an example of what happens when a 3D output file is specified. In this case, the Front View (Window 3) was exported and Hidden Edges were included. The resulting 3D elements look just fine in the Front View but if you look at the Right Isometric you can see that they are three-dimensional. Please remember that these are the **exported** elements, not the solid model itself.

We have exported the edges to a 3D file (named kcb3dexample.hln), which is shown here.
- Visible and Hidden Edges were both specified.
- Overrides as discussed previously.

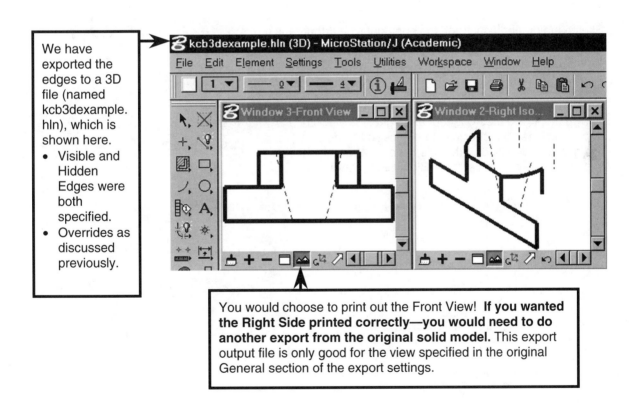

You would choose to print out the Front View! **If you wanted the Right Side printed correctly—you would need to do another export from the original solid model.** This export output file is only good for the view specified in the original General section of the export settings.

QUESTIONS

① State the four tool boxes that are included in the 3D Main tool box.

② Name the basic primitives available. For each primitive, state a real object that it could model.

③ Give the specific settings required to get a pointed Cone primitive.

④ What is the direction that the Angle setting of a Wedge is swept through—clockwise or counterclockwise?

⑤ Sketch the proportional front, top, and right side orthographic views that would result from this Place Slab setting. Label the Length, Width and Height values.

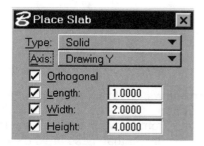

⑥ Sketch the proportional front, top and right side orthographic views that could result from this Place Cone. Label the Top Radius, Base Radius, and Height values.

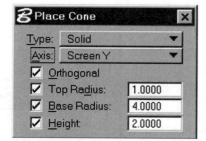

⑦ Shown below are the front, top, and right side views of a slab. State the correct values for Fields A, B, and C in this Place Slab dialog box for the orientation of the slab shown. Assume you are working in the Right Side View window.

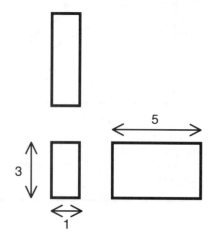

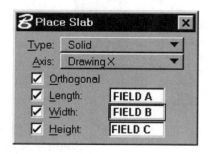

EXERCISE 19-1

Basic Primitives Using Drawing Z Axis

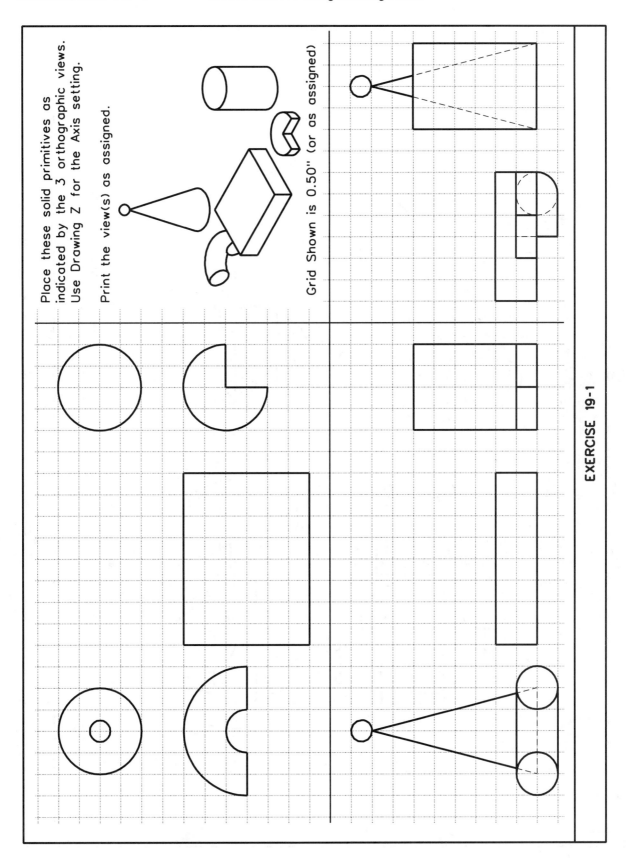

EXERCISE 19-2

Rough Model of a Pull-Cart

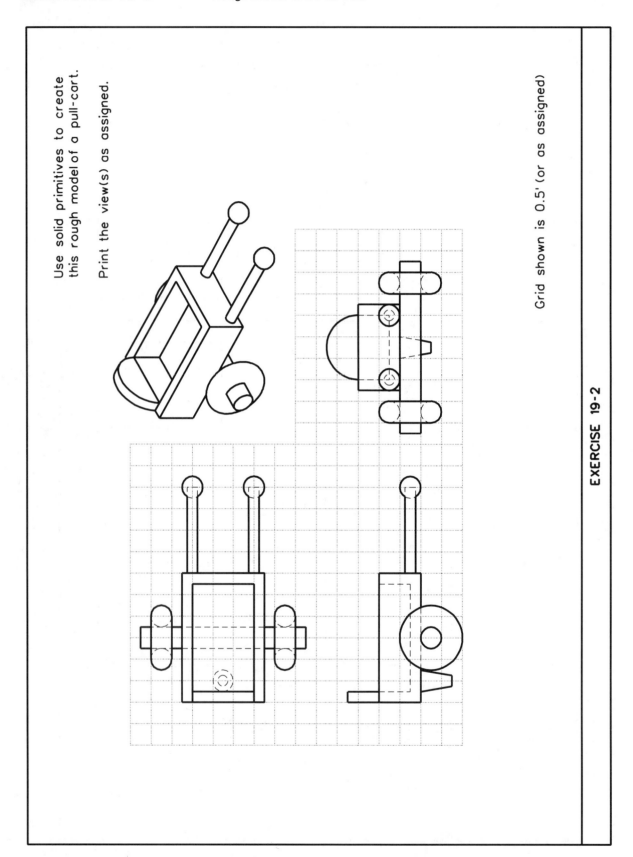

EXERCISE 19-2

EXERCISE 19-3 Fixture Brace

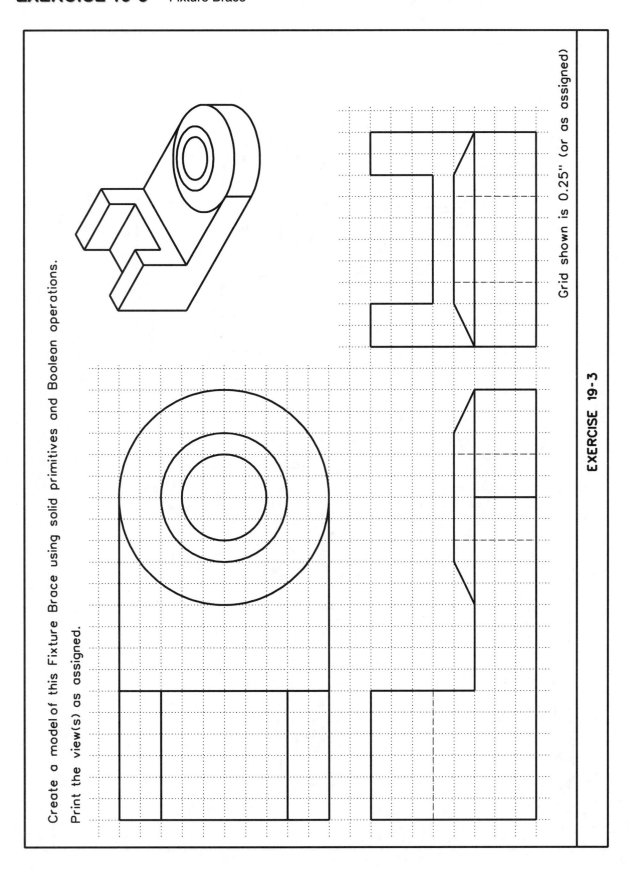

EXERCISE 19-4 Box Culvert

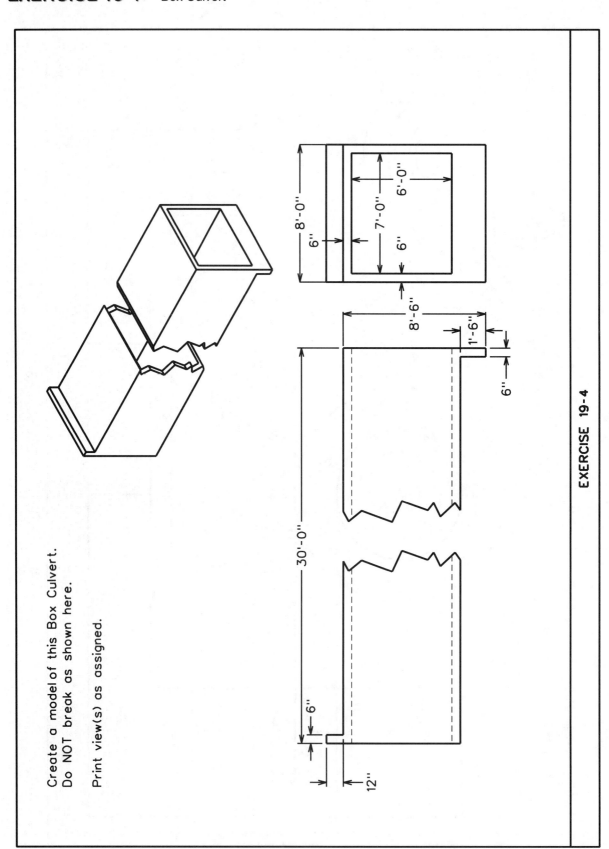

Create a model of this Box Culvert. Do NOT break as shown here.

Print view(s) as assigned.

Chapter 20: Advanced Model Construction and Modification

SUBJECTS COVERED:

- 3D Construct Tool Box
- Extrude
- Construct Revolution
- Extrude Along Path
- Shell Solid
- Thicken to Solid
- 3D Modify Tool Box
- Modify Solid
- Remove Face and Heal
- Cut Solid
- Fillet Edges
- Chamfer Edges

Using primitives and Boolean operations can only get you so far in modeling. Using a unique profile or "cross-section" in conjunction with modeling tools such as Extrude and Revolve can often do more complex models quicker and easier. These tools found in the 3D Construct tool box require a closed shape for a profile. The profile can be extruded in a straight line (a procedure sometimes referred to as projection) or revolved about an axis. There is also a tool that will project the profile along a three-dimensional path.

After making the basic form of the model, it may need to be modified. Whether you make a mistake in size or shape or you want to chamfer or fillet at edge, there are tools in the 3D Modify tool box that you'll want to use. This tool box was discussed in the previous chapter because it also contains the Boolean operation tools. First let's look at constructing some more complex models and then we'll work on using the 3D Modify tools that are available.

3D CONSTRUCT TOOL BOX

The 3D Construct tool box is found in the upper right corner of the 3D Main tool box as illustrated below. In order to use these tools a closed profile is usually required. It is important to note that **where you select the profile** will affect the result, so make sure you pay attention.

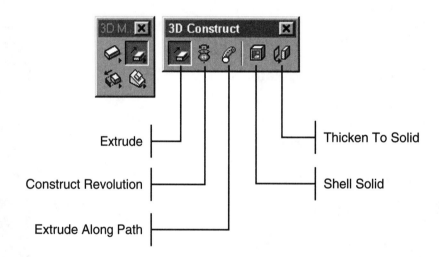

Closed Shape Profile

A closed shape is needed for the profile that is to be extruded or revolved. Using SmartLine and being sure that both settings of Join Elements and Close Element are checked ON is probably the easiest way to get this closed shape. If you already have individual elements, then use the Create Complex Shape tool found in the regular Groups tool box in the Main tool box.

If you make a mistake in your profile, then it is easier to undo your extrusion and make a new profile. There is a Modify Profile tool, but it is for a different set of circumstances. It is used when using dimension-driven techniques, which is an advanced topic.

 While we're talking about SmartLine, it is a good time to remind you about AccuDraw. Don't forget how handy it is in 3D. Sometimes the examples don't show using AccuDraw because the compass can really clutter up the illustration, but you should take advantage of it.

EXTRUDE

The Extrude tool takes the profile and projects it in a straight line perpendicular to the profile. Imagine a cross-section of a beam: Construct Projection takes that cross-section and pushes it through the length of the beam. There are numerous settings that allow you to spin and or change the scale of the cross-section as it extrudes.

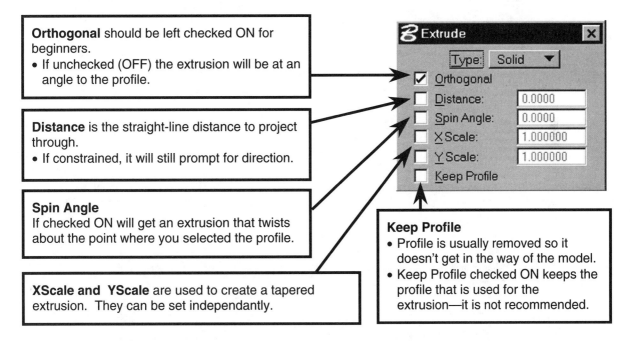

Orthogonal should be left checked ON for beginners.
- If unchecked (OFF) the extrusion will be at an angle to the profile.

Distance is the straight-line distance to project through.
- If constrained, it will still prompt for direction.

Spin Angle
If checked ON will get an extrusion that twists about the point where you selected the profile.

XScale and YScale are used to create a tapered extrusion. They can be set independantly.

Keep Profile
- Profile is usually removed so it doesn't get in the way of the model.
- Keep Profile checked ON keeps the profile that is used for the extrusion—it is not recommended.

For Example: Here is a profile created with a closed shape. It was drawn in the Right View, so it will be located at the Active Depth for that view. It appears on edge in the Front View. The steps for creating a extrusion will be to identify the profile and then specify the direction. If the Distance setting hasn't been constrained, then the data point that specifies the direction will also specify the distance. Since this distance is perpendicular to the profile, it should be specified in a view that shows the profile on edge.

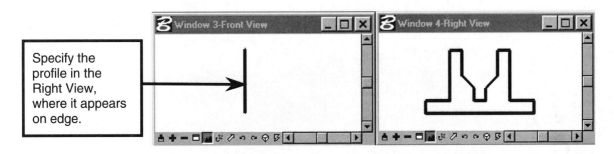

Specify the profile in the Right View, where it appears on edge.

Here the profile has been identified with a data point, the profile highlights, and you can dynamically see the distance being defined when you move the cursor. Once a data point defines the distance, the solid will be made.

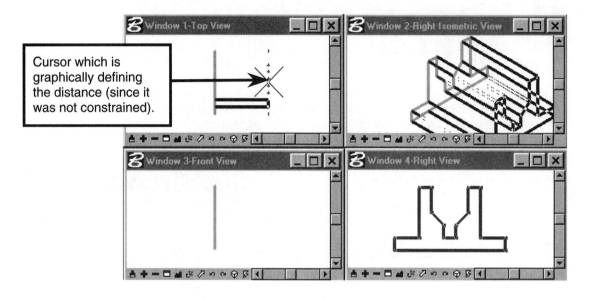

Cursor which is graphically defining the distance (since it was not constrained).

Here is the solid that results from using the Extrude tool with only the Orthogonal setting checked ON. This model would require a lot of slabs and Boolean operations but is simple to do with Extrude. The view display has been set so that you can see the hidden lines.

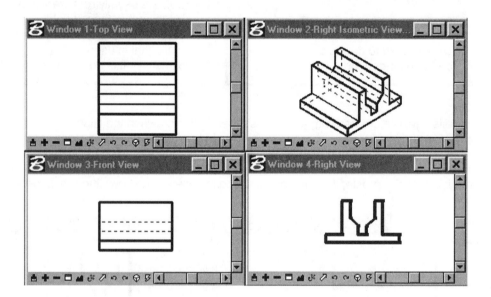

Here's another example using the same original closed shape profile but this time the Distance and Spin Angle are constrained as indicated. Distance was selected to go in the **negative Drawing X direction** from the profile by just a data point in the Front View.

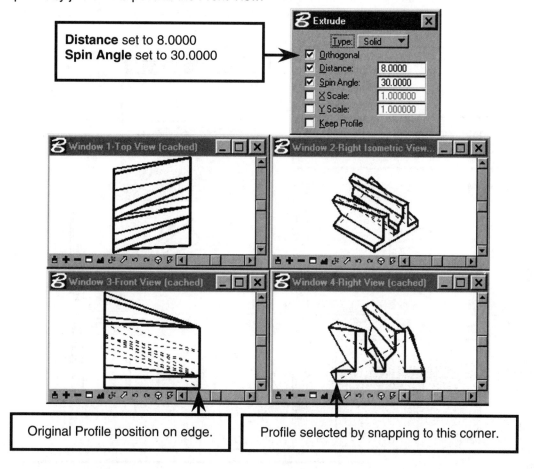

Shown below is a different model that can be made just by specifying the profile in a different location and extruding in a different direction—still using the same profile and constraints.

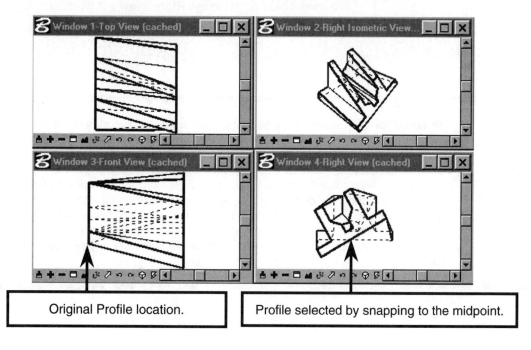

Illustrated below is an example of using Extrude with the settings XScale and YScale constrained as shown. The resulting model is tapered (in this case larger) as it goes through the distance. Once again, where you identify the profile makes a difference. Now it acts like a "scale about" point. It is important to note that the X and Y here refer to the **view** coordinates of the view in which we see the profile in true size.

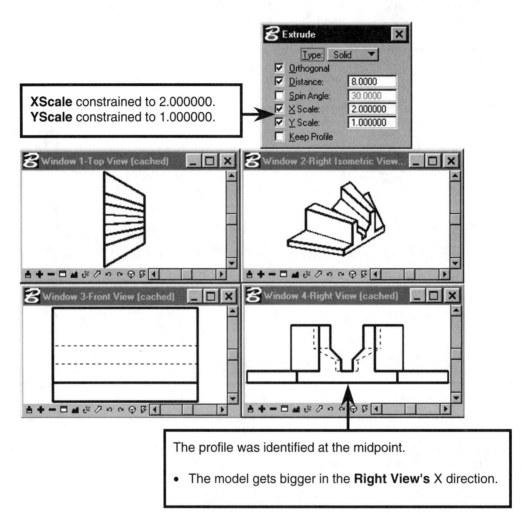

The profile was identified at the midpoint.

- The model gets bigger in the **Right View's** X direction.

The illustrations shown here will be using different display modes for the view windows. You may have noticed the (cached) notation in the view window title bar. This means that QuickvisionGL is using stored information in order to display that view quicker. Sometimes this "cached" view doesn't look correctly. First, try changing the view's display mode to Wireframe (remember you click on the lightning bolt in the view controls at the bottom of the view to get to the setting). If that doesn't work, exit MicroStation and come back into the design file. If that doesn't work—cuss—it doesn't help but it makes you feel better! No, seriously, if you can't get the views to look correct—see the software documentation for more information.

CONSTRUCT REVOLUTION

The Construct Revolution tool also uses a closed shape to construct the solid. In this case, the profile is revolved around an Axis so that you get a radially symmetric object. The orientation of the Axis is important, and you have some choices—so pay special attention to that setting.

Angle
This is the sweep angle of revolution that will need to be set.
- Goes in a positive direction.
- Can be 360° or any part of a circle.

Keep Profile
Same settings as the Extrude tool.

Axis orientation can be based on:
- Points—not recommended for beginners.
- Drawing—X, Y, or Z
- Screen (View)—X, Y, or Z

For Example: Here's our closed profile being used with Construct Revolution with a Drawing Y Axis and using an Angle setting of 360°. The profile was identified at the midpoint of the bottom edge. It highlights as shown. The prompt will read Define Axis of Revolution. This actually means to specify the **distance** away from the profile that you want the axis of revolution to be located. The **orientation** of the axis of revolution has already been determined by the Axis setting (in this case, Drawing Y).

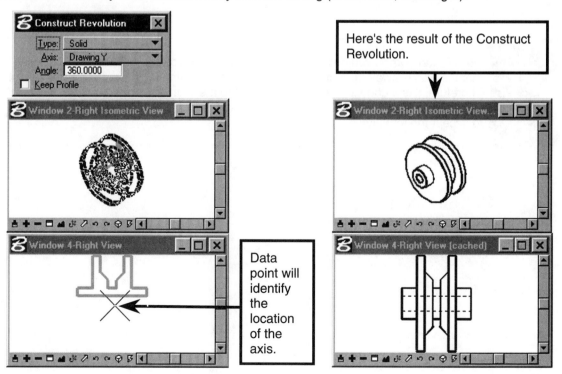

Data point will identify the location of the axis.

Here's the result of the Construct Revolution.

How about a different Axis setting? Shown below is the same closed profile, but the **Axis was set to Drawing Z**. Once again, by just changing a setting, you can get a whole different model from the same profile shape!

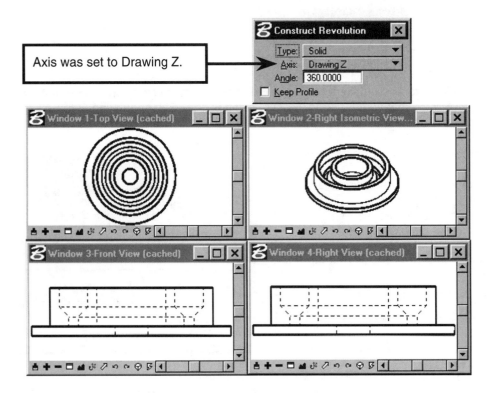

 Don't forget—when you define the axis of revolution, you are defining the distance that the axis is away from the profile. This of it as setting up the radius of revolution. Therefore, it is a good idea to tentative point to a **specific point** on the profile when identifying it to use in the revolution—that way you can be precise. Just don't forget to accept the tentative point.

Advanced Model Construction and Modification 471

EXTRUDE ALONG PATH

The Extrude Along Path uses a linestring to define a path for the cross-section to be extruded along. This path can be 3D and so that makes this tool very useful for laying out pipes. That is why the tool has a Define By setting. You can use a closed profile or you can use the Circular setting, which lets you constrain the outside and inside radii—perfect for circular pipes!

Define By Circular

Shown below is the 3D path that will be used to extrude along. There isn't a profile because it will be established by the Inside Radius and Outside Radius constraints as specified. You'll be asked to identify the path; it will preview as highlighted and then you'll need to accept it. Please read the prompts and be patient, it may take a while for the preview to show up. The linestring that makes up the path will not be deleted.

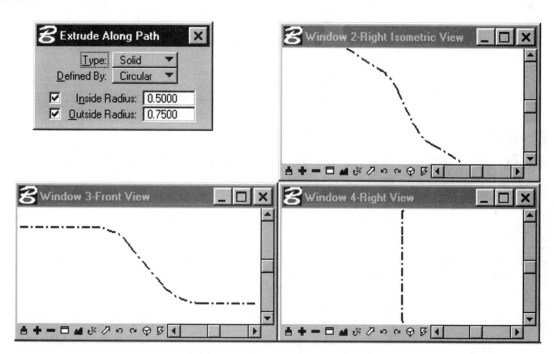

Here's the result. There will be visible lines at the curved portion of the pipe. Also notice that the original linestring may be displayed only in the straight sections of the pipe. Don't worry: The whole linestring is still there. You can easily tell from the end of the pipe in the Right Isometric View that it is a hollow pipe.

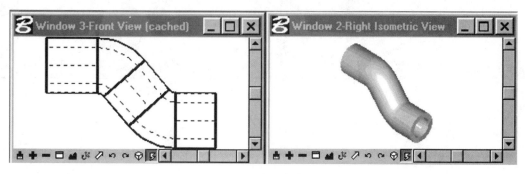

But what happens if you don't want a circular extrusion? Then you need to use the Profile option of the Define By setting!

Define By Profile

When you use Extrude Along Path with the Define By Profile setting you need to have an existing closed profile. Illustrated below is the path that has already identified to be used and a rectangular block that has been placed for the profile.

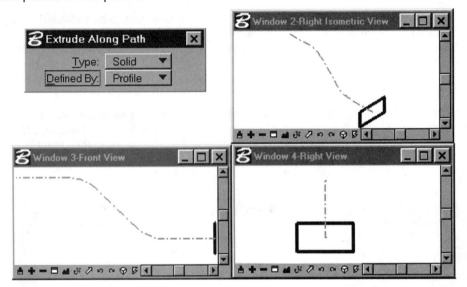

 It may help to use the Set Active Depth tool found in the 3D View Control tool box to help establish the profile at the end of the 3D linestring.

Illustrated below is the result of the Extrude Along Path. This is a better illustration of how the original linestring is not deleted. This solid is not hollowed out, but the next tool to be covered from the 3D Construct tool box can help do just that.

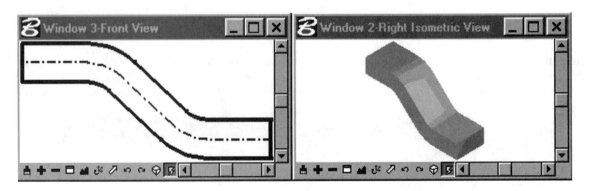

SHELL SOLID

The Shell Solid tool takes an existing model and hollows it out. The Shell Thickness setting determines the thickness of the wall. If Shell Outward is unchecked (OFF) then the outside wall of the model stays the same size. If Shell Outward is checked ON then the existing outside wall of the model becomes the new inside wall.

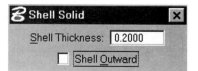

Once you've identified the solid, it will highlight and you will be prompted to Identify face to open. This is illustrated below. As you move the cursor around, the different faces of the model will be highlighted as a thicker dashed line. Shown below is the face at the original profile's position being located. You can continue to select different faces (once selected it will stay highlighted). When you're done selecting faces use a datapoint to accept and finish the process—just read the prompts.

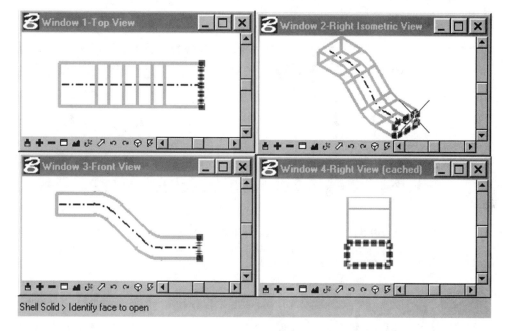

For this example, only the one face was selected. The resulting shell is illustrated below. Note in the Front View that there is a wall thickness on the end of the model. In order to have it hollow all the way through, the rectangular face on that end would have also needed to be identified as a face to open.

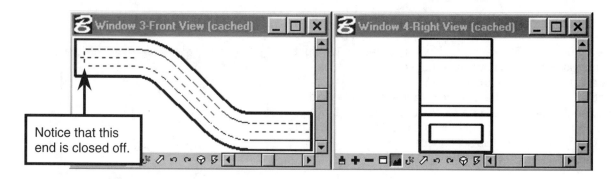

THICKEN TO SOLID

The last tool in the 3D Construct tool box is the Thicken to Solid tool. It will take an existing surface and add thickness to it to make it a solid. When you use Thicken to Solid with a profile and the Add to Both Sides setting checked ON it may work better than Extrude in some situations. Let's look at an example of when this tool is especially useful—constructing a support rib.

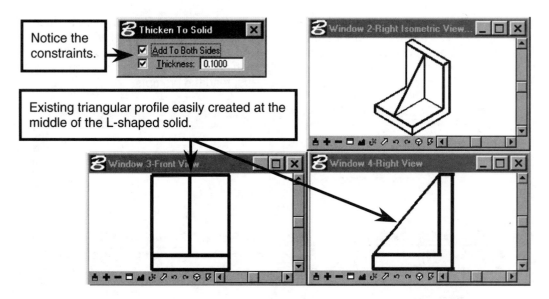

Illustrated below is the highlighted profile that has been identified to be thickened. Arrows appear showing that it will be in both directions (because Add To Both Sides was checked ON), but a data point is still needed to be used to complete the process.

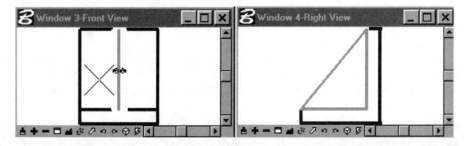

The result of the Thicken to Solid is illustrated on the right. Using Extrude would have meant offsetting the profile; instead this tool made the construction of the rib much simpler. The rib can then be unionized with the L-shaped solid so the model is all one piece. Let's assume that has been done and now look at how to use 3D Modify tools to make adjustments to this solid model.

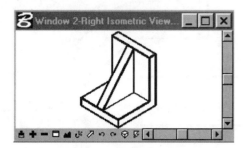

3D MODIFY TOOL BOX

This 3D Modify tool box shows up in the bottom left corner of the 3D Main tool box as illustrated below. We've already discussed the Boolean operation tools that are in the middle of the tool box. The other tools allow you to modify or remove faces, cut the model, or create basic fillets and chamfers on your model. We'll discuss all of them.

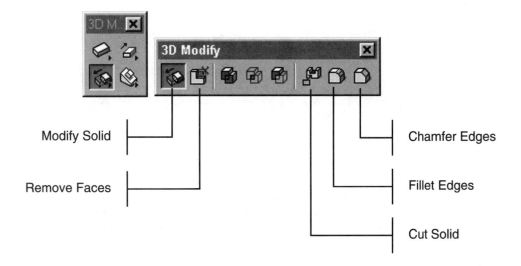

MODIFY SOLID

Modify Solid is used to change the size of the model by changing the position of a face. The face will be moved a specified distance in the direction normal (perpendicular) to the face. You can do only one face at a time. If the modification is not possible (due to construction limitations) then the status bar will read "Unable to modify solid".

Distance
- If unchecked (OFF) then a single data point will define both the distance and direction.
- If checked ON then a single data point will still be needed to define the direction.

476   Chapter 20

For Example: Let's say that our L-shaped brace needs to be longer in the Drawing X direction (on the right side of the rib). First you need to identify the target solid. This has been done and the entire solid highlights; then you're prompted to Select face to modify as illustrated below.

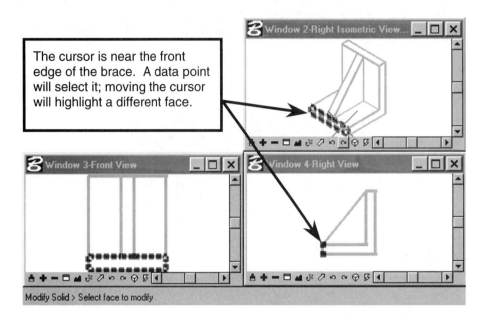

The cursor is near the front edge of the brace. A data point will select it; moving the cursor will highlight a different face.

The front face isn't what we wanted to modify. Instead we need to select the face that is illustrated below. A data point will select the face, and it will highlight as a solid line rather than the dashed as shown here.

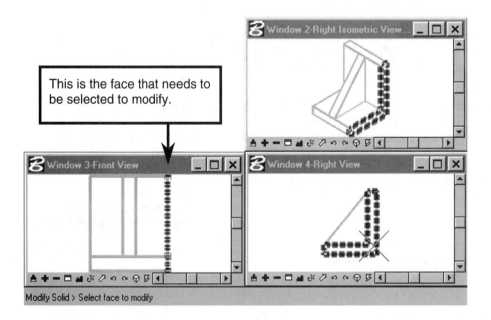

This is the face that needs to be selected to modify.

Once the face has been selected, then you need to Define distance. Even if the Distance was constrained, a data point is still needed to specify the direction (move the face outward or inward). Since we want the brace larger, the data point will be located as illustrated below.

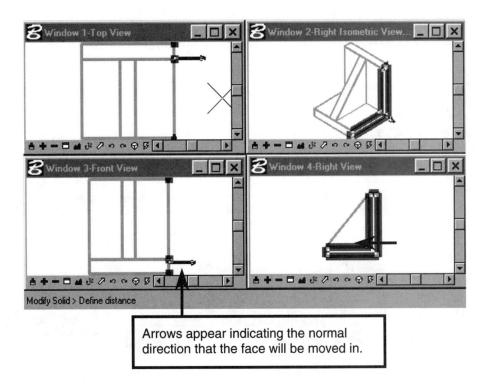

Arrows appear indicating the normal direction that the face will be moved in.

Here's the end result of modifying the solid brace. You can tell from the Top View that it has gotten large in the Drawing X direction.

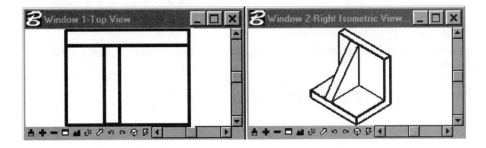

REMOVE FACE AND HEAL

The Remove Face tool allows you to modify the solid by removing one or more faces. The "and Heal" portion means that the opening that is left will be closed or "healed". Once again, after identifying the target solid, move the cursor and a face will highlight. A data point selects that face and you can go on to select another face if you want. Reset will deselect a face that was incorrectly selected (last one selected is first one deselected) or you can just data point on the selected face again; that will also deselect it.

There are two Method options: Faces and Logical. If set to Faces only selected faces will be removed. Logical Groups will allow you to pick one face, and other faces that may also be associated with the selected face will also be selected. We'll look at the Logical Groups Method after discussing the Cut Solid tool; it'll make more sense then.

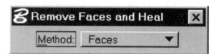

Shown at the right is an example of highlighting the inclined surface to be removed. Right now it is still dashed so it hasn't been officially selected with a data point. Once the data point selects this inclined plane you have the choice to identify additional faces or data point (off the model) to Accept.

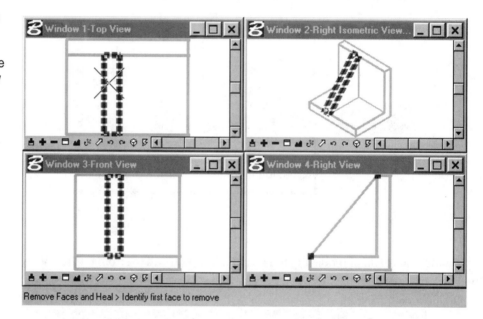

Shown below is the resulting model that has had a single face removed and then been "healed."

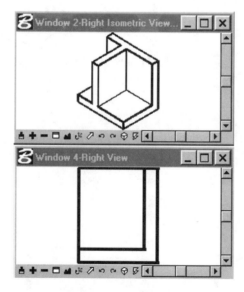

CUT SOLID

The Cut Solid uses a profile to cut into a model. This profile can be a closed shape or it can be a linestring. The cut can go all the way through the solid or you can specify a specific depth. Just adjust the settings as necessary.

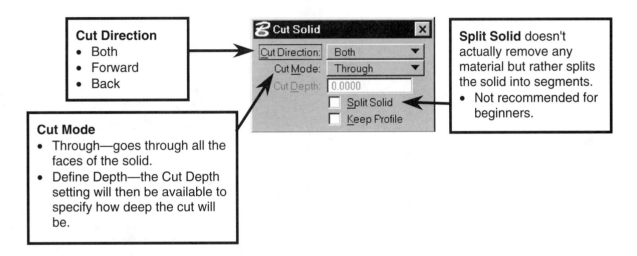

Cut Direction
- Both
- Forward
- Back

Cut Mode
- Through—goes through all the faces of the solid.
- Define Depth—the Cut Depth setting will then be available to specify how deep the cut will be.

Split Solid doesn't actually remove any material but rather splits the solid into segments.
- Not recommended for beginners.

For Example: Let's make a hexagon-shaped cut through the base of the brace. A hexagon polygon was created on the top face of the brace as shown below. The solid is already highlighted because it has already been identified. Now the cutting profile needs to be identified as shown by the prompt in the status bar.

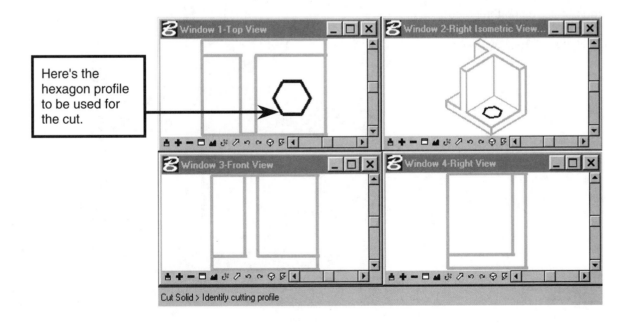

Here's the hexagon profile to be used for the cut.

Once the cutting profile has been identified, arrows appear indicating the direction of the cuts. If you weren't cutting all the way through the model then the cut depth would need to be defined at this time.

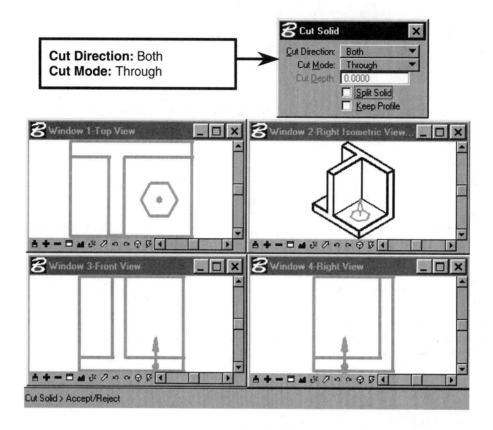

Shown at the right is the result from this Cut Solid operation. There is now a hexagon "hole" going through the base. The hidden lines in the Right View show it has been cut all the way through.

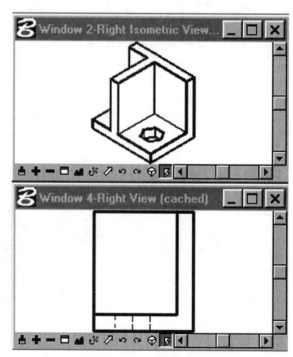

Now is a good time to go back and look at removing the cut just made. This is a prime example of the Cut and Heal Solid tool's Method **Logical Groups** setting.

This is the individual face that highlights (dashed) when the cursor moves in that area.

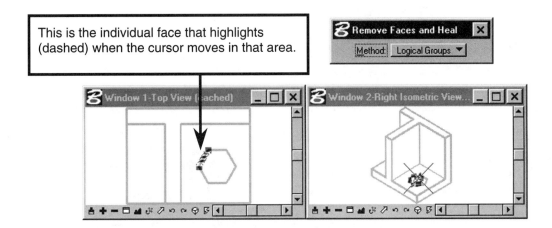

When a data point is actually done to identify the single face—**ALL** of the faces associated with the hexagon cut are selected.
- This is because the Method is set to Logical Groups.
- If the Method was set to Faces then each face would need to be selected—very time consuming.

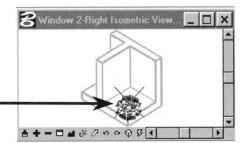

Here's the result of using the Remove Faces and Heal tool on the cut we made earlier.

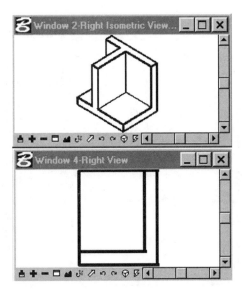

FILLET EDGES

 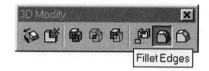

The Fillet Edges tool allows you to round off edges of the solid. You can do more than one edge at a time. If you specify an improper Radius the prompt will let you know that you need to try again!

Radius set to 0.2500
Select Tangent Edges checked ON—if edge selected curves, then all segments will be filleted.

Three edges have already been selected.

Here's the result of the Fillet Edges tool.

Lines at tangent points will be displayed (even if it doesn't follow standard graphics convention).

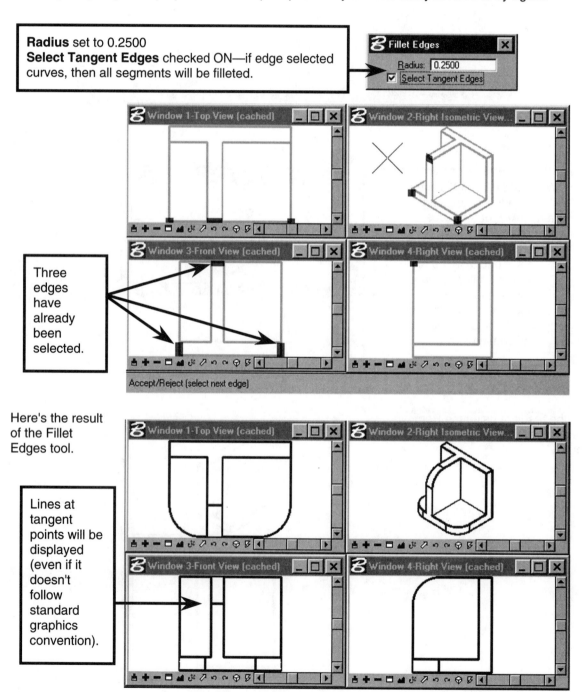

CHAMFER EDGES

The Chamfer Edges tool cuts off the edges selected with a straight cut rather than curved. There are two distances involved so if they aren't equal then there is a bit of trial and error involved. First, you try it with Flip Direction unchecked (OFF). If it isn't the right way, then use Undo, make Flip Direction checked ON, and try it again!

- **Distance 1** set to 0.5000
- **Distance 2** set to 2.000
- **Flip Direction** unchecked (OFF)

Single Edge selected.

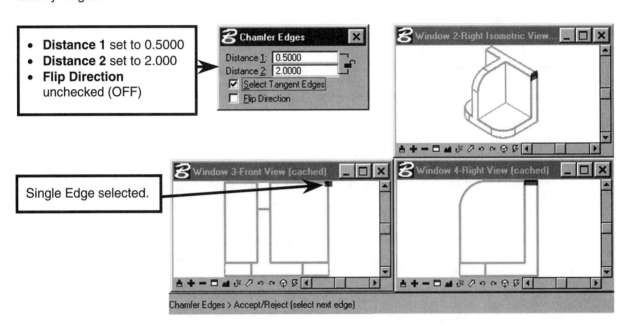

Here's the result of the Chamfer Edges tool with Flip Direction unchecked (OFF)

If you don't like it, use the Undo tool and...

Set Flip Direction to checked ON and do the Chamfer Edges tool again!

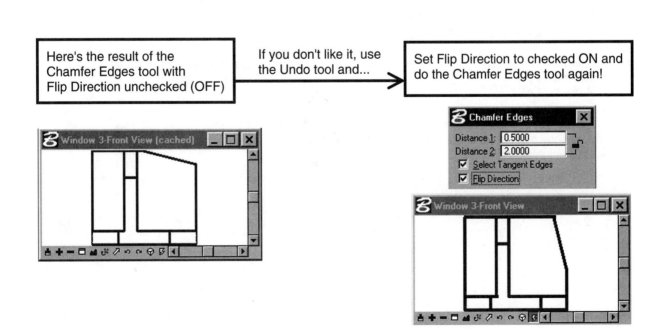

QUESTIONS

① Explain the difference between an Extrude and a Revolution.

② What type of element works well for the profile needed to construct an extrusion or revolution?

③ Given this Front View profile, sketch a pictorial that would illustrate the solid made by the following:
   a) Extrude.
   b) Revolution with Axis set to Drawing Z.
   c) Revolution with Axis set to Drawing X.

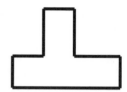

④ Sketch the Front and Right side views that result from the Extrude tool on this equilateral triangular Front View profile with the settings as shown on the right.

   a) With the data point at position A and going in the positive Drawing Y direction.
   b) With the data point at position B and going in the positive Drawing Y direction.

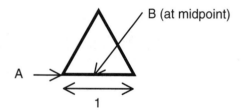

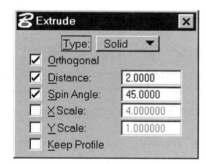

⑤ What does it mean to "heal" a solid?

⑥ Explain why there is a Flip Direction setting for the Chamfer Edges tool and use a sketch to illustrate your answer.

⑦ Sketch the resulting right side view of using the Fillet Edges tool on this model shown in a right isometric. The edges have already been selected and are highlighted.

EXERCISE 20-1   Pipe Support

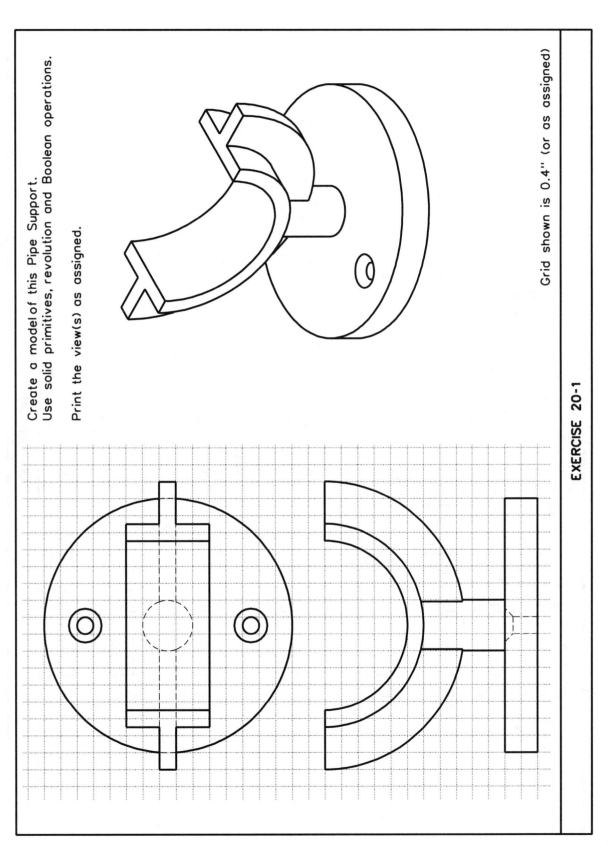

EXERCISE 20-2 Pipe Holder

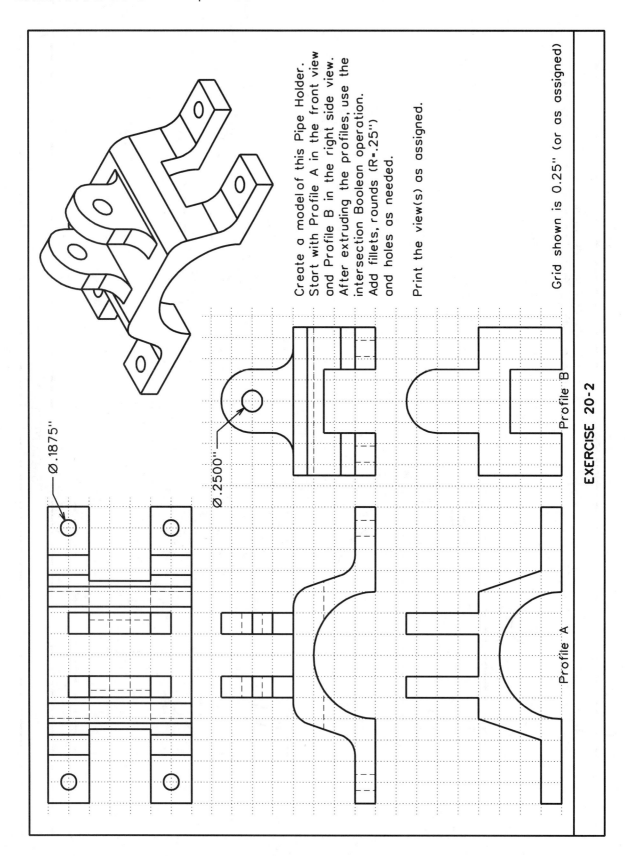

EXERCISE 20-3   L-Bracket

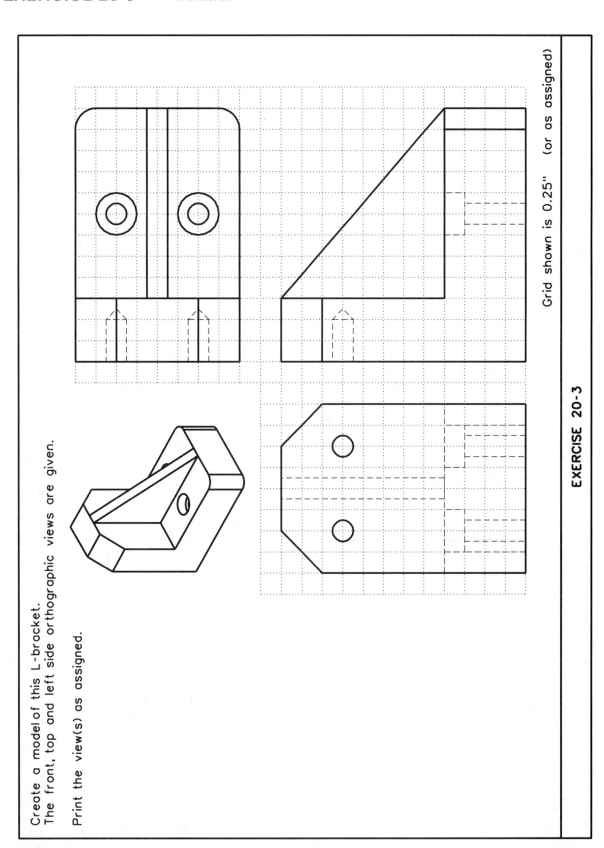

EXERCISE 20-3

EXERCISE 20-4   Slotted Object

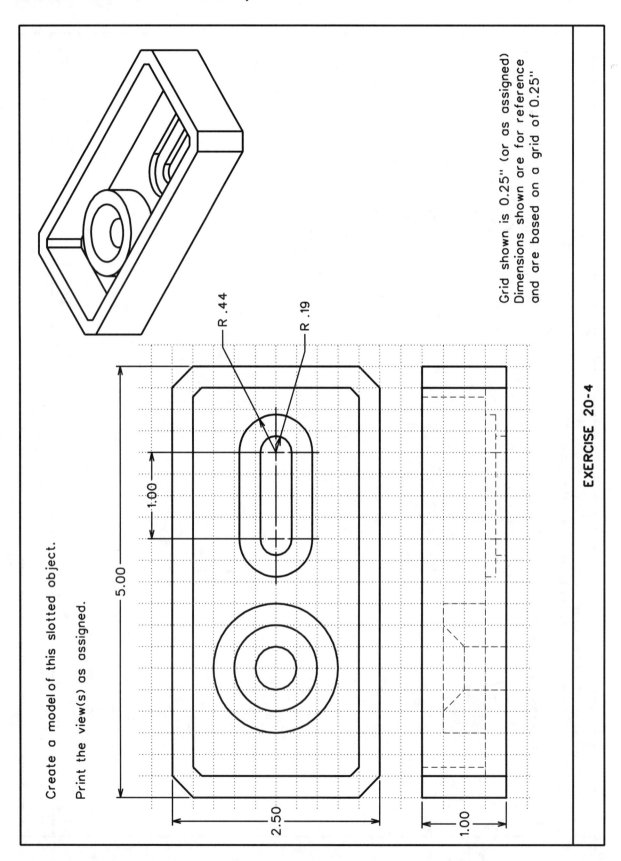

EXERCISE 20-5 Light Base and Globe

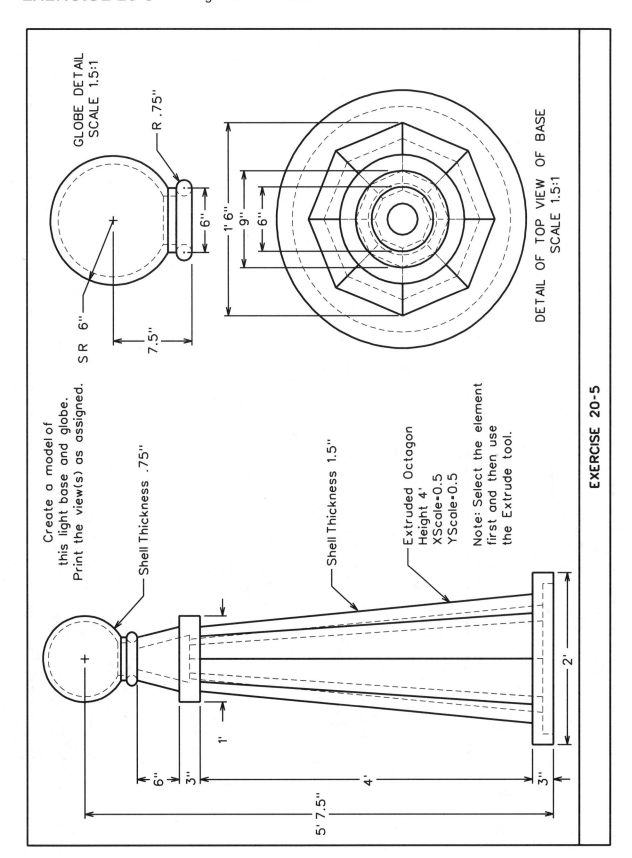

EXERCISE 20-6   T-Beam Bridge and Railing

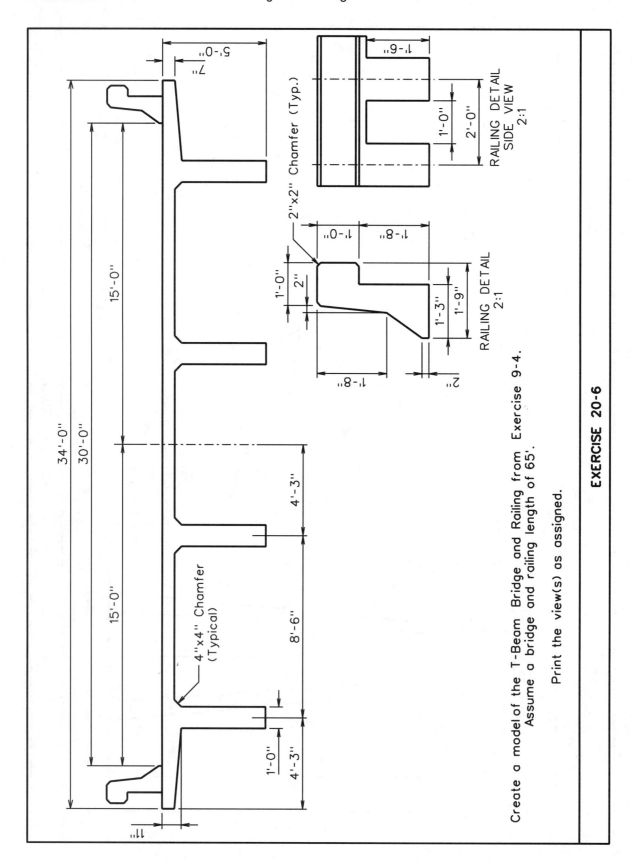

Chapter 21: Drawing Composition

SUBJECTS COVERED:

- General Sequence
- Drawing Composition Dialog Box
- Sheet View
- Attach Border
- Attaching Views
- Attach Saved View
- Attach Standard
- Attach Folded
- Example of Establishing a Sheet View Layout
- Modification of Existing Attached Views
- Example of Modifying a Sheet View Layout
- Dimensions and Text
- Printing
- Sectioned View
- Generate Section Dialog Box
- Section by View
- Sectioned View in a Sheet Layout

Up to this point, we've been limited to printing just what we see in one view. You've been able to print a top, right, or front view or a pictorial but not at the same time on the same sheet. Drawing Composition uses reference files and the ability of a design file to reference itself to get more than one view onto a single sheet view. For 2D drawings, Drawing Composition is generally used to lay out saved views onto a single sheet view. For 3D models, Drawing Composition is really a great advantage. It allows you to show the standard orthographic views plus a pictorial on a sheet view **in proper alignment with consistent dimensions.** This sheet view can also contain text or dimensions to annotate your model. Since Drawing Composition is so advantageous when used in conjunction with 3D models, the following discussion will focus on that area. Its use with 2D design files will be pointed out, but not with as much detail. Because Drawing Composition uses existing elements and models, we need to discuss not only its tools but also the sequence in which to do things. You can't attach a saved view unless you've already saved it! Planning ahead and doing things in a sequential manner will make using Drawing Composition easier.

GENERAL SEQUENCE

Sequence for 2D Design Files

Here is a general guide to the steps to follow when using Drawing Composition with a 2D file. As stated previously, we'll be focusing on using Drawing Composition with a 3D model. However, material that is important to 2D files will be pointed out as needed. The steps of opening a sheet view and attaching a border are the same for both 2D and 3D. It is during the step of attaching views that 2D and 3D will distinctly vary. Some of these deal with saved views and you may think, "I don't even know what a saved view is!" Don't panic—we'll cover how to create a saved view right before learning how to attach a saved view, just in the nick of time. The other steps will be discussed while we're looking at Drawing Composition and a 3D model.

Here's the sequence for 2D design files:
1. Make saved views that have display and attributes as needed.
2. Open a sheet view.
3. Attach an existing file to use as a border (optional).
4. Attach saved views.
5. Add text or dimensions.
6. Print the sheet view.

Sequence for 3D Models

The steps for using Drawing Composition in conjunction with a 3D model are similar to those used with 2D design files. However, when it comes to attaching views of the model, the Standard views and Folded view options are well suited for getting orthographic and pictorial view layouts directly from the model. Since Front, Top, Back, Right Isometric, etc., are already established as Standard views, you don't have to go through the step of setting up and saving these views.

Here's the sequence for 3D models:
1. Create model.
2. Open a sheet view.
3. Attach an existing file to use as a border (optional).
4. Attach views of the model.
5. Add text or dimensions.
6. Print the sheet view.

DRAWING COMPOSITION DIALOG BOX

The Drawing Composition dialog box is opened by going to the main pull-down menu under FILE>DRAWING COMPOSITION. It has different Parameters sections. These contain the settings that need to be adjusted according to your needs. The Drawing Composition dialog box illustrated below on the right also has its very own pull-down menus of File, Tools, and Settings. There are quite a few tools and parameters to be considered here.

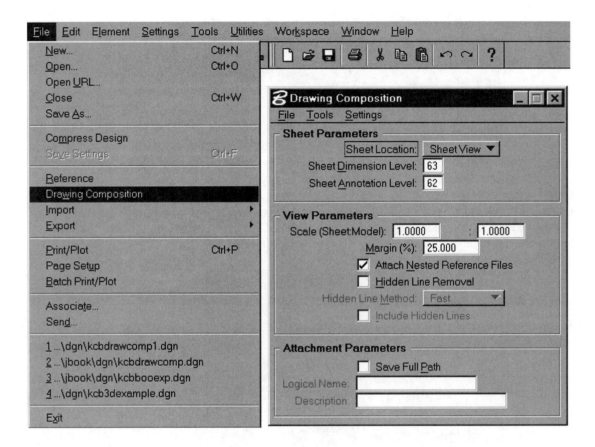

Here are some of the pull-down tools available in the Drawing Composition dialog box. The tools should look similar to those from the Reference Files dialog box. That's because Drawing Composition is just a handy method for using reference files techniques to attach various views of the same model onto a single sheet view. We won't delve deeply into all of these tools. Instead we'll discuss the ones necessary to get started, and then you can experiment with some of the others on your own.

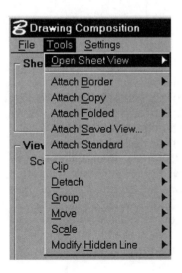

SHEET VIEW

A sheet view isn't the same as a normal window view. Drawing Composition first needs a sheet view before you can attach views (a process similar to attaching reference files). When a sheet view is opened, its special characteristics will be based on the Sheet Parameters that have been set in the Drawing Composition dialog box. The default Sheet Parameters that are shown below are fine for beginners working with either 2D or 3D. If you choose to change the parameters, you should do so **before** opening the Sheet View.

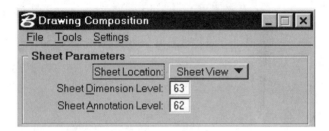

Open Sheet View

Once the Sheet Parameters have been set, then the sheet view is ready to be opened. This is done by going to the Drawing Composition dialog box's pull-down TOOLS menu. Use Open Sheet View as illustrated below. By habit, usually the sheet view is opened in Window 8.

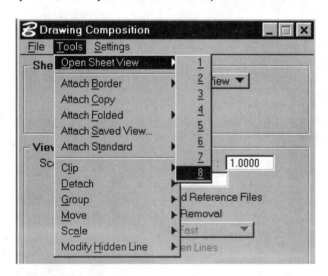

Before the window is actually opened as a sheet view, there will be an Alert as shown below. It is highly recommended that you **ANSWER YES TO THIS ALERT.** Selecting **Yes** to this alert will automatically turn the levels off as noted. Otherwise you can get unwanted results when you go to dimension or put notes in the sheet view. If you say Yes to the Alert, then dimensions will appear only in the view they refer to. They won't show up in the other views. For instance, a dimension placed in a Top view of a 3D model won't show up on edge in the Front view.

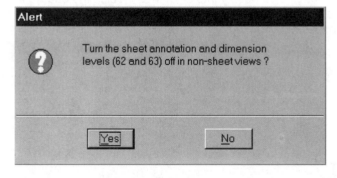

Drawing Composition 495

Once you've opened the sheet view, look around in the graphic area for Window 8 with the title Sheet View. There won't be anything displayed in the window (until you attach a border or views), but it should be there. Usually you will want to maximize the window so that you can see a larger area.

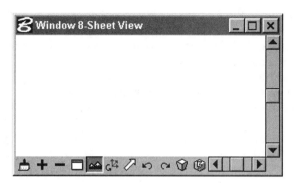

ATTACH BORDER

If you have a standard border that you used for a printed hard copy, then that can be attached now. Attaching a border is not required—you can skip this step and still attach views to the sheet view. For instance, models are created full size, but attached views of the model can come in at different scales. An attached border would allow you to easily lay out the attached views so that the sheet view prints out according to your specifications. If you have a border for an 8½" x 11" piece of paper, then by attaching it first, you'd be able to adjust your attached views to fit appropriately onto the actual paper.

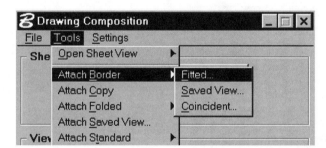

Attach Border Fitted is easiest for beginners. The Saved View and Coincident are for advanced users or when you do this process many times.

Once you've clicked on Fitted, you will get the Attach Border File dialog box as shown at the right. Some standard borders are found in the directory shown. You may use one of these or your own.

Clicking on OK will close out this dialog box. You will still need to data point in the sheet view to actually define where the border should be placed in the view. Remember to read the status bar prompt.

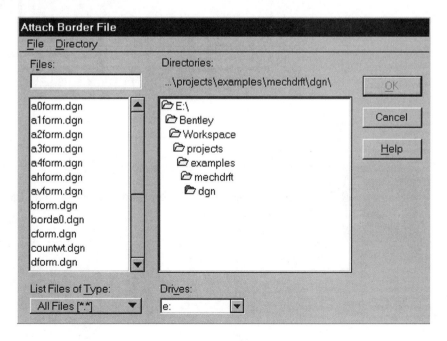

Shown at the right is how it looks after you've specified which border to attach. It is prompting you to Identify view center. The crosshairs indicate where the center of the border is (you won't see the actual border until the attachment is complete). This center can be specified with a data point somewhere within your sheet view; we're not going to be too picky right now.

If you data point to identify the center but still don't see the border—do a Fit View. Since we're working with reference files, make sure you use the Files:All setting as shown below.

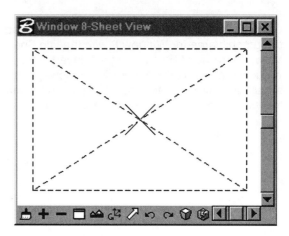

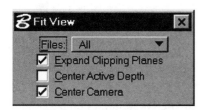

Shown below is the border that will be used for our examples. It fills the view because a Fit View was done. Now when the views are attached, it will be easier to adjust their scale to be sure that the layout fits within the necessary border, which can then be printed to scale for the hard copy.

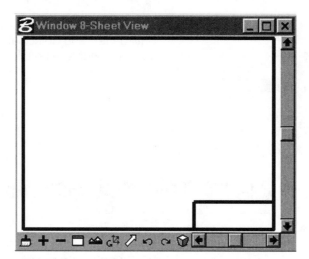

 The border needs to be in the drawing X-Y plane. You can use a 2D design file border for a 3D model.

ATTACHING VIEWS

Now we're almost ready to start attaching views. It is important to remember that each attachment will be made according to the View Parameters that are set in the Drawing Composition dialog box. Make any necessary changes to these parameters **before** attaching the views. Some characteristics of the view attachment can be changed after the attachment, but it is more difficult. Each view can have its own set of parameters—they don't have to be the same for all views. This is a great feature, especially for 3D models, because it allows you to show pictorials without hidden lines on the same sheet as standard orthographic views that do display hidden lines.

<u>View Parameters</u>

These are the default View Parameters. The one that is usually changed for 3D models is the Hidden Line Removal. If left unchecked (OFF) as shown then the views of the 3D model will come in as wireframe. If checked ON then there are different options for Hidden Line Method to choose from and you can choose to Include Hidden Lines or not.

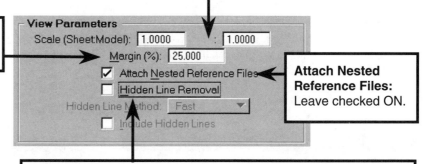

Scale (Sheet:Model): This ratio will need to be changed if you want the attached view to come in scaled up or down from the real-size elements.
• Not limited to integer values.
• 1:2 <u>or</u> 0.5:1 would have the attached views at half-size.

Margin (%): This can be left set to 25.000.

Attach Nested Reference Files: Leave checked ON.

Hidden Line Removal
• **Unchecked (OFF)** means that the model will not undergo an evaluation for visible and hidden edges. This default is usually changed for 3D models so that it doesn't appear as a wireframe.
• **Checked ON** will make the settings Hidden Line Method and Include Hidden Lines available.

For 2D design files, note that the Scale (Sheet:Model) could be thought of as (Sheet:Saved View) since there isn't necessarily a model being used.

It is also recommended that for **2D** design files, you leave the Hidden Line Removal unchecked (OFF). You can then skip the following section covering Incremental Hidden Line Removal and jump right to Attach Saved Views.

Hidden Line Removal and Method

Hidden Line Removal is an evaluation of the visible and hidden surfaces of the model. This evaluation can be turned on or off for each attached view (by toggling the Hidden Line Removal setting). There are different algorithms that can be used in this evaluation process. Some are more extensive than others and give better results (but take more hardware and time). The Hidden Line Method setting refers to which algorithm will be used. There are three options: Exact, Fast, and SmartSolids. Each attached view can have hidden lines included or not on an individual basis—pictorials don't have to include hidden lines but orthographics can. When Hidden Line Removal is checked ON the Hidden Line Method and Include Hidden Lines settings are available as shown below. Just a reminder that it is easier to set the View Parameters <u>before</u> making the view attachment than it is to modify the settings later.

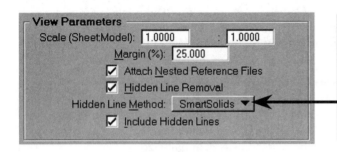

Hidden Line Method:
- **Exact**—midrange evaluation.
- **Fast**—default and fastest but not as good as Exact.
- **SmartSolids**—preferred and will be used for examples in this text.

Hidden Line Settings Dialog Box

Under the Drawing Composition pull-down menu SETTING>HIDDEN LINE SETTINGS you can bring up the Hidden Line Settings dialog box as illustrated below on the right. Most of the settings here can be left as the defaults (especially for beginners) but you may want to change the Hidden Edge Overrides. These control the attributes of the hidden lines shown (**if** Hidden Line Removal is checked ON **and** Include Hidden Lines is checked ON).

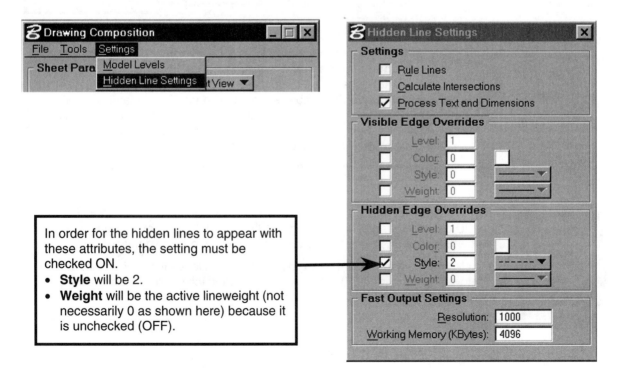

In order for the hidden lines to appear with these attributes, the setting must be checked ON.
- **Style** will be 2.
- **Weight** will be the active lineweight (not necessarily 0 as shown here) because it is unchecked (OFF).

Attachment Parameters

The Attachment Parameters section of the Drawing Composition dialog box deals with the reference file attachments that are being used to attach views. It will automatically assign the reference file a Logical Name. Illustrated below is the Logical Name setting that was assigned when an Attach Standard Top view was done. It named the attachment Top for you—isn't that nice? These parameters can left alone for now. As you gain more experience, then you may want to start controlling the Attachment parameters, but for now we'll let them set automatically.

Save Full Path It is recommended that this setting be checked ON.

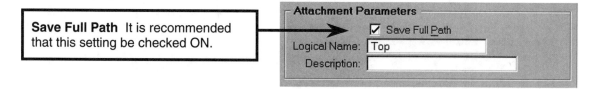

ATTACH SAVED VIEW

Using Drawing Composition with 2D design files generally requires knowing how to use the utility called Saved Views. Saving a view allows you to name a view and save its display and attributes. These saved views can then be attached in order to form a layout in a sheet view. These views can come in at different scales, but the elements in the views retain their precise size and location. Before getting into Drawing Composition, we need to learn how to use the Saved Views utility.

Saved Views

In looking at how to use the Saved Views utility, we'll revisit the columns 2D design file from Chapter 15 dealing with reference files. It is illustrated below. Window 1 on the left is the layout of the columns. Window 2 on the right is a view that results from zooming in on one of the columns so that the detail of the reinforcement is shown more clearly. Each of these views will become a saved view.

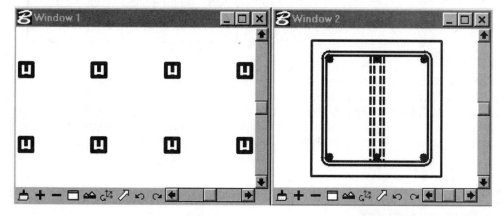

The Saved Views utility is accessed under the main pull-down menu UTILITIES>SAVED VIEWS as shown on the right.

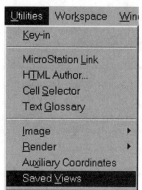

This will bring up the Saved Views dialog box illustrated below. This is where you specify the view that is to be saved (source). The other tools allow you to use the saved views in different ways other than Drawing Composition. We'll just look at how to get the view saved, which is all we need for Drawing Composition.

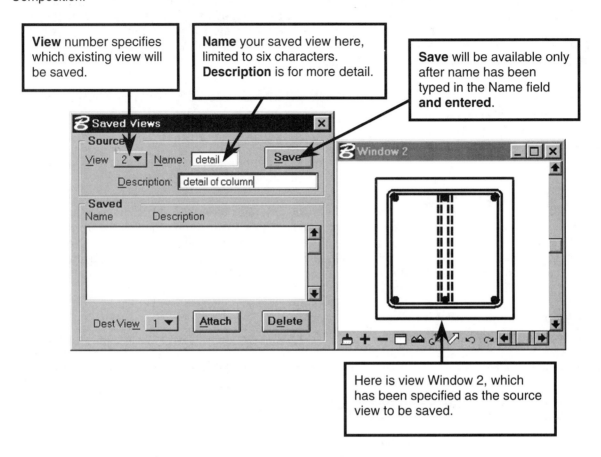

View number specifies which existing view will be saved.

Name your saved view here, limited to six characters. **Description** is for more detail.

Save will be available only after name has been typed in the Name field **and entered**.

Here is view Window 2, which has been specified as the source view to be saved.

Beginners are often frustrated because they will type in the Name field, and the Save button still won't be available. The Save button will be available only once the name has actually been entered. If you still have a blinking cursor in the Name field, it is still waiting for more input. Use the ENTER key to actually finish the naming process. Bringing focus to a different field (by clicking in a different field or using the TAB to move to the next field) will also complete the entering of the name. This concept is illustrated below.

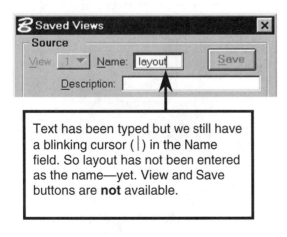

Text has been typed but we still have a blinking cursor (|) in the Name field. So layout has not been entered as the name—yet. View and Save buttons are **not** available.

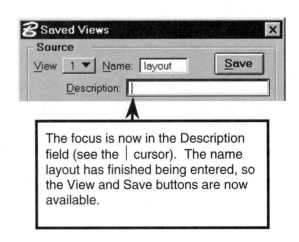

The focus is now in the Description field (see the | cursor). The name layout has finished being entered, so the View and Save buttons are now available.

Once the Save button is available, then clicking on it will complete the task of making a saved view. Shown at the right are the two saved views that have been made and will be available for use in Drawing Composition.

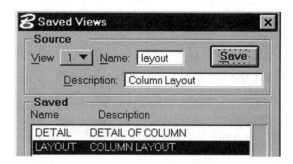

Accessing Attach Saved View

Now that we have some saved views, let's look at how to attach them using Drawing Composition. The Attach Saved View is accessed from the Drawing Composition dialog box's Tools pull-down menu as illustrated below on the left. This will open the Select Saved View box illustrated on the right.

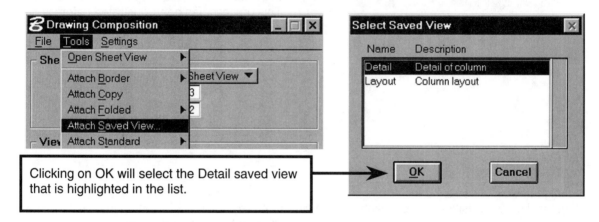

Clicking on OK will select the Detail saved view that is highlighted in the list.

Once the Detail saved view has been selected, you'll be prompted to Identify view center as illustrated below on the left. The view center is identified with a data point **in the sheet view**, and the Detail saved view appears as illustrated below on the right.

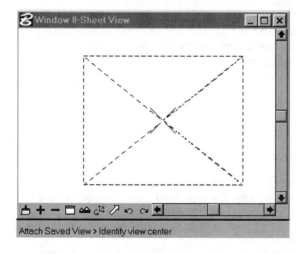
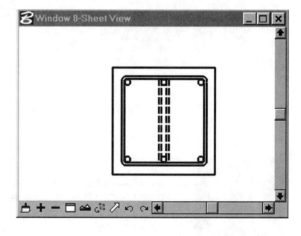

That's all we'll look at for attaching saved views of a 2D design file. The sheet view can be annotated and printed out. This will be discussed later in regard to a 3D model, but it can also be applied to 2D.

The Attach Standard and Attach Folded methods of attaching views are related to engineering graphics concepts of orthographic, auxiliary, and isometric views. Using Drawing Composition and these methods can establish a proper sheet view layout from a 3D model. The methods will be discussed and then an example will be done that illustrates their use in establishing a proper sheet view layout of a 3D model.

ATTACH STANDARD

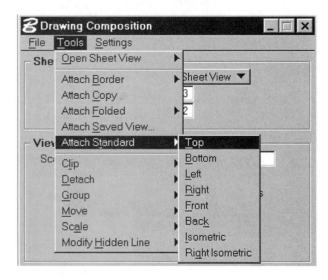

Here are the Standard views available for attachment. They are the same as the options available for View Rotation. You have the standard orthographic views as well as the isometric views. Click on the view you want and then you will be asked to identify the view center. This process has been illustrated before when we were attaching a border or saved view.

ATTACH FOLDED

This option utilizes the engineering graphics concept that any two adjacent views are perpendicular to each other. There is a folding line between the two views. If this folding line is orthogonal on the paper (horizontal or vertical), then you are working between the standard orthographic views. If the folding line isn't orthogonal but instead lies at an angle, then it is referred to as an auxiliary view. The two Attach Folded options, Orthogonal and About Line, allow you to achieve both orthographic and auxiliary views. **By using Attach Folded, the alignment between the two adjacent views is maintained. This is very important.**

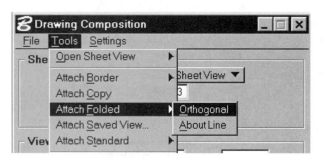

 In order to use Attach Folded, there first needs to be a view to fold from! It is recommended that you use Attach Standard for the first view placed in your sheet view. The rest of the orthographic or auxiliary views can then be done using Attach Folded.

The process of attaching a folded view requires that you first identify the principal attachment. This is the **existing** attached view that you want to fold from. This identification is done with a data point. Once the view has been identified, it needs to know where the folding line is. For the Orthogonal option, you data point on one of the four existing horizontal or vertical edges of the view selected. Then you will be prompted to specify the view center; this will determine the spacing between the two adjacent views.

When using the About Line option of Attach Folded, after identifying the principal attachment, the two endpoints of the folding line will each need to be given (usually graphically with a data point). The location of this folding line will determine the orientation of the auxiliary view. It also determines the spacing between the two adjacent views, so you will not be asked to identify the view center.

EXAMPLE OF ESTABLISHING A SHEET VIEW LAYOUT

Let's look at getting a top and front view of an object along with a pictorial view. We've already seen how to Attach Border, so we'll use that border and put our attached views within its boundaries. The plan of attack is to Attach Standard Front and then Attach Folded Orthogonal for the top view. After modifying the View Parameters so that Hidden Lines are not included, an Attach Standard Right Isometric will then be done. Here's a look at the model we'll be using. After illustrating how to establish the front, top, and isometric pictorial layout, we'll look at getting an auxiliary view showing the true size of the inclined plane.

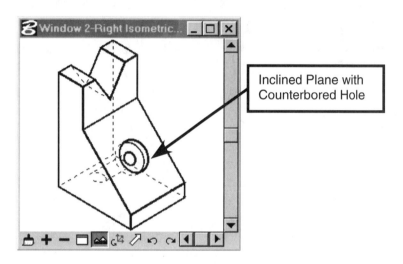

Inclined Plane with Counterbored Hole

Here is the first sheet layout that we want to establish.

- Aligned front and top views with hidden lines displayed.

- Half-size right isometric pictorial without hidden lines.

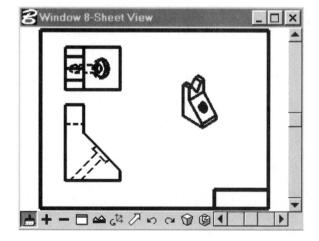

The Drawing Composition parameters need to be set before attaching the views. Settings in the Drawing Composition parameters handle the evaluation of the model and whether or not to display hidden lines in the attached views. How hidden lines are displayed is determined by the settings in the Hidden Lines dialog box (accessed by the Drawing Composition's pull-down menu SETTINGS>HIDDEN LINE SETTINGS.)

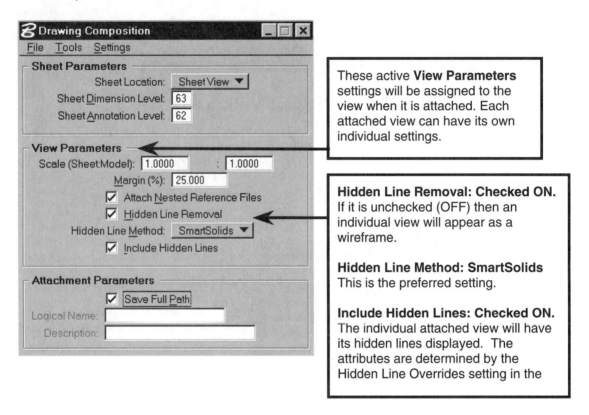

These active **View Parameters** settings will be assigned to the view when it is attached. Each attached view can have its own individual settings.

Hidden Line Removal: Checked ON. If it is unchecked (OFF) then an individual view will appear as a wireframe.

Hidden Line Method: SmartSolids This is the preferred setting.

Include Hidden Lines: Checked ON. The individual attached view will have its hidden lines displayed. The attributes are determined by the Hidden Line Overrides setting in the

What happens if you forget to change the settings dealing with hidden lines before you make the view attachment? Don't worry—there's a Modify Hidden Line tool that will save the day. It's covered later on in this chapter.

Attach Standard View—Front

With the View Parameters set as illustrated on the previous page, we can now attach our front view. Using the Tools pull-down menu from the Drawing Composition dialog box, select the Attach Standard Front option as illustrated at the right. You will be prompted to Identify view center as illustrated below.

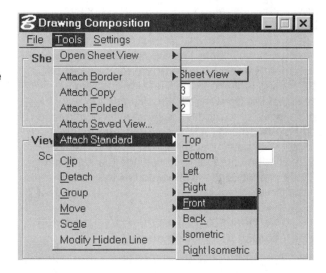

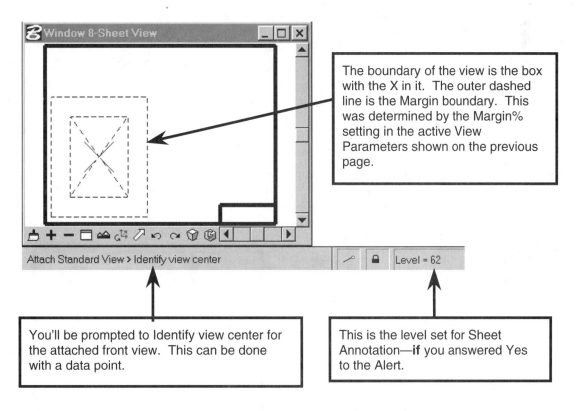

The boundary of the view is the box with the X in it. The outer dashed line is the Margin boundary. This was determined by the Margin% setting in the active View Parameters shown on the previous page.

You'll be prompted to Identify view center for the attached front view. This can be done with a data point.

This is the level set for Sheet Annotation—**if** you answered Yes to the Alert.

Once the data point has identified the placement of the view center, the front view of the model will show up as illustrated to the right. The Logical Name of the attached view will be Front.

If your front view looks like a wireframe, you forgot to change **Hidden Line Removal to checked ON.** This setting is found in the View Parameters section of the Drawing Composition dialog box. There is a way to modify the hidden line evaluation and display setting. It'll be covered later in the chapter. For now, use Undo, make the corrections to the Hidden Line Removal settings (be sure that Include Hidden Lines is checked ON too) and attach the view again.

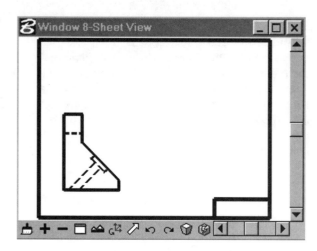

Attach Folded View—Orthogonal

Using the same Drawing Composition parameters and the same Hidden Line settings, the Top view is going to be folded off of the Front view. If at this point you just attach another standard view, it would be tough to get them aligned correctly.

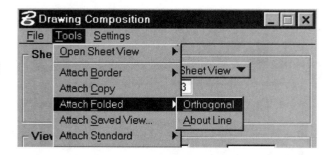

The Front View was identified (with a data point) as the principal attachment. Now it needs to be accepted with a data point at this location for a **Top** View to be folded.

Once the fold line has been specified, the Top view spacing can be set. The cursor can move anywhere, but the view (indicated by dashed lines) maintains its alignment with the Front View.

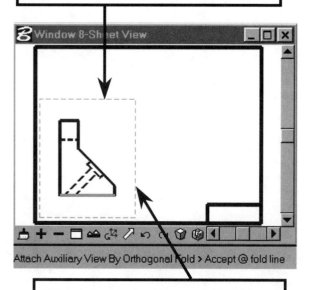

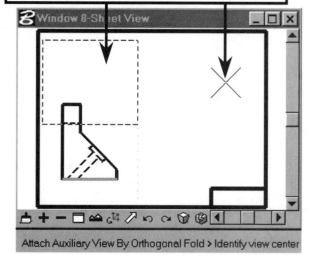

If the data point had been on this fold line, it would fold a Right view.

Drawing Composition 507

Once the top has been folded from the front, you can still specify another view to fold off of the Front view. Illustrated below on the left is the Top view already attached but the fold lines still shown. A data point on the right fold line will result in the illustration below on the right—dashed lines indicating the placement of the Right view (still maintaining alignment.)

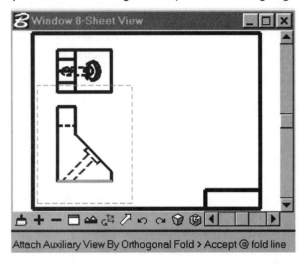
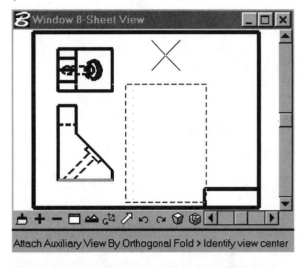

Here is the resulting front and top orthographic view layout in the sheet view. There are options to move an attachment after it has been placed. These will be discussed later on in this chapter.

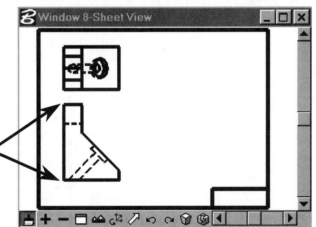

Notice that the attached Top and Front views **maintain the conventions of consistent sizes and mutual alignment.** This is due to folding one view from the other.

Attach Standard View—Right Isometric

With this attachment, the **View Parameters** of Drawing Composition will be changed—before we make the attachment! Pictorials usually don't include hidden lines, so the Include Hidden Lines setting will be unchecked (OFF). **The Hidden Line Removal will remain checked ON.** The ratio value for the Scale (Sheet:Model) of the attachment will be changed so the Right Isometric view is attached as half-size.

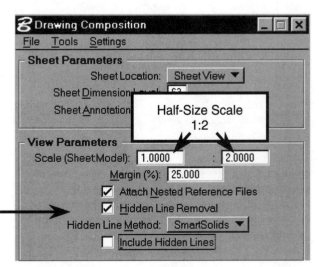

Hidden Line Removal: Checked ON

Include Hidden Lines: Unchecked (OFF)

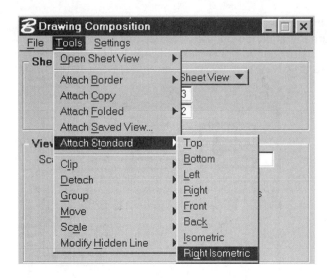

This time we'll use Attach Standard but choose the Right Isometric option as illustrated on the left.

The result of the attachment of the Right Isometric view is as shown on the right. The sheet view in Window 8 could now be printed. You would have the orthographic and pictorial layout as shown.

Shown below on the left is the Right Isometric view (indicated by the dashes) being placed with a data point. The result of the attachment of the Right Isometric view is shown below on the right. The Window 8-Sheet View could now be printed. You would have the orthographic views (with hidden lines) and the pictorial (without hidden lines) layout as shown.

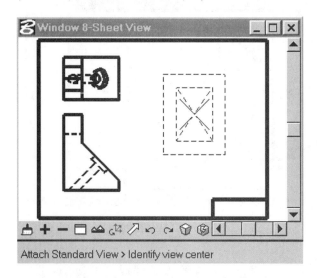
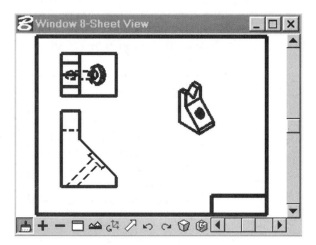

Editing the Model in the Sheet View

The attached views that you see are directly linked to the model itself. This means that you can make changes to the model in the sheet view—you don't have to go back to the regular view windows. Any changes made to the model will be reflected in all the views. Very handy indeed, except it is a bit slower to work in the attached views of the sheet view layout.

Drawing in the Sheet View

You can also draw 2D elements (such as lines) in the sheet view. Text can also be placed just by drawing in the sheet view. Dimensioning the model in the sheet view will be covered in detail later.

MODIFICATION OF EXISTING ATTACHED VIEWS

There will be times when you need to modify the existing view attachments. The view placement doesn't always fit nicely, so it may need to be moved. It may also be necessary to get rid of an attached view. Just like reference file attachments, the attached views can be modified and manipulated. These options are found in the Drawing Composition's pull-down menu Tools. The tools dealing with attached views are grouped together and illustrated below. The ones to consider for now for our 3D model are Detach, Move, and Modify Hidden Line.

Each of these options has the choices of Single, Group, or All:
- **Single** will do just one view.
- Attach Folded will **Group** the two views involved together.
- **All** will affect all of the attached views.

After specifying what you want to choose, you will need to identify the single view or group with a data point. If you choose All, then it already knows what views to modify—all of them!

Use these tools to modify the attachments. These views are reference files attachments and must be treated as such. **Do not treat the attached views as elements. Do not use the regular tools in the modification and manipulation tool boxes on an attached view.**

Detach

This will detach any or all of the views. If you've chosen Detach All, an Alert will pop up asking you if you're sure about the detachment. If you are doing Detach Single or Group, then you will be prompted to accept the selection.

 You must **Detach** an unwanted attached view. If you use the Delete Element tool, you will delete your entire model—not wise.

Move

This will move the attachment. Some things to remember when doing a move of an attachment:
- Be careful to maintain alignment when moving an attachment. Use Axis Lock or AccuDraw to be sure that the views stay aligned.
- Move can also be used to get the attached views to be spaced evenly. Again AccuDraw or use of relative coordinate key-ins can be useful here.
- Move is often necessary after using Attach Folded. It may be necessary to specify the fold line by snapping to the model. This results in a very close spacing between the two adjacent views.

Modify Hidden Line

This is for those times (rare as they may be) when you forget to set the Drawing Composition's View Parameters before doing the attachment.

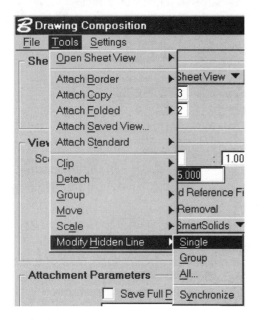

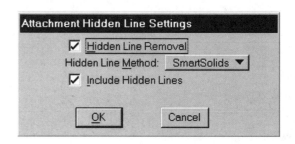

Once the attached views to be modified have been selected, the Attachment Hidden Line Settings dialog box illustrated below will open. Make your changes here, and then use the OK button to accept them.

The Attachment Hidden Line Setting dialog box doesn't help if you've forgotten to set the Hidden Edges Overrides settings the way you wanted them! Remember, they were accessed by the Drawing Composition's pull-down menu SETTINGS>HIDDEN LINE SETTINGS as illustrated at the right. You would need to go in and make the changes to this dialog box first.

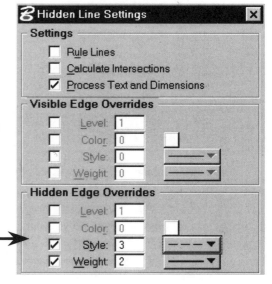

Make the necessary changes to these settings.
- Style changed to 3 and checked ON.
- Weight changed to 2 and checked ON.

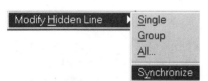

Then you would need to use the TOOLS>MODIFY HIDDEN LINE>SYNCHRONIZE. This will update the hidden lines display to the new settings.

The Synchronize tool may also come in handy if the attached views do not reflect the current state of the model (such as after a modification to the model). This is because the evaluation of what's hidden or not takes time—so the information is calculated once and stored (like a shortcut). Synchronize reevaluates the hidden lines!

Drawing Composition

EXAMPLE OF MODIFYING A SHEET VIEW LAYOUT

We're going to go back and make some modifications to our existing sheet view layout. First we'll detach the existing attached right isometric pictorial. Then an auxiliary view will be established using Attach Folded. This new attached view will then be moved and its hidden line settings modified.

Detach—Single

In the illustration below, the attached view (Right Isometric) has been identified, and the status bar informs you to Accept to detach. This is a good time to read the status bar. **If it says Delete Element rather than Detach Single Attachment, then you're not detaching a reference file; you're deleting the model!** Pay attention to the status bar and what you are doing. To actually complete the detachment, data point within the graphic area.

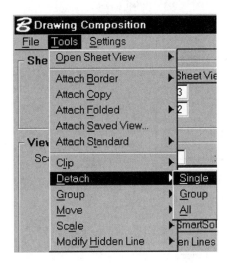 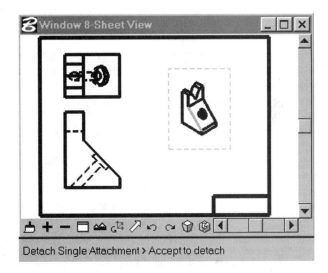

Attach Folded View—About Line

Now an auxiliary view will been done using the About Line option of Attach Folded. In order to see the slanted surface true size, the folding line must be parallel to the slanted surface. In order to achieve this, the model will be snapped to when specifying the endpoints of the fold line.

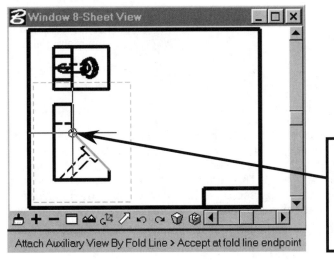 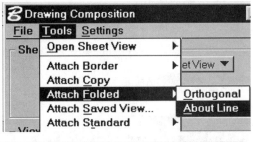

The Front view has already been identified (highlighted gray) and the prompt is to Accept at fold line endpoint. Shown here is the fold line endpoint being specified by snapping to the corner of the slanted surface.

Once the tentative point has been accepted, the second endpoint of the folding line needs to be specified. You can see the folding view's orientation change dynamically by moving the cursor. The first endpoint stays fixed.

Snapping to the endpoint of the slanted line will identify the other fold line endpoint.

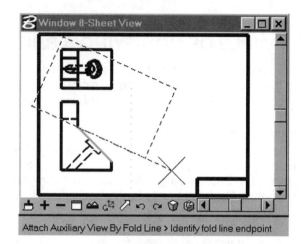

Illustrated on the right is the auxiliary view established by using Attach Folded. It is necessary to modify the attachment by moving it. In order to maintain alignment, the point to move from will be specified on an edge of the model (in the folded view).

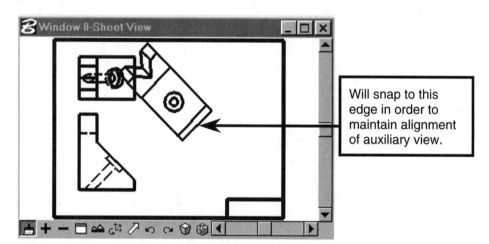

Will snap to this edge in order to maintain alignment of auxiliary view.

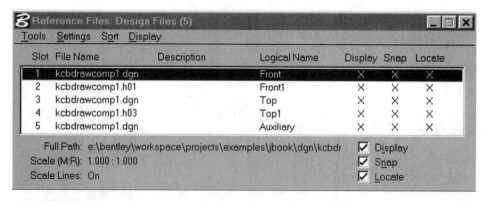

If you are unable to snap to the model in the attached views, then you will need to check the Snap and Locate settings for the reference files. This dialog box is obtained from the main pull-down menu FILE>REFERENCE FILES.

In the Reference Files dialog box, some of the File Name listings have .h01 or .h03 extensions. These files are where the hidden line geometry is stored (their creation, attachment, and detachment are done "behind the scenes" for us). The files have the same name as the main design file. The numbers in the extension represent the attachment number of the view they're associated to. Don't worry about the hidden line geometry files—but if you run across them in your directory, don't delete them.

Move—Single

To initiate the moving of a single attached view, go to the Drawing Composition's pull-down menu TOOLS>MOVE>SINGLE as illustrated on the right.

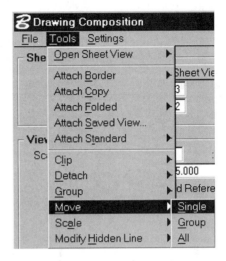

The data point that identifies the view to move also sets the point to move from. In order to maintain alignment, the attached view will be identified by snapping to one end of the edge. The result is illustrated below on the left. The next data point will do two things: Accept the view's identification **and** define the distance as shown below on the right.

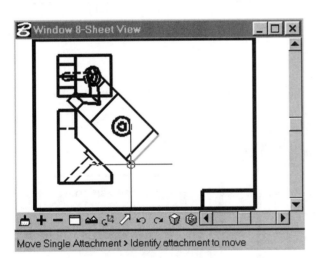
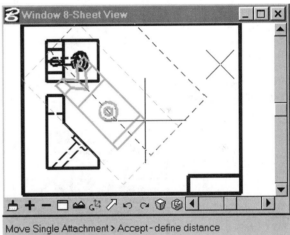

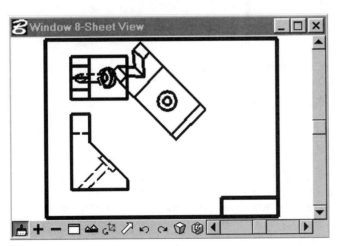

Here's the result of the move—the alignment has been maintained. Unfortunately it wasn't moved far enough, so another move would be needed. Upon closer inspection, you will also notice that hidden lines are not included in the attached folded view (look for the bottom edge of the model). This gives us the opportunity to use Modify Hidden Line.

Modify Hidden Line

This is also initiated by using the Drawing Composition pull-down menu TOOLS>MODIFY HIDDEN LINE. When Single is clicked on (as in this case), the status bar will prompt you to Identify element—this means to data point on the attached view(s). The selected view (or part of it) will highlight as shown and the Attachment Hidden Line Settings dialog box will appear.

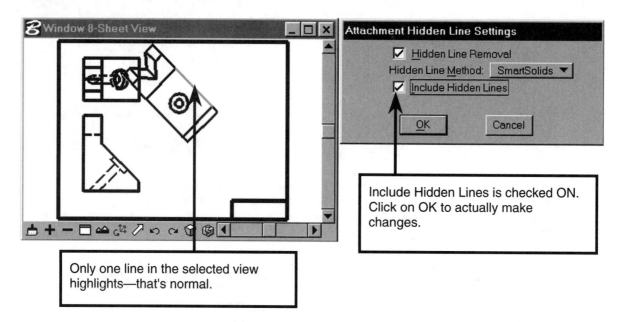

Only one line in the selected view highlights—that's normal.

Include Hidden Lines is checked ON. Click on OK to actually make changes.

After clicking on the OK button, the attached auxiliary view now shows the hidden line for the back corner as illustrated below.

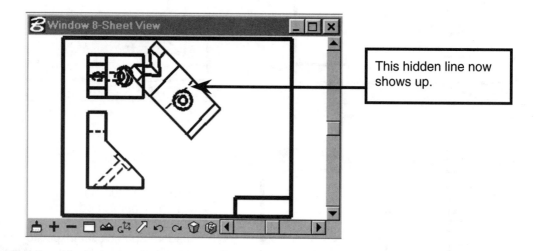

This hidden line now shows up.

It is suggested that Hidden Line Removal be turned off before dimensioning and annotating the sheet view layout. You can use Modify Hidden Line and the Attachment Hidden Lines Settings to do that—use the Group or All option of Modify Hidden Lines to do more than one view at a time! Just make sure you modify them back to Hidden Line Removal checked ON before printing.

DIMENSIONS AND TEXT

You can add dimensions directly to the attached views in the sheet views, in both 2D and 3D cases. You use the regular dimensioning tools found in the Main tool box. For instance, when using a 3D model, by placing dimensions in the sheet view, the software can control their display so that front view dimensions don't show up on edge in the other views. If you place dimensions in the regular view windows, then they are with the model and will end up showing in all the views that the model shows up. Since the attached views are a direct link to the model, associated dimensions placed in the sheet view will reflect changes made to the model. Text elements can also be placed in the sheet view too. Just like with 2D dimensioning, you may need to adjust the dimension setting before doing any dimensioning. The seed files for 3D may have dimension settings that may differ from the seed files for 2D.

Dimensions will come in stated as the actual size of the elements, even if the views were attached at a different scale. No need to adjust the Scale Factor dimension setting.

Levels

When opening the sheet view, the Sheet Parameters in the Drawing Composition dialog box were that Sheet Dimension Level was set to 63 and Sheet Annotation set to 62. You were alerted that sheet annotation and dimension levels (62 and 63) were being turned off in non-sheet views—and it was recommended that you answer Yes to this Alert. These choices result in the software doing some automatic changes to the levels that will be displayed in different views. Pay close attention to the Active Level setting either in Primary tools (recommend that you have it open and docked so you can see the Active Level setting) or in the status bar display. **You will want to have the Active Level be 62 or 63 when placing elements in the sheet view.**

Previous versions of MicroStation had the Active Level change to 62 **automatically** when you worked in the sheet view. Now that doesn't happen, so the Active Level in the sheet view may be Level 1. This can cause a problem because the sheet view is set up to **not** display Level 1. Therefore, you can be drawing an element in the sheet view but it won't show up because its attribute is Level 1 (which is not being displayed). **You should check the Active Level setting before drawing in the sheet view!**

Once again, you can always go to the regular Reference Files dialog box and use the settings there to look at the attachment's level display control.

Here is a dimension that has been placed in the Top view. It has been selected to Change Dimension to Active Settings so that you can see in the status bar that it is being placed on Level 63.

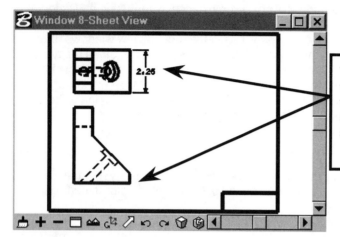

This dimension shows up in the Top view but doesn't show up in the Front view. This is a great feature of the software: it automatically figures out the settings to control to obtain this result.

If you are seeing a single dimension in all of the views, it is usually because you dimensioned in the regular view window with the model and not in the sheet view like you should have.

PRINTING

Once the sheet view layout is what you want, you can print out a hard copy. You print out of sheet view using the same method that you would with a regular view—except this time you'll get all the necessary views of a 3D model all on one sheet. Yes!

SECTIONED VIEW

A sectioned view of an object shows the internal features that would be hidden in standard orthographic views. A planar cut is made through the object, part of the object is removed, and the remaining portion is viewed. Section lining indicates where solid material was cut through, and features that were shown as hidden lines will become visible. The cutting plane runs parallel to the view that shows the sectioning. For instance, a horizontal sectioned view results from a horizontal cutting plane. Once again, there are standard engineering graphics conventions that should be followed, such as including surface limits and excluding hidden lines. MicroStation/J has the ability to take the solid model and generate a section. Using this capability in conjunction with Drawing Composition, you can layout a sheet view that shows a sectioned view along with the orthographics. To get a sectioned view and all the standard orthographics on one sheet, you must be familiar with the ability of a design file to reference itself and adjusting each reference's own individual level display settings. First let's look at the Section Generation dialog box.

GENERATE SECTION DIALOG BOX

When you generate a section, a "cutting plane" orientation is specified that determines where the solid should be "cut." The solid is evaluated and new lines are created at the boundary of the solid material that has been cut. By reorienting the "cutting plane" you can generate different sectioned views. It is then necessary to use Patterning tools to create section lining within these boundary lines.

The Section Generation dialog box is accessed by the main pull-down menu UTILITIES>GENERATE SECTION as illustrated at the right. We won't discuss all of these settings in detail because there are many different ways to utilize Section Generation. Instead we'll focus on the ones that are necessary in order to layout a sheet view with sectioned views and orthographic views. You can experiment with the other options on your own.

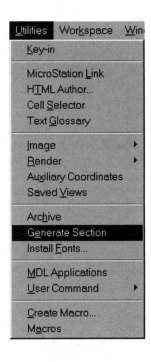

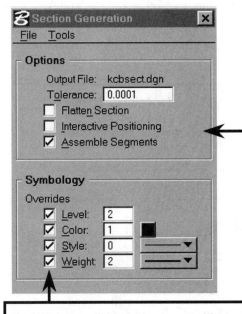

Symbology will have these overrides set as shown.
- Level set to 2 will put the section on a different level (2) than the model (level 1). This is important because level manipulation will be necessary to get the correct sheet view layout.
- Color is set so that the section generated is easier to see—not necessary but helpful.

Output File will default to the active design.
- If you want a different file then you'll need to use the Section Generation File pull-down menu.
- We'll put our section in our active design file—kcbsect.dgn

Tolerance will be left as the default shown.

Flatten Section will be unchecked (OFF). This leaves the section boundary created parallel to the cutting plane.

Interactive Positioning will be unchecked (OFF). The section boundary elements will automatically be located at the position of the cutting plane—"within" the model.

Assemble Segments will be checked ON. The elements that make up the section will be joined together as a complex shape.

SECTION BY VIEW

There are different ways to section the model. These are available In the Section Generation pull-down menu TOOLS as shown at the right. These options can be thought of as different ways to specify the "cutting plane." You can use an Element, Fence, Plane, Projection, or View. Lots of choices, but for beginners the Section by View tool may prove to be the easiest to use. **Section by View lets you cut the model with a plane running Horizontal or Vertical in the view. Therefore, it is important to pay attention to what view you are working in.** Once chosen then all you need to do is specify the **location** of the cut—the Horizontal or Vertical setting already determines its orientation. The Section by View Depth choice isn't as intuitive, so we'll leave that alone.

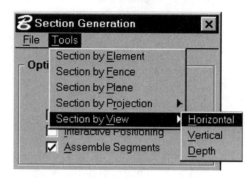

Horizontal

With Section by View–Horizontal, the cutting plane will be running horizontal (straight right or left) **in the view selected**. Working in different views will give you different results. A cutting plane that is horizontal in the Top View has a different orientation from a cutting plane that is horizontal in the Front View.

For Example: With the settings as shown previously (Assemble Segments checked ON and Symbology overrides checked ON) let's select the model to be Sectioned by View–Horizontal in the **Top** View.

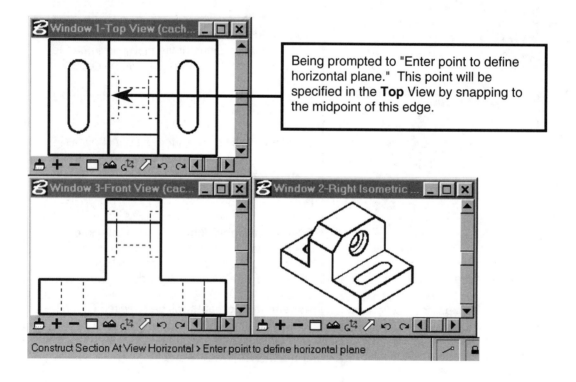

Being prompted to "Enter point to define horizontal plane." This point will be specified in the **Top** View by snapping to the midpoint of this edge.

Once the point to define the location of the cutting plane has been specified, the location of the cut will highlight and you'll need to enter a data point to accept or a RESET to reject as illustrated below.

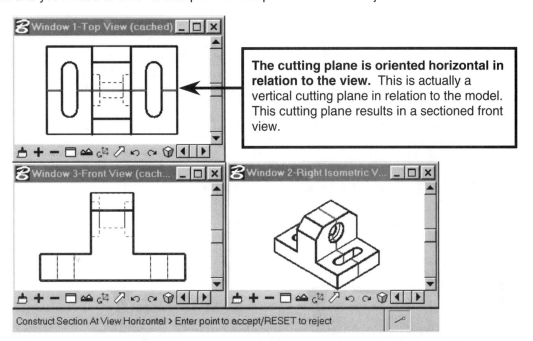

The cutting plane is oriented horizontal in relation to the view. This is actually a vertical cutting plane in relation to the model. This cutting plane results in a sectioned front view.

With the display mode set to Hidden Lines, it may be hard to see the elements that were automatically placed to represent to boundaries of the solid. Also the model itself may get in the way of seeing what section generation actually created. Because we had the Level set to 2 before we generated our section, level control can help us see the section a bit better.

By turning off Level 1 in View Number 3, the Front View now shows the elements (closed shapes) created by the section generation.

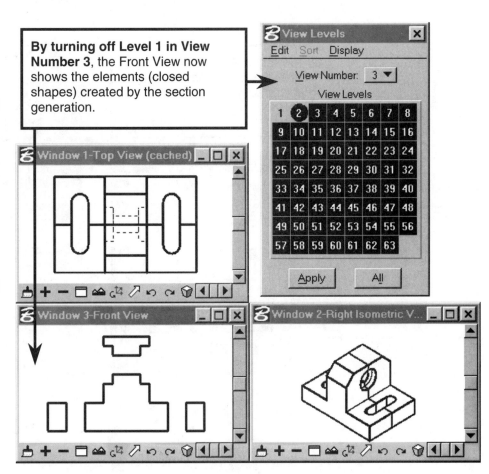

Now using the same Section by View–Horizontal, let's see what section you get by working in the **Front View** instead of the Top View.

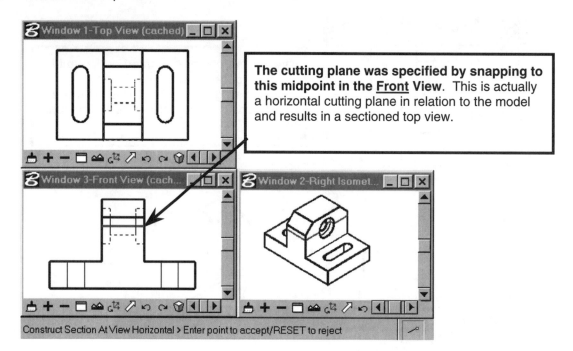

The cutting plane was specified by snapping to this midpoint in the <u>Front</u> View. This is actually a horizontal cutting plane in relation to the model and results in a sectioned top view.

Once again it may be difficult to see the elements that were created. By manipulating the levels on and off, we can see the section without the model getting in the way. This is illustrated on the right.

By turning off Level 1 in View Number 1, the Top View now shows the elements created by the section generation.

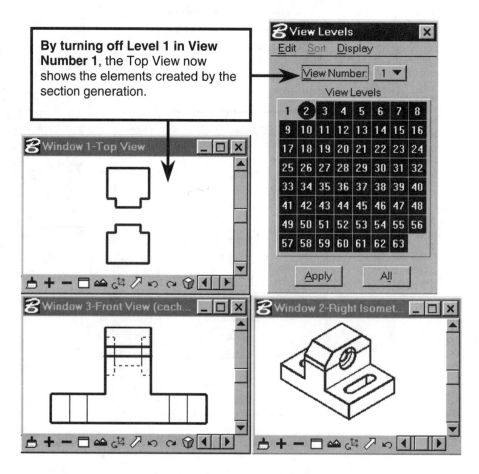

Vertical

With Section by View–Vertical, the cutting plane will be running vertical (straight up or down) **in the view selected.** Once again, working in different views will give you different results so we need to look at an example.

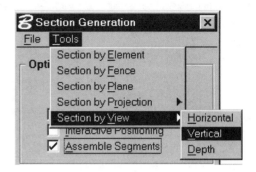

For Example: The settings will remain the same (Assemble Segments checked ON and the section elements assigned to Level 2) and we'll start by working in the **Top** View.

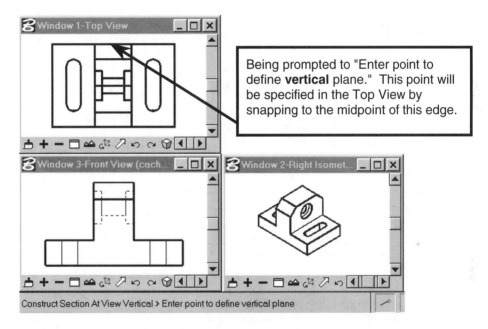

Being prompted to "Enter point to define **vertical** plane." This point will be specified in the Top View by snapping to the midpoint of this edge.

The elements created from the Section by View–Vertical and working in the Top View are shown below.

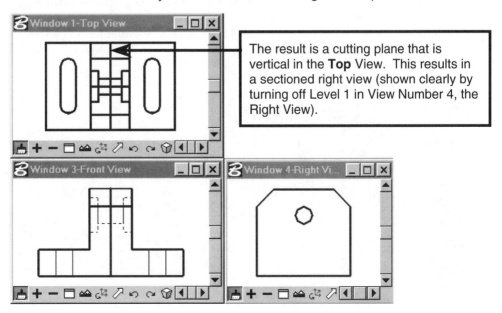

The result is a cutting plane that is vertical in the **Top** View. This results in a sectioned right view (shown clearly by turning off Level 1 in View Number 4, the Right View).

How about using the Section by View–Vertical and working in the **Right** View? Well, if you pick in the middle of the model, you'll get a sectioned front view as illustrated below. This is just like what we got using Section by View–Horizontal and working in the Top View. This is because the cutting planes would be at similar orientations.

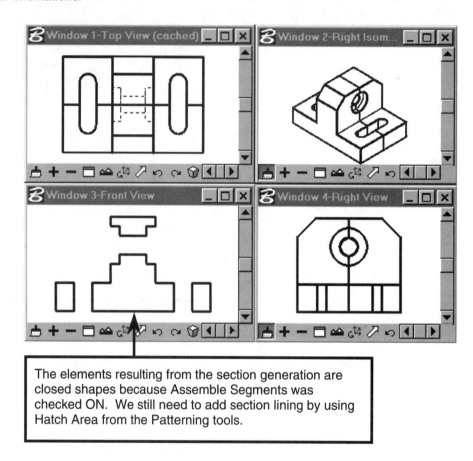

The elements resulting from the section generation are closed shapes because Assemble Segments was checked ON. We still need to add section lining by using Hatch Area from the Patterning tools.

SECTIONED VIEW IN A SHEET LAYOUT

So we're back to where we started and can now continue on with laying out our sheet view! First, let's review how to get to Hatch Area in the Patterning tools so that we can add section lining to indicate the solid material within our section. **It is important to have the Active Level set to the same level as the section elements (in our case Level 2) before using Hatch Area.** That way, all of the elements for the section are on the same level and can be manipulated together.

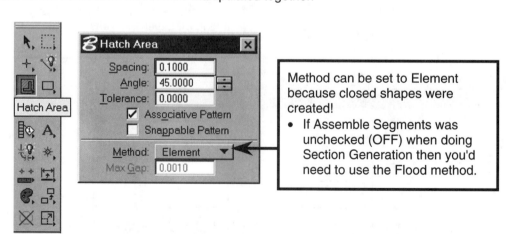

Method can be set to Element because closed shapes were created!
- If Assemble Segments was unchecked (OFF) when doing Section Generation then you'd need to use the Flood method.

Shown on the right is the result of using Hatch Area on the section. Display mode for Window 2 was set to wireframe so that it can be seen there also.

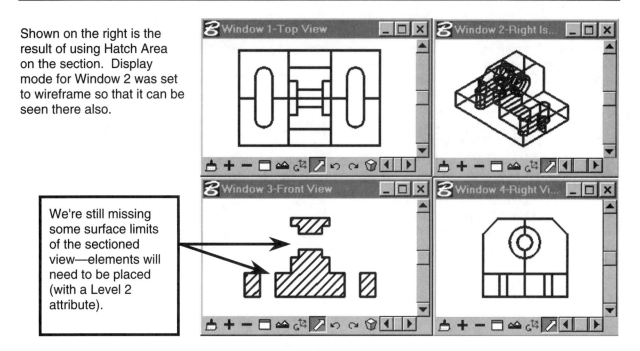

We're still missing some surface limits of the sectioned view—elements will need to be placed (with a Level 2 attribute).

Here are the regular window views that resulted from our section generation and adding the appropriate section lining and elements to represent the surface limits. The model is on Level 1 and the entire section is on Level 2. We're ready to use Drawing Composition to layout the sheet view.

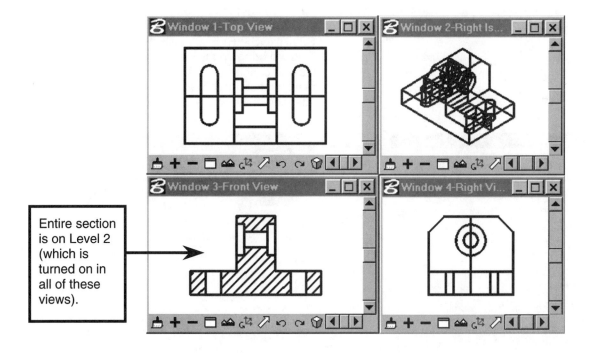

Entire section is on Level 2 (which is turned on in all of these views).

By using Drawing Composition, the views were attached to give the layout in the sheet view as shown on the right. All views attached had Hidden Line Removal checked ON, the Hidden Line Method set to SmartSolids, and Include Hidden Lines checked ON. The section appears as an edge in the top and right side views. The front view doesn't show the section correctly and the right isometric has some of the outline of the section. Some of this unusual representation is caused by the Hidden Line Removal. By using Level manipulation of the reference files everything can be adjusted to appear correctly.

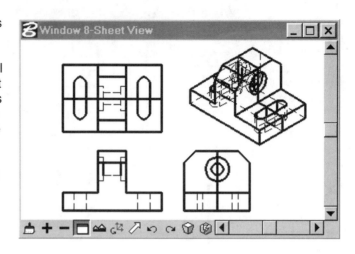

Reference Levels

In order to get the attached views to correctly display either section or model, we must manipulate the levels displayed in the reference files. This means we need to use the Reference Files dialog box. It is accessed as illustrated below on the left. You use the main pull-down menu FILE>REFERENCE.

When you get the Reference Files dialog box, use its pull-down menu of SETTINGS>LEVELS as illustrated below. This will bring up the Ref (erence) Levels dialog box that we need in order to do level manipulation for the reference files (our attached views in the sheet layout.)

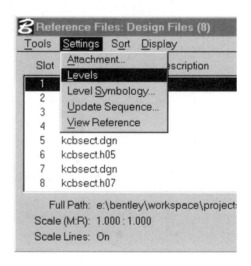

Shown below is the Reference Files dialog box. The reference file in Slot 1 that is the Front view attachment is highlighted. The Ref Levels dialog box shown below on the right shows the level settings for the Front view attachment in the Sheet View. Each different attachment has its own level controls.

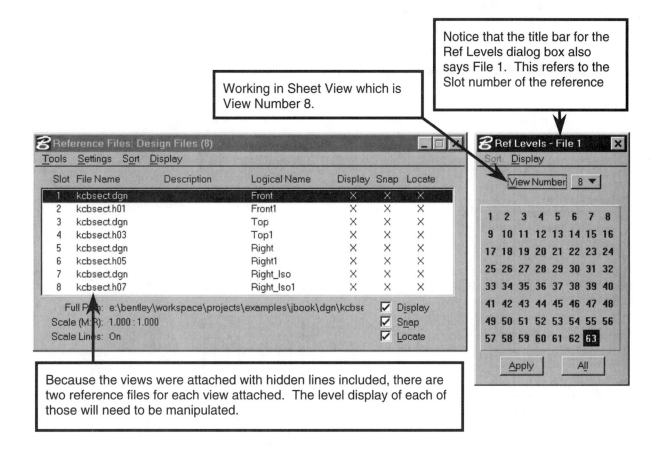

Because the views were attached with hidden lines included, there are two reference files for each view attached. The level display of each of those will need to be manipulated.

Illustrated below is the reference file in Slot 2 highlighted. It is the .h01 file that is the hidden lines for the Front view. Notice that its Ref Levels dialog box has different level settings. These will also need to be manipulated.

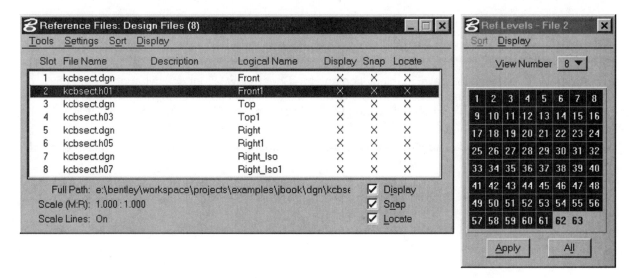

For the hidden lines file of the Front view, the model level (Level 1) will be toggled off as illustrated below. Remember to use the Apply button to actually apply the level display change.

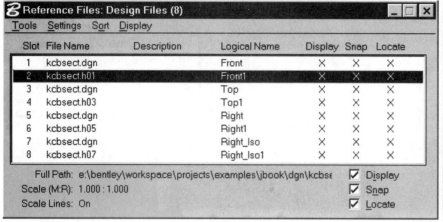

Here's the Sheet View layout that results from this level manipulation. The sectioned front view looks better but we're still missing section lines. That needs to be fixed by going to the other reference file attachment that is associated with the Front view.

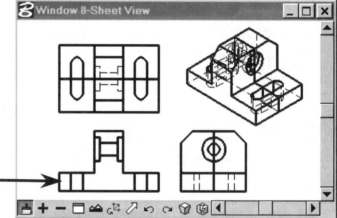

Still need to see the section lining—more level manipulation is needed.

WARNING: If you do a level manipulation and the Sheet View layout doesn't appear correctly, it may be necessary to do a Synchronize. This is accomplished by going to the Drawing Composition dialog box pull-down menu TOOLS>MODIFY HIDDEN LINE>SYNCHRONIZE.

Shown below is the reference file in Slot 1 highlighted. It is also associated with the Front view attachment and so Level 2 needs to be turned on as shown below on the right. Once again, be sure to use the Apply button to actually make the manipulation.

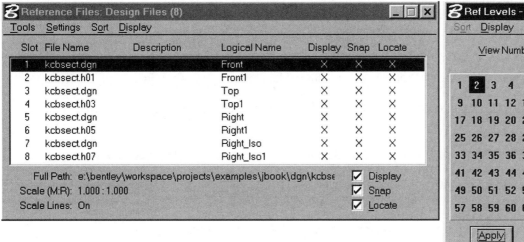
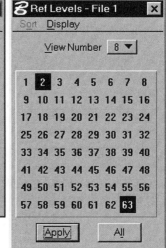

With Level 2 turned on in the Front view, the section lining now appears correctly as illustrated on the right. The other attached views will need to have level manipulation done so that the section does not appear in those views (turn Level 2 off in all of those associated reference files). This takes a bit of time and finesse. Remember that "synchronization" may be necessary in order to get the layout in the Sheet View to correctly reflect the changes.

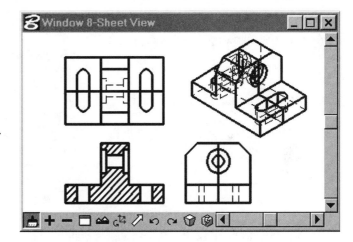

Shown here on the right is the final layout in the Sheet View. The level display for each attached view has been manipulated in the associated reference file(s). The right isometric had its hidden lines modified so that they are not displayed. Lines were drawn in the Sheet View to indicate the cutting plane in the top view. By printing the Sheet View (or a fence in the Sheet View) a hard copy of this layout of orthographic and sectioned views can be obtained.

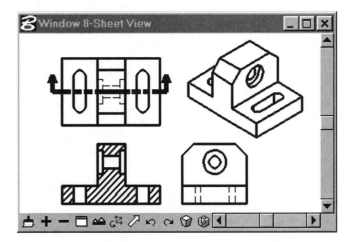

QUESTIONS

① Explain the relationship between Reference Files and Drawing Composition.

② Explain why the Hidden Line Removal setting is important and the difference between having it checked ON and having it unchecked (OFF)?

③ When attaching a border, what drawing plane should its elements be drawn in?

④ State the dialog box that contains the Hidden Edge Overrides settings and explain how you access the dialog box.

⑤ List the **three** different Hidden Line Method options, and give a brief explanation of each.

⑥ Explain the recommended order of view attachments for a 3D model that will result in the standard conventions of orthographic views. Why is it recommended?

⑦ Name the specific method of view attachment to use for placing auxiliary views.

⑧ What would cause the inability to snap to elements when dimensioning on attached views in a sheet view layout? State how to correct this problem.

⑨ A front view of a model was attached with the Drawing Composition View Parameters set as shown. Answer the following questions.

a) What would be the scale size of the attached front view?
b) Assume Scale Factor = 1. If the actual width of the model was 8 inches, what would be the width's dimensioned value shown in the front view?

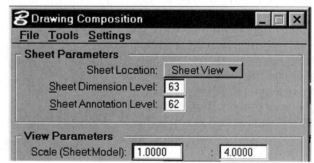

⑩ Explain how to open up the Saved Views dialog box.

EXERCISE 21-1 U-Bracket

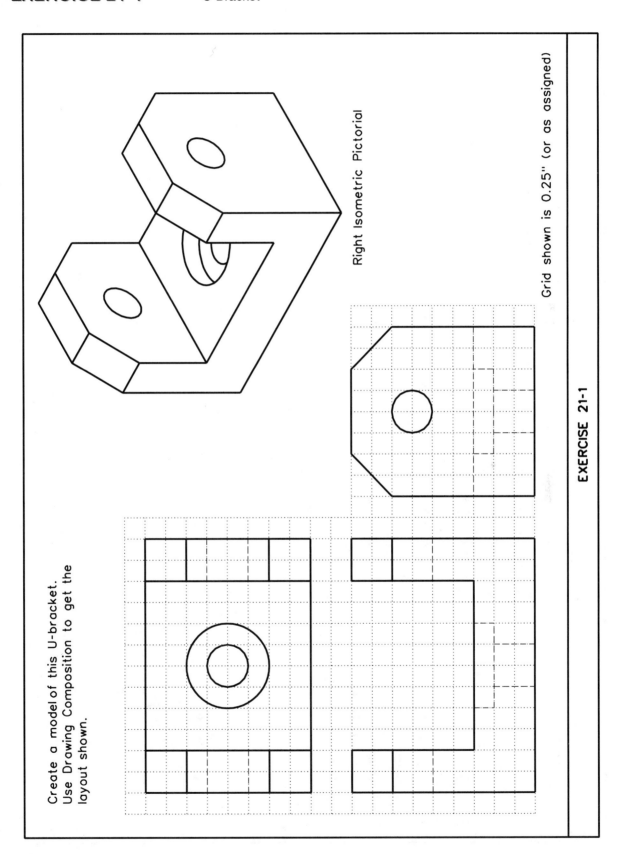

EXERCISE 21-1

EXERCISE 21-2   Pole Support

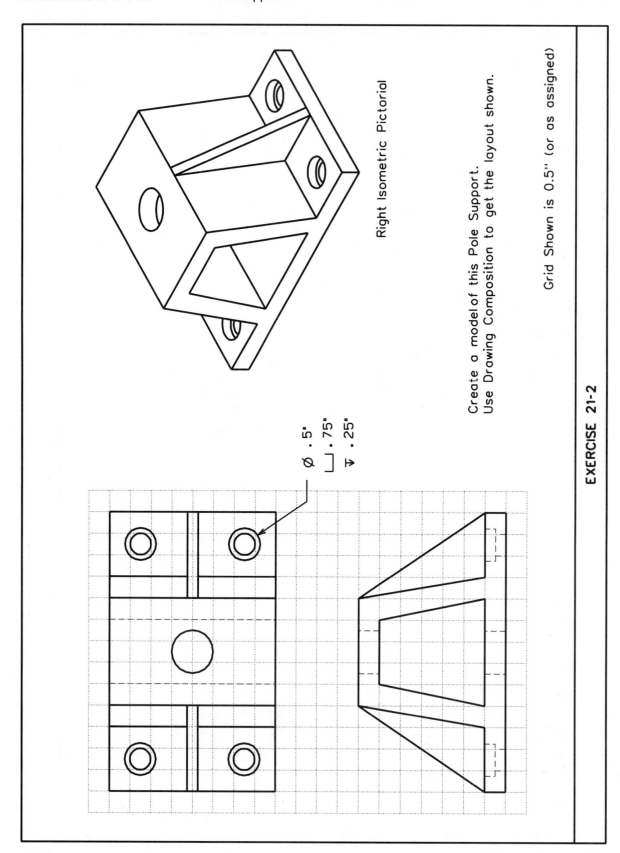

EXERCISE 21-2

Drawing Composition

EXERCISE 21-3   Dimensioning a Bracket Arm

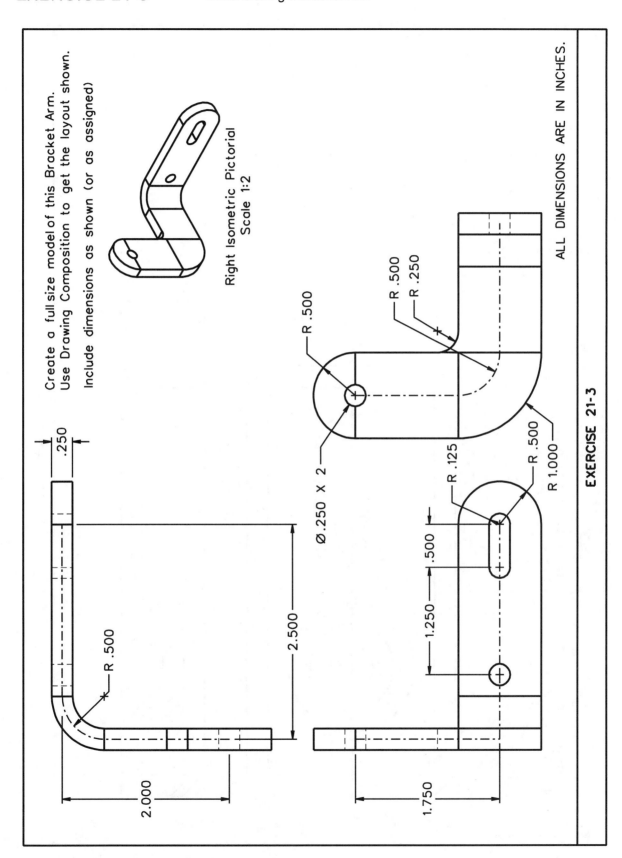

EXERCISE 21-3

EXERCISE 21-4

Slotted Object (Sectioned)

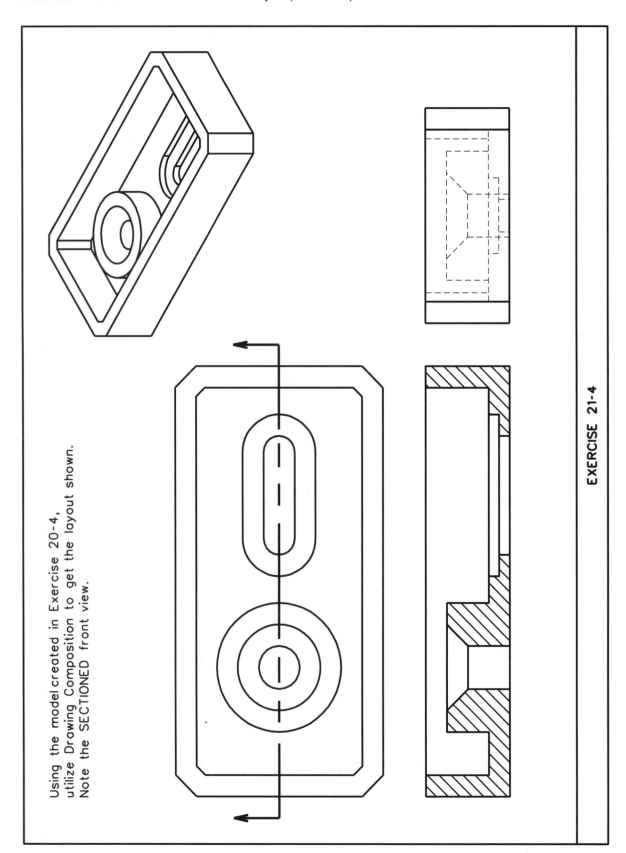

Using the model created in Exercise 20-4, utilize Drawing Composition to get the layout shown. Note the SECTIONED front view.

EXERCISE 21-5 Angled Bracket (Sectioned)

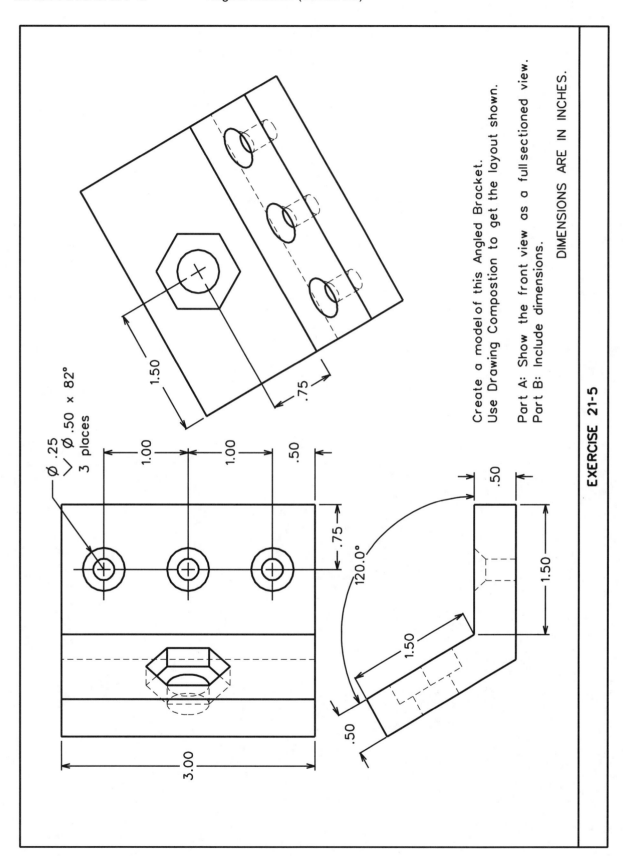

EXERCISE 21-5

Chapter 22: Engineering Configurations, ProjectBank, and Java

SUBJECTS COVERED:

- Engineering Configurations
- ProjectBank
- Java
- Future Direction of MicroStation

MicroStation/J is a great CAD software program all by itself. However, using additional applications that are customized for the various fields of engineering, architecture and building can really increase functionality and power. These specialized applications are called Engineering Configurations. These Engineering Configurations are advanced programs that utilize the MicroStation core product, extend its capabilities by adding a whole lot of new tools, and functions, and then customize its working environment. Therefore, when you start the special engineering configuration, it will have a specialized graphical interface with its own menus, tools and options. MicroStation/J is still involved, but because of the application it may not look exactly like what you've been seeing. For instance, the Main tool box may have some of the same tool boxes as MicroStation/J (but not necessarily all of them) and there will be some new tool boxes available. For a beginner, this can be a bit bewildering. You'll say, "Where is the Cells tool box?" and "What's this tool box?" That's okay—the stuff you've seen in the core product is still there but you may not need it right at your fingertips anymore. Standard MicroStation tools can still accessed by other means. Remember, by first learning MicroStation/J, you've gained the knowledge needed in order to branch out into the special Engineering Configurations. It is beyond the scope of this text to explain how to run all of these configurations. Instead we'll look at each configuration and its specialty. Then you'll know what is best suited for your needs and can take it from there.

ENGINEERING CONFIGURATIONS

We'll look at the Engineering Configurations that are available with the academic version of MicroStation/J. They include at this time: MicroStation Modeler, MicroStation TriForma®, MicroStation Geographics®, CivilPAK™, and MicroStation Schematics™. Each of these configurations is an application developed for a specific area. What an architect needs differs from what a manufacturing engineer needs. By using a different Engineering Configuration, they can each have the tools and functions they need to get the job done.

MicroStation Modeler

This application addresses the needs of mechanical and manufacturing engineers. It expands on the SmartSolids technology and enhances the 3D modeling capabilities. The Feature tool box allows you to add features such as counterbored or countersink holes, ribs, and bosses to a 3D model. These features can easily be arrayed, mirrored, and rotated. There are also tools that allow you to create a Sheet View layout quickly and easily with a click of the mouse. It is much more streamlined than Drawing Composition, but, in fact, still uses the concepts involved there. They've just automated things at bit more for you!

MicroStation Modeler also has the ability to create assemblies of parts, analyze them, and easily make modifications. It uses the ParaSolid® solid-modeling kernel and is definitely the way to go if you want to do extensive 3D modeling of mechanical parts.

MicroStation Triforma

This engineering configuration packs a lot of power into it. It is tailored especially for architects and has a whole range of functionality just for building and plant design. This is an extremely detailed piece of programming that takes a project through the modeling, design, reporting and documentation stages. It has the ability to work on a combination of both raster data (such as scanned images) and vector data (such as line elements from MicroStation) in the design process and to utilize this for presentations.

Triforma does involve a great deal of 3D modeling, and it too uses the SmartSolids technology. Rather than just primitives, it has a broader range of 3D modeling tools such as forms. Forms represent a quick way to model standard building components such as walls and columns. In order to handle the reporting and documenting stages of a project like bill of materials and quantities, part information is associated to a form. The information can be graphical (such as dimensions) but can also contain a non-graphical component such as name or material. This part information can be sorted, calculated, and extracted. Parts can be grouped into families, which also helps in the organizational aspect of the design process.

Triforma not only makes models of buildings and plants, but the pieces that make up the models contain a vast amount of information that can also be utilized.

Another strength of Triforma is its compound cells. A compound cell consists of both 2D and 3D portions that allow the cell to behave differently in rendering 3D model than in a 2D document derived from the model. Compound cells that are available with Triforma include typical items found in a building such as doors, windows, and even toilets. So you don't have to model a door each time; instead you can just pick out one from the "catalog" of compound cells. Some compound cells contain "perforators" that actually cuts a hole in a solid when needed. For instance, you don't need to subtract a portion out of a wall and then place a door in the open spot. Instead you have a solid wall and place a door compound cell. The doorframe actually cuts the solid wall at the proper size so that the doorframe fits. It even knows whether to show up as a left-hand door or a right-hand door on the plans. Pretty slick stuff.

Triforma is not for the beginner. It requires some existing building and architectural knowledge in addition to MicroStation experience. But with a good background, it is an extremely powerful and productive piece of software.

MicroStation Geographics

MicroStation Geographics is suited for people using MicroStation as the graphics for a GIS system. GIS is short for Graphical Information System. A GIS system links your graphical information to a database. The graphical information can consist of vector and raster maps that are "real-world" representations. The sizes are precise to actual geography. This graphical information can be directly linked to a database that maintains information about the "real world" represented. This database information can be queried, sorted, and extracted for reports. Since the graphics and data are linked together, you have the ability to manipulate one and have the other respond. For instance, say you have a map of a subdivision that has a database with lot size, cost, owner, taxes, and school affiliation. If you wanted to know which lots are larger than 1 acre, you could query the database. It would then sort out the lots that fall into the criteria. Consequently, these lots can then be indicated graphically on the map. MicroStation Geographics streamlines the management of the maps and information in a process suitable for this specialty.

CivilPAK

Civil engineers—this one's for you. The CivilPAK engineering configuration uses GEOPAK, an application suited for roadway design projects. CivilPAK deals with COGO (coordinate geometry) necessary in laying out horizontal and vertical alignments and creating cross-sections, plans, and profiles, all of which are important concepts in civil engineering and the transportation field. Digital terrain modeling is also a significant 3D aspect of this engineering configuration.

MicroStation Schematics

This engineering configuration is for creating schematics with intelligence. Instead of graphics elements such as line and circles representing real-life piping layouts, you actually place components such as check-valves, reducers, and pumps. These components are "intelligent" even if they are represented as a schematic. There are a large number of different configurations within MicroStation Schematics itself. They require an extensive knowledge base in this specialized field.

Future of Engineering Configurations

The Engineering Configurations will also figure heavily into the future direction of MicroStation, which is Engineering Component Modeling (ECM). ECM utilizes an object-oriented environment (which is now possible because of Java) that creates "smart" objects rather than graphic elements. More on objects later; just realize that Engineering Configurations will someday evolve into different Engineering Component Modeling software.

PROJECTBANK

ProjectBank is a new concept that is being introduced in conjunction with MicroStation/J. Its intention is to eliminate the file management headaches that often accompany large engineering projects. ProjectBank acts like a large database that serves as one big storage unit for the elements contained in the design files associated with a single project. For each project, there would be a ProjectBank database set up that would store all of the project's graphic elements. Currently, there is ProjectBank DGN that stores and manages the graphic elements from **dgn** files. The graphic elements are equivalent to the current MicroStation/J elements. That will change in the future with the introduction of Engineering Component Modeling, but we'll talk about that later. Right now let's look at some of the new concepts that are involved in ProjectBank: Server-based Usage, Database Functionality, and History Information.

Server-based Usage

ProjectBank is a central storage so its usage is more like a server-based network. The information is in one location (server) and more than one user can access it at this location. This "Server-based Usage" is especially suited for a large company with lots of people working on one project together. More than one person can be working on the same graphic elements simultaneously—without the use of reference files. Now what if you are a single user? Well, ProjectBank just sets up on your local machine and only you have access to this "local" server.

Database Functionality

If the one huge file that stores all the graphic elements can be accessed by more than one person at a time, how does ProjectBank keep everything straight? Well, since it is just one big database, a lot of the jargon and concepts of a database can be applied to ProjectBank. Terms such as commit, transaction, synchronize, and conflict are common in ProjectBank because it functions like a database. Let's look at a very simple scenario of how two users would work on a single project in ProjectBank.

User A checks out the Kiara Ranch project first. User A plans on making modifications to the existing line element that represents a fence-line by adding a gate. Meanwhile User B also checks out the Kiara Ranch project with the intent of placing a complex shape to represent a pond that the owner requested. Each user has a briefcase that is storing all the changes and modifications that that user makes. **Both of the users are working in an interface that looks and behaves just like MicroStation/J.** The only difference is that the elements are not in a design file but rather in the ProjectBank database for the Kiara Ranch project. This is illustrated below.

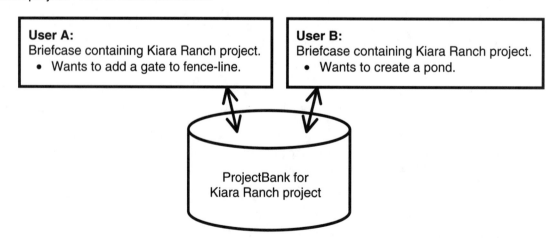

Now User B gets done making the pond. Since it is still just stored in User B's briefcase, a transaction is necessary to take the pond elements and commit them to the database. User B must **commit** the changes to the database. This is illustrated below.

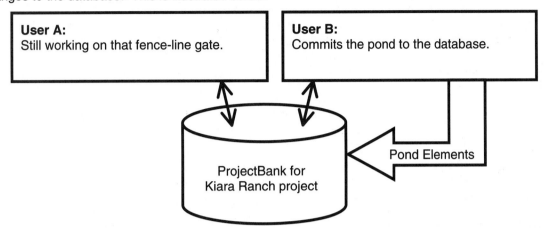

Now User A is alerted that a transaction has been made at the database. User A must **synchronize** in order to be up-to-date with the new stuff in ProjectBank. Meanwhile, after committing the pond, User B realizes that there is a fence-line running through the pond! User B now starts moving the fence-line. This is the same fence-line that User A is modifying by putting in a gate. There's trouble a brewin'.

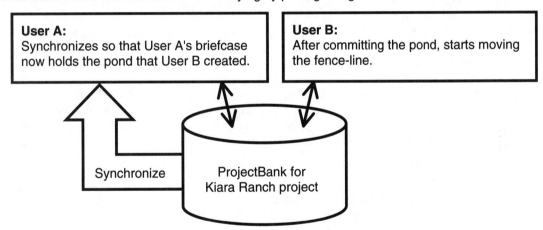

The link between the briefcases that each user has and the database is a smart one! If ProjectBank sees that an element or elements are being modified at the same time by different people—a red flag goes up. This is not true literally, but what happens is that a **conflict** is posted. This is illustrated below.

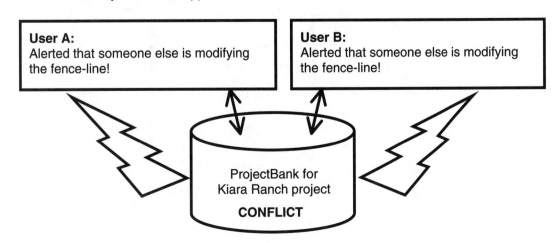

Now, just like in real life, the conflict between the two users must be resolved. In the case of the Kiara ranch project, User A agreed that moving the fence-line was acceptable. Upon resolution of the conflict, the fence-line was moved, and the changes that resulted in the gate were committed too. When the users are done with their work, they empty out their briefcases and close them by checking the database information back into ProjectBank. The database has grown and it contains the new elements as well as the changes to the existing elements.

ProjectBank for Kiara Ranch project
- New Pond
- Relocated Fence
- Gate in Fence

History Information

One of the things that you can do with ProjectBank that isn't possible with a .dgn file is track the history of the project. ProjectBank stores all of the history of each element in its database. Because all of this information is stored in the database, it can be accessed. So you are not limited to seeing the current elements. You can go back to any point in the history of the project, and see what's there and where it came from. Not only does the ProjectBank for the project store the graphical elements but it also stores other information such as which user was involved, any comments that were entered, and when the changes to the database were committed. All of this information is available: You just need to query the database. The amount of history information available is a major characteristic of ProjectBank that is not available in a regular .dgn file.

Let's revisit our simple scenario. A couple of weeks after User A and User B made their changes, suppose User C (in charge of User A and User B) wants to check on the progress of the Kiara Ranch project. User C opens up the ProjectBank for the project. **Again, User C has a briefcase that contains the information from the database but the interface looks and behaves like MicroStation/J.** User C notices the new complex shape representing the pond and the new gate in the fence-line. User C doesn't like the fact that the fence-line was moved and wants to know what happened! User C can go back in the history of the database and see when the fence was moved and any comments. The fence-line can also be moved back to its original location without much trouble. This ability to track the changes to each graphic element and review the history of the project's design is a primary feature of ProjectBank. Without it—User C would need to go around asking, "Why was this fence-line moved?"

User C:
Briefcase containing the Kiara Ranch project.
- Reviews the history to see why the fence-line was relocated.

ProjectBank for the Kiara Ranch project
- New Pond
- Relocated Fence
- Gate in Fence

JAVA

Java is a programming language that has the advantages of being platform independent. It can run on any computer that has the Java Virtual Machine. It is truly object-oriented and is an industry standard. Java is also Internet-ready. MicroStation/J has incorporated the Java Virtual Machine directly into the software. This allows MicroStation to run Java applets and applications. An applet runs within a browser. It is more secure and safer across the Internet. An applet has no local access to your hardware and software. MicroStation's ability to run applets increases its capacity for sharing data and programs over the Internet. More importantly, with Java MicroStation/J steps into the realm of an object-oriented environment for CAD. It introduces the concepts of "objects" that combine data and programming code. Currently CAD elements contain data only. **True objects contain data and programming logic**, which makes them smart. This object-oriented environment leads to a new way of solving traditional design situations referred to as Engineering Component Modeling (ECM). By having the ability to utilize Java in MicroStation/J, the way is paved for component modeling.

So what exactly is a "smart" object? Imagine you have an object that represents a free-flow pipe. As a graphic element, it can have a length and other attributes but doesn't really "know" it's a pipe. As an object, the pipe can "know" that when it reaches over 20 feet, it will need a joint (and a joint object will be placed). The pipe also knows that a clean-out will be needed at every 100 feet. Again, the pipe is smart enough to adjust and place a clean-out. Because there is programming code combined with the data, the object is "smart".

Those are not the only strong points of an object! One object can be used to "evolve" into another object. For instance, that free-flow pipe object could also be utilized as a pressurized pipe. The object would evolve so that instead of a clean-out, check valves are placed. Additional programming could be added that allows for capped ends and gages. The two types of pipes are related, but one has additional programming. By using an object, the process is more efficient because of the reuse and modification of data.

FUTURE DIRECTION OF MICROSTATION

MicroStation/J is just a stepping-stone to implementation of true object-oriented engineering design—it isn't there yet. Currently, the elements that make up the design file are still made up of just data, so they aren't full-fledged objects yet. If you work with MicroStation/J in the traditional mode, you will still have a .dgn file and the workflow will be the same as in the previous versions of MicroStation. If you work with MicroStation in conjunction with the ProjectBankDGN, your workflow will change to more of a client/server type application as discussed earlier. Instead of opening a .dgn file, the elements will be stored in the database and you will "check out" what you want to work on. These elements will be assembled into typical .dgn format that can be worked with in the traditional manner. When you are done working, you can commit your changes to the database. ProjectBankDGN will then store the new elements in the database.

If MicroStation is just a stepping-stone, then what's the destination? Looking down the road, there will be a new version of ProjectBank (tentatively referred to as ProjectBankECM) that will implement only true object-oriented engineering design. This means that elements (objects with just data) will be replaced by **true** objects combining data and programming code. These objects will require the implementation of a database and the workflow that goes along with it. This workflow includes incorporating new design methods based on Engineering Component Modeling, an evolution from the current Engineering Configurations.

Appendix

Seed File Information: View Attributes and Active Settings

SEED FILE INFORMATION—VIEW ATTRIBUTES

Defaults for View Attributes settings for View Window 1

X checked ON
– unchecked (OFF)

View Attribute	ARCHSEED.DGN	SDARCH2D.DGN	SDARCH3D.DGN	CIV2D.DGN	CIV3D.DGN	MAP2D.DGN	MAP3D.DGN	SDMAP2D.DGN	SDMAP3D.DGN	SDMAPM2D.DGN	SDMAPM3D.DGN	MECHDET.DGN	MECHDETM.DGN	MECHLAY.DGN	MECHLAYM.DGN	SDMECH2D.DGN	SDMECH3D.DGN	SDMENG2D.DGN	SDMENG3D.DGN	2DM.DGN	3DM.DGN	SCHEM2D.DGN	SCHEM3D.DGN	SDSCH2D.DGN	SDSCH3D.DGN	SEED2D.DGN	SEED3D.DGN	SEEDZ.DGN	SHEETSD.DGN	TRANSEED.DGN
Text Nodes	X	X	X	X	X	X	X	X	X	X	X	·	X	·	X	·	X	·	X	·	·	X	X	X	X	·	X	X	X	X
Text	X	X	X	X	X	X	X	X	X	X	X	X	X	X	X	X	X	X	X	X	X	X	X	X	X	X	X	X	X	X
Tags	X	X	X	X	X	X	X	X	X	X	X	X	X	X	X	X	X	X	X	X	X	X	X	X	X	X	X	X	X	X
Ref Boundaries	·	·	·	·	·	·	·	·	·	·	·	·	·	·	·	·	·	·	·	·	·	·	·	·	·	·	·	·	X	·
Patterns	X	X	X	X	X	X	X	X	X	X	X	X	X	X	X	X	X	X	X	X	X	X	X	X	X	X	X	X	X	X
Line Weights	X	X	X	X	X	X	X	X	X	X	X	X	X	X	X	X	X	X	X	X	X	X	X	X	X	X	X	X	X	X
Line Styles	X	X	X	X	X	X	X	X	X	X	X	X	X	X	X	X	X	X	X	X	X	X	X	X	X	X	X	X	X	X
Level Symbology	·	·	·	·	·	·	·	·	·	·	·	·	·	·	·	·	·	·	·	·	·	·	·	·	·	·	·	·	·	·
Grid	X	X	·	X	·	X	X	·	·	·	·	·	·	·	·	·	·	·	·	·	·	·	X	·	·	X	·	·	·	·
Fill	X	·	·	·	·	·	·	·	·	·	·	·	·	·	·	·	·	·	·	X	X	·	·	·	·	·	·	·	·	·
Fast Ref Clipping	·	·	·	·	·	·	·	·	·	·	·	·	·	·	·	·	·	·	·	·	·	·	·	·	·	·	·	·	·	·
Fast Font	·	·	·	·	X	·	·	·	·	·	·	·	·	·	·	·	·	·	·	·	·	·	·	·	·	·	·	·	·	·
Fast Curves	·	·	·	·	X	·	·	·	·	·	·	·	·	·	·	·	·	·	·	·	·	·	·	·	·	·	·	·	·	·
Fast Cells	·	·	·	·	·	·	·	·	·	·	·	·	·	·	·	·	·	·	·	·	·	·	·	·	·	·	·	·	·	·
Data Fields	X	X	X	X	X	X	X	X	X	X	X	X	X	X	X	X	X	X	X	X	X	X	X	X	X	X	X	X	X	X
Dynamics	X	X	X	X	X	X	X	X	X	X	X	X	X	X	X	X	X	X	X	X	X	X	X	X	X	X	X	X	X	X
Dimensions	X	X	X	X	X	X	X	X	X	X	X	X	X	X	X	X	X	X	X	X	X	X	X	X	X	X	X	X	X	X
Constructions	X	X	X	X	X	X	X	X	X	X	X	X	X	X	X	X	X	X	X	X	X	X	X	X	X	X	X	X	X	X
Camera	·	·	·	·	X	·	·	·	·	·	·	·	·	·	·	·	·	·	·	·	·	·	·	·	·	·	·	·	·	·
Background	·	·	·	·	·	·	·	·	·	·	·	·	·	·	·	X	X	·	·	·	·	·	·	·	·	·	·	·	·	·
ACS Triad	·	·	·	·	·	·	·	·	·	·	·	·	·	·	·	·	·	·	·	·	·	·	·	·	·	·	·	·	·	·

Path prefixes (all under BENTLEY\WORKSPACE\):
- PROJECTS\EXAMPLES\ARCH\SEED\: ARCHSEED.DGN, SDARCH2D.DGN, SDARCH3D.DGN
- PROJECTS\EXAMPLES\CIVIL\SEED\: CIV2D.DGN, CIV3D.DGN
- PROJECTS\EXAMPLES\MAPPING\SEED\: MAP2D.DGN, MAP3D.DGN, SDMAP2D.DGN, SDMAP3D.DGN, SDMAPM2D.DGN, SDMAPM3D.DGN
- PROJECTS\EXAMPLES\MECHDRFT\SEED\: MECHDET.DGN, MECHDETM.DGN, MECHLAY.DGN, MECHLAYM.DGN, SDMECH2D.DGN, SDMECH3D.DGN, SDMENG2D.DGN, SDMENG3D.DGN
- SYSTEM\SEED\: 2DM.DGN, 3DM.DGN, SCHEM2D.DGN, SCHEM3D.DGN, SDSCH2D.DGN, SDSCH3D.DGN, SEED2D.DGN, SEED3D.DGN, SEEDZ.DGN, SHEETSD.DGN, TRANSEED.DGN

SEED FILE INFORMATION—ACTIVE SETTINGS

File (BENTLEY\WORKSPACE\)	MU Unit	SU Unit	Sub Units per MU	PU per SU	Working Area (MU²)	Grid Ref	Grid Master (PU)	Active Level	Active Color	Active Style	Active Weight	Active Font	Active Text Height (PU)	Active Text Width (PU)	Active Line Spacing (PU)	Snap Lock	Snap Divisor
PROJECTS\EXAMPLES\ARCH\SEED																	
ARCHSEED.DGN	-	-	12	8000	44739	12	8000	1	0	0	0	0	96000	96000	48000	on	1
SDARCH2D.DGN	-	-	12	8000	44739	12	8000	1	0	0	0	41	72000	48000	24000	on	2
SDARCH3D.DGN	-	-	12	8000	44739	12	8000	1	0	0	0	41	96000	96000	24000	on	2
PROJECTS\EXAMPLES\CIVIL\SEED																	
CIV2D.DGN	FT	TH	10	100	4294967	10	1000	1	3	0	6	3	1500	1500	5000	on	1
CIV3D.DGN	FT	TH	10	100	4294967	10	12700	1	4	0	0	0	25400	31750	25400	on	1
PROJECTS\EXAMPLES\MAPPING\SEED																	
MAP2D.DGN	FT	TH	10	100	4294967	10	1000	1	0	0	0	3	2500	2500	5000	on	1
MAP3D.DGN	FT	TH	10	100	4294967	10	1000	1	0	0	0	3	2500	2500	5000	on	1
SDMAP2D.DGN	ft	th	10	100	4294967	10	100	1	0	0	0	3	1000	1000	500	on	2
SDMAP3D.DGN	ft	th	10	100	4294967	10	100	1	0	0	0	3	1000	1000	5000	on	2
SDMAPM2D.DGN	m	mm	1000	10	429496	1000	10	1	0	0	0	3	20000	20000	30002	on	2
SDMAPM3D.DGN	m	mm	1000	10	429496	1000	10	1	0	0	0	3	60000	60000	30001	on	2
PROJECTS\EXAMPLES\MECHDRFT\SEED																	
MECHDET.DGN	in	th	1000	254	16909	1000	254	1	0	0	0	3	1200	1200	7500	on	1
MECHDETM.DGN	mm	mm	1000	100	42949	100	100	1	1	0	0	3	3000	3000	1500	on	1
MECHLAY.DGN	in	th	100	254	16909	1000	254	1	1	0	0	3	1200	1200	7500	on	1
MECHLAYM.DGN	mm	mm	100	100	42949	100	100	1	1	0	0	3	3000	3000	1500	on	1
SDMECH2D.DGN	in	th	10	254	16909	1000	254	1	0	0	0	3	508000	508000	127000	on	2
SDMECH3D.DGN	in	th	10	254	42949	1000	254	1	0	0	0	3	25400	25400	7500	on	2
SDMENG2D.DGN	mm	su	1000	100	42949	100	100	1	0	0	0	3	200000	200000	5000	on	2
SDMENG3D.DGN	mm	su	1000	100	429496	100	100	1	0	0	0	3	1000	1000	5000	on	2
SYSTEM\SEED																	
2DM.DGN	m	mm	1000	100	42949	10	10000	1	2	0	0	3	10000	10000	15000	on	2
3DM.DGN	m	mm	1000	100	42949	10	10000	1	2	0	0	3	10000	10000	15000	on	2
SCHEM2D.DGN	mm	th	100	1000	42949	100	1000	1	0	0	0	3	50000	50000	5000	on	2
SCHEM3D.DGN	mm	th	100	1000	42949	10	1000	1	0	0	0	3	1000	1000	5000	on	2
SDSCH2D.DGN	in	th	10	1000	42949	10	1000	1	0	0	0	3	60000	60000	5000	on	2
SDSCH3D.DGN	in	th	10	1000	42949	10	1000	1	0	0	0	3	1250	1250	5000	on	2
SEED2D.DGN	mu	su	10	10	429496	10	1000	1	2	0	0	3	10000	10000	5000	on	2
SEED3D.DGN	mu	su	10	1000	429496	10	1000	1	0	0	0	3	1250	1250	5000	on	2
SEEDZ.DGN	mu	su	10	1000	429496	10	1000	1	0	0	0	3	1250	1250	5000	on	2
SHEETSD.DGN	mu	su	10	1000	429496	10	1000	1	0	0	0	3	10000	10000	5000	on	1
TRANSEED.DGN	mu	su	10	1000	429496	10	1000	1	0	0	0	3	10000	10000	5000	on	1

INDEX

A
absolute coordinates, 29, 33-34, 112
AccuDraw
  and ACS, 392-93
    Get ACS, 393
    Rotate ACS, 392
    Write to ACS, 393
  applications, 235-37
  compass, 229-32
  Context Sensitivity setting, 234
  coordinate system, 229-232
  Coordinate System Rotation setting, 390-91
  isometrics, 235-37
  keyboard shortcuts, 3, 249
  key-ins, 34
  placing text, 235
  and precision input, 34
  quitting, 228-29
  rotating view, 390-91
  settings, 233-34
  starting, 228-29
  in 3D, 390-91
  Unit Roundoff setting, 234
AccuDraw dialog box, 228-29
AccuDraw Settings dialog box, 233-34
ACS (Auxiliary Coordinate Systems), 382
  and AccuDraw, 392-93
    Get ACS, 393
    Rotate ACS, 392
    Write to ACS, 393
  ACS Plane Lock, 385
  ACS Plane Snap, 385
  ACS tool box, 384
  ACS triad, 382
  Auxiliary Coordinate Systems dialog box, 384
  defining by element, 386
  defining by points, 387
  defining by view, 388-89
  key-ins, 383
  naming, 393
  orientation, 388-89
  recalling, 393
  returning to drawing plane, 388-89
  rotating views, 392
  saving, 388
  uses of, 382-83
ACS Plane Lock, 385
ACS Plane Snap, 385
ACS tool box, 384
Active Angle Method, 212
Active Angle setting, 143
Active Attributes settings, 176
Active Scale Method, 211
Active Snap, 74
Align Edges tool, 216-17

alignment, dimensioning, 309, 320, 406
Analyze Element tool, 168
angle, measuring, 163-64
angle constraint, 114
angle dimensions, 324-26
animation, 396, 418-19
  FlyThrough, 418-425
  Keyframe, 418
  Parametric Motion Control, 418-19
  Path, 418
antialias shading, 399
Apply Material tool, 412-15
arcs
  Arcs tool box, 122
  Center Method, 123-24
  Construct Circular Fillet tool, 195-96
  Edge Method, 125
  example, 126
  extending, 186
  modifying, 181
  Place Arc tool, 123-26
  with SmartLine, 243
Arcs tool box, 122
area, measuring, 165-67
  and Boolean operations, 165-67
  Difference Method, 167
  Element Method, 165
  Flood Method, 167
  Intersection Method, 166
  Union Method, 166
Area Source Lighting, 408
Area type, 282
arrays, 218-20
  polar, 219-20
  rectangular, 218-19
arrowheads, 311
arrow symbols, 7
associative dimension, 319
associative pattern, 283
Attach Border, 495-96
Attach Folded View, 502-3, 506-7
  About Line, 511-12
Attach Reference File tool, 341-42
Attach Saved View, 499-502
Attach Standard View, 502
  Front, 505-6
  Orthogonal, 506-7
  Right Isometric, 507-8
Attach URL tool, 351
attaching views, 497-99
Attachment Parameters, 499
attributes
  Active Attributes settings, 176
  elements. *See* element attributes
  patterning, 295
  text, 138-42

view attributes, for rendering, 416
views, 63
Auxiliary Coordinate Systems. *See* ACS
Auxiliary Coordinate Systems dialog box, 384
Axis Lock, 47, 78
Axis setting, SmartSolids solid modeling, 434-35

B

background color, 34
base radius, 441
baseline dimensioning, 323
block fence, 87
blocks, modifying, 178-80
BMP format, 401
Boolean operations
  area measurement, 165-67
  Construct Difference, 451
  Construct Intersection, 452
  Construct Union, 449-450
  solids for, 445-48
borders
  attaching, 495-96
  plots, 101
  windows, 6
boundary of fence, 89, 101
bump map, 411
buttons
  command buttons, 7
  options buttons, 7
  tool buttons, 6, 8-9
By Edge Method, placing polygons, 130

C

calculator, 248
Cast Shadows setting, Define Materials dialog
    box, 408
.cel file extension, 256
Cell Library dialog box, 256, 271
Cell Selector utility, 271
cells, 256
  active angle, 264
  active cell, 263-65
  cell libraries, 256-59
    attaching an existing library, 257-58
    creating, 259
    detaching, 259
  Cell Library dialog box, 256, 272
  Create button, 272
  Delete button, 272
  deleting cell definition, 272
  Edit button, 272
  editing name and description, 272
  example of active cell placement, 265
  Graphic cell type, 256
  Interactive setting, 264
  nesting, 261
  Pattern button, 263
  Place Active Cell tool, 263-65
  Placement button, 263
  Point button, 263
  Point cell type, 256
  Relative setting, 264
  Share button, 272
  shared cells, 271
  Terminator button, 263
  types, 256
  XScale setting, 264
  YScale setting, 264
Cells tool box, 260
Center Method
  placing arcs, 123-24
  placing circles, 119-20
center size, dimensioning, 310
Chamfer Edges tool, 483
chamfers, 195-96, 242, 483
Change Attributes tool box, 174
Change Element to Active Area tool, 282
Change Elements Attributes tool, 175
Change Text Attributes tool, 148
Change View Display Mode, 379, 448
Change View Perspective tool, 369
Change View Rotation tool, 376
check boxes, 7
circles
  Arcs tool box, 122
  Center Method, 119-20
  diameter, 119
  Diameter Method, 121
  Edge Method, 120
  isometric circle, 237
  modifying, 180-81
  Place Circle dialog box, 119-21
  Place Circle tool, 118
  Place Ellipse tool, 122
  radius size, 119
Circumscribed Method, placing polygons, 129
Class attribute, 57
clicking, 5
closed complex shape, 238
closed element, 42, 238
closed shape, 238
closing windows, 6, 34
Color attribute, 56
command buttons, 7
compass, AccuDraw, 229-32
complex chain, 238
complex shapes, 247
cones, 441, 450
Constant Render Mode, 399
constrained values, 114-15
Construct Array tool, 218-20
Construct Chamfer tool, 196
Construct Circular Fillet tool, 195-96
Construct Revolution tool, 469-70
Construct Difference, 451
Construct Intersection, 452
Construct Union, 449-50
Construction class, 57
Context Sensitivity setting, 234
conventions used in text, 3
coordinate system, 29-31. *See also* ACS
  (Auxiliary Coordinate Systems)
  AccuDraw, 229-32
  positive right-hand coordinate system, 362
  for three-dimensional space, 363-64
Coordinate System Rotation setting, 390-91

coordinates, 29, 33-34, 112
  absolute, 29, 33, 112, 363
  information about, 168
  key-ins, 29, 34, 112
  polar, 29, 33
  rectangular, 29, 33
  relative, 29, 33, 363
Copy Attachment tool, 349
Copy tool, 205-7
copying
  elements, 205-7
  reference files, 351-52
correcting mistakes, 39-40, 57-58
Create Complex Shape tool, 247
Create Design File dialog box, 20-21
Create New Cell dialog box, 262
Create Region tool, 247
Crosshatch Area tool, 291
cursor, 91
Cut Solid tool, 479
cylinders, 439-40

D
data field, 149
  Fill In Single Enter-Data Field tool, 150-51
data field character, 149
data point, 5
Define By Circular, 471
Define By Profile, 472
Define Light dialog box, 408
  Shadow setting, 408
Define Light tool, 404-8
Define Materials dialog box, 408
Define Path tool, 421
Delete Pattern tool, 284
deleting. *See also* detaching
  cell definitions, 271
  Delete Fence Contents tool, 89
  Delete Pattern tool, 284
  Delete tool, 40
  Partial Delete tool, 182
Design File Settings dialog box, 28
design files
  coordinate system, 29-31
  creating, 20, 41
  display of, 40
  examples, 402
  saving, 36
  scaling down, 152-53
  working units, 29-31
Detach Reference File tool, 350
detaching. *See also* deleting
  cells, 259
  reference files, 350-51
  views, 509, 511
.dgn file extension, 12
diagonal lines, 48-49
dialog boxes, 6-7
  AccuDraw, 228-29
  AccuDraw Settings, 233-34
  Auxiliary Coordinate Systems, 384
  Cell Library, 256
  Create Design File, 20-21

  Create New Cell, 262
  Define Light, 408
  Define Materials, 408
  Design File Settings, 28
  Dimension Settings, 306-7
  Drawing Composition, 493
  Element Attributes, 54
  Export Visible Edges, 455
  FlyThrough Producer, 420
  Font, 139
  Generate Section, 475
  Hidden Line Settings, 498
  Level Manager, 64
  Level Names, 68
  Locks, 78
  MicroStation Manager, 7, 12-13
  Movies, 425-26
  Page Setup, 99, 103
  Place Regular Polygon, 129-30
  Place SmartLine, 238
  Place Text, 143
  Plot, 94-95
  Plot Layout, 100
  Plot Options, 101-2
  Print Setup, 99
  Reference File, 353
  Reference Files, 341
  Reference Levels, 345
  Rendering Setups, 417
  Rendering View Attributes, 408, 416
  Save Image, 400-401
  Set View Display Mode, 379
  Text, 138
  Text Editor, 143
  View Attributes, 63
  View Levels, 64-67
diameter, 119
Diameter Method, placing circles, 121
Difference Method
  area measurement, 167
  Crosshatch Area tool, 291
Dimension Angle Between Lines tool, 326
Dimension Angle Location tool, 325
Dimension Angle Size tool, 324
Dimension Element tool, 321-22
Dimension Lines category, 308
Dimension Location (Stacked) tool, 323
Dimension Radial tool, 327
Dimension Settings dialog box, 306-7
Dimension Size with Arrows tool, 322-23
Dimension tool box, 318-21
dimensioning, 304
  angle dimensions, 324-26
  baseline dimensioning, 323
  Dimension Angle Between Lines tool, 326
  Dimension Angle Location tool, 325
  Dimension Angle Size tool, 324
  Dimension Element tool, 321-22
  Dimension Location (Stacked) tool, 323
  Dimension Radial tool, 327
    Center Mark Mode, 327
    Diameter Extended Mode, 329
    Diameter Mode, 328

Radius Extended Mode, 328
Radius Mode, 327
Dimension Size with Arrows tool, 322-23
Dimension tool box, 318-321
  alignment, 320
  associative dimension, 319
  length of extension line, 320
  Geometric Tolerance tool, 330
  alignment, 309
  arrow head, 311
  center size, 310
  common settings, 307
  Dimension Lines category, 308
  Dimension Settings dialog box, 306-7
  Extension Lines category, 308
  minimum leader, 311
  orientation, 313
  Placement category, 309-10
  show secondary units, 316
  Terminators category, 310-12
  Text category, 313
  text height units, 305
  Tolerance category, 314
  type, 314
  Unit Format category, 317
  Units category, 315-16
  terminology, 304
  Update Dimension tool, 329
  vertices, 318
dimensions, Drawing Composition, 515-16
Display Level Names, 66-67
Display Level Numbers, 64
Distant Source lighting, 407
docking tool boxes, 8
double-clicking, 5
dragging, 5
  dynamic drag, 178
  elements, 178
  tool boxes, 8
  windows, 6
Drawing Composition, 353-54, 492
  Attach Border, 495-96
  Attach Folded View, 502-3, 511-12
  Attach Saved View, 499-502
  Attach Standard View, 502, 505-8
  attaching views, 497-99
  dimensions, 515-16
  Drawing Composition dialog box, 493
  levels, 515-16
  margins, 516
  modification of existing attached views, 509-10
  modifying a sheet view layout, example, 511-14
  printing, 516
  sequence, 492
    3D models, 492
    2D design files, 492
  Sheet View, 494-95
  sheet view layout, example, 503-8
  dialog box, 493
drawings
  adding text, 50-51
  creating, 37-52
  diagonal lines, 48-49
  editing, 51-52
  elements, 42
  knobs, 45-47
  parts of, 38
  printing, 52
  rectangular elements, 42-44
  standardizing. *See* cells
drivers, plotter, 103
Drop Element tool, 246
dynamic drag, 178

E

Edge Method
  placing arcs, 123-24
  placing circles, 120
Edit Light Source tool, 410-11
Edit Text tool, 145-46
editing
  cell names and descriptions, 271
  drawings, 51-52
  light source, 410-11
  sheet view, 508
  text, 145
element attributes, 54-57
  accessing active attributes, 54-55
  Change Element Attributes tool, 175
  changing, 81, 175
  Class attribute, 57
  Color attribute, 56
  Level attribute, 55
  Match Element Attributes tool, 176
  Style attributes, 56
  Weight attribute, 57
Element Attributes dialog box, 54
Element Method
  area measurement, 165
  Hatch Area tool, 286-89
Element Selection tool, 81
Element Selection tool box, 79-80
elements, 42
  Area type, 282
  changing, 174
  closed, 42, 238
  copying, 205-8
  dimension elements, 321-22
  dragging, 178
  dropping status, 246
  Extend Element to Intersection tool, 186
  Extend Elements to Intersection tool, 184-86
  grabbing, 178
  information about, 168
  IntelliTrim tool, 189-94
  joining, 245
  Manipulate tool box, 204
  Modify Element tool, 178-81
  modifying, 174
  modifying with SmartLine, 244
  moving, 208
  moving, parallel, 209
  open, 42
  rotating, 212-13
  scaling, 210-11
  selecting, 81-85

symmetric objects, 214-15
Trim tool, 187-88
ellipses
Ellipses tool box, 118
Place Ellipse tool, 122
Ellipses tool box, 118
Engineering Configurations, 536-37
CivilPAK, 537
future of, 537
MicroStation Geographics, 537
MicroStation Modeler, 536
MicroStation Triforma, 536
enhanced precision, 33
enter data field. *See* data field
Entity setting, 95-97
errors, correcting, 39-40, 57-58
example design files, 402
examples folder, 257
Export, 453-55
General, 454
Hidden, 454
Output, 455-56
Visible, 455
Extend Element to Intersection tool, 186
Extend Elements to Intersection tool, 184-86
Extend Line tool, 183-84
Extension Lines category, 308
Extrude tool, 465
Extrude along Path tool, 471-472

F
fence, 86
boundary of, 89, 101
copying elements, 205-7
Delete Fence Contents tool, 89
Fence Mode, 88
Inside, 88
Overlap, 88
Fence tool box, 86
Fence Type, 87
block, 87
shape, 87
placing, 86-88
rotating elements, 213
Fence tool box, 86
file extensions
.cel, 256
.dgn, 12
file management, 12-13
FILE menu, MicroStation Manager, 13
files. *See also* design files; reference files; seed files
image files, 401
naming, 20
palette files, 411, 413
plot files, 103
seed3d.dgn, 366
3dm.dgn, 366
Fill In Single Enter-Data Field tool, 150-51
fill pattern. *See* patterning
Filled Hidden Line Render Mode, 398
fillet, 195-96, 482
Fillet Edges, 482

Fit View tool
reference files, 351
three-dimensional space, 373
Fit View view control, 62
Flashbulb Global Lighting, 406
Flood Method
area measurement, 167
Hatch Area tool, 289-90
FlyThrough animation, 418-25
FlyThrough dialog box, 420
folded view, 502-3
fonts, 139
Fonts dialog box, 139
Format options, rendered images, 402

G
Generate Section dialog box, 517
Geometric Tolerance tool, 330
Get ACS, 393
Global Lighting, 405-6
global origin, 31-32
default, 32
grabbing elements, 178
Graphic cell type, 256
graphical user interface (GUI), 4
dialog boxes, 6-7
Help, 26-27
key-ins, 24
mouse actions, 5
pull-down menus, 10
status bar, 11
tool boxes, 8-9
tool buttons, 6, 8-9
windows, 6-7
grid, 75-76
example, 76
Grid Master setting, 75
Grid Reference setting, 75
locking, 78
parameters, 75
zooming, 76
grouping tool boxes, 10
Groups tool box, 245
GUI. *See* graphical user interface

H
hardcopies. *See* printing
Hatch Area tool, 285-91
Difference Method, 291
Element Method, 286-89
Flood Method, 289-90
Union Method, 291
height of text, 140
Help, 26-27
hidden line removal, 498
Hidden Line Render Mode, 398
highlight color, 34
holding, 5
Hole Area type, 282

I
icons, 8
borderless, 37

image files, 401
Imaging feature, 401
Immediately Save Design Changes, 36
Input category, 36
Inscribed Method, placing polygons, 129
Inside Fence Mode, 88
IntelliTrim tool, 189-94
   Advanced Mode, 192-94
   Quick Mode, 189-91
Interface setting, 19
Intersection Method, area measurement, 166
Intersection snap, 72
isometric circle, 237
isometrics, 235-37

J
Java, 541
joining elements, 245
JPEG format, 402
justification, 140

K
Keep Originals option, 444
keyboard shortcuts, AccuDraw, 232
Keyframe animation, 418
key-ins, 3, 24
   AccuDraw, 3
   ACS, 383
   for coordinates, 29, 112
Keypoint snap, 71
knobs, 45-47

L
leaders, 312
length, measuring, 164
length constraint, 114, 115
Level attribute, 55
levels
   Drawing Composition, 515-16
   names, 66-69
   numbers, 65
   reference levels, 345
   view levels, 64
lighting
   Ambient Global Lighting, 406
   Area Source Lighting, 408
   Define Light dialog box, 408
   Define Light tool, 404-8
   Distant Source lighting, 407
   Edit Light Source tool, 410-11
   Flashbulb Global Lighting, 406
   Global Lighting, 405-7
   Place Light Source tool, 408-9
   Point Source Lighting, 407
   shadowing, 399, 408
   Solar and Sky Global Lighting, 406
   Source Lighting, 407-8
   Spot Source Lighting, 407
line spacing, 140
line string, 238, 246
line weights, 57
Linear Elements tool box, 113
Linear Pattern tool, 294

lines. *See also* SmartLine
   chamfers, 196, 242
   diagonal, 48-49
   example, 116-18
   Extend Line tool, 183-84
   extending, 183-84, 186
   modifying, 180-81
   Place Line dialog box, 114-18
   shortening, 183-84
locks, 77-78
   accessing, 77
   ACS Plane Lock, 385
   axis lock, 47, 78
   grid lock, 78
   snap lock, 78
Locks dialog box, 78

M
main pull-down menu, 10
Main tool box, 10
Manipulate tool box, 204
margins, 100
   Drawing Composition, 516
Master Unit (MU), 29-31
Match Element Attributes tool, 176
Match Pattern Attributes tool, 295
Match Text Attributes tool, 147
Material Assignment Table, 411
materials, 411-12
   Apply Material tool, 412-15
   assigning, 414-15
   bump, 411
   definitions, 414
   Material Assignment Table, 411
   palette files, 411, 413-14
   pattern, 411
   transparent materials, 415-16
maximizing windows, 6
Measure Angle tool, 163-64
Measure Area tool, 165-67
Measure Distance tool, 161-63
Measure Length tool, 164
Measure tool box, 160
measurement
   Analyze Element tool, 168
   Measure Angle tool, 163-64
   Measure Area tool, 165-67
   Measure Distance tool, 161-63
   Measure Length tool, 164
   Measure Radius tool, 163
   Measure tool box, 160
   perimeter of closed shapes, 164
   perpendicular distance, 162-63
   between points, 161-62
menu bars, pull-down, 7
menus
   main pull-down menu, 10
   pull-down menus, 10
   Settings pull-down menu (Drawing Composition), 514
Merge Into Master tool, 349
messages, 23
MicroStation, 2, 541

MicroStation/J, 2, 541
  Active Angle field, 212
  Active Angle setting, 143
  animation, 418
  Element Selection tool, 79-80
  icons, 8
  IntelliTrim tool, 189-94
  MicroStation MasterPiece, 396
  Modify Element dialog box, 181
  Place SmartLine tool, 234
  Popup Calculator, 232
  PowerSelector tool, 82-85, 329
  Preview Reference, 297
  Reference File dialog box, 341
  Rendering Setup dialog box, 417
  Rendering Tools tool box, 403
  scaling method, 211
  SmartLine Modification Settings, 244
MicroStation Manager, 12-13
  FILE menu, 13
  opening files, 12-13
minimizing windows, 6
minimum leader setting, 312
Mirror tool, 214-15
  Mirror About Horizontal, 215
  Mirror About Line, 215
  Mirror About Vertical, 214
mistakes, correcting, 39
modeling. See SmartSolids solid modeling
modes
  PowerSelector tool, 85
  Render modes, 398-99
Modify Element dialog box, 181
Modify Element tool, 178-81, 244
  arcs, 181
  blocks, 178-80
  circles, 180-81
  lines, 180-81
Modify Solid tool, 475-77
Modify tool box, 177
mouse actions, 5
mouse buttons, 39
Move Parallel tool, 209
Move Reference File tool, 347
Move tool, 208
Movies animation, 425-26
Movies dialog box, 425-26
moving
  elements, 208
  parallel movement, 209
  views, 509, 513
MU (Master Unit), 29-31

N
naming conventions, 20
naming levels, 68-69
nesting cells, 261
noun/verb order, 188

O
open element, 42
Operation category, 35, 36
options buttons, 7

orientation
  ACS, 388-89
  dimension text, 313
  for printing, 99
Overlap Fence Mode, 88

P
Page Setup dialog box, 99, 103
palette files, 411, 413
Pan View view control, 62
parallel line patterning, 285-91
parallel movement of elements, 209
Parallel snap, 74
Parametric Motion Control animation, 418-19
Partial Delete tool, 182
Path animation, 418
Pattern Area tool, 292-93
Pattern button, 263
pattern map, 411
patterning, 282
  Area type, 282
  associative pattern, 283
  Crosshatch Area tool, 291
  Delete Pattern tool, 284
  displaying patterns, 282
  Hatch Area tool, 285-91
  Linear Pattern tool, 294
  Match Pattern Attributes tool, 295
  parallel lines, 285-91
  Pattern Area tool, 292-93
  Patterns tool box, 283
  Show Pattern Attributes tool, 295
Patterns tool box, 283
perimeter, measuring, 164
perpendicular distance, measuring, 162-63
Perpendicular snap, 73-74
Phong Render Mode, 399
Place Active Cell tool, 263-65
Place Arc tool, 123-26
Place Block tool, 42-44, 127-28
Place Circle dialog box, 119-21
Place Circle tool, 118
Place Cone tool, 441, 450
Place Cylinder tool, 439-440, 445-47
Place Ellipse tool, 122
Place Fence tool, 86-88
Place Light Source tool, 408-9
Place Line dialog box, 114-18
  constrained values, 114-15
Place Regular Polygon dialog box, 129-30
Place Slab tool, 436-38, 445
Place SmartLine dialog box, 238
Place SmartLine tool, 234, 238-44
Place Sphere tool, 438-39
Place Text dialog box, 143
Place Text tool, 143-45
Place Torus tool, 442
Place Wedge tool, 443
Placement button, 263
Placement category, 309
plot borders, 101
Plot dialog box, 94-95
plot files, 103

Plot Layout dialog box, 100
Plot Options dialog box, 101-2
Plot tool, 98
plotter drivers, 103
Point button, 263
Point cell type, 256
Point Source Lighting, 407
pointing, 5
polar arrays, 219-20
polar coordinates, 33
polygons
  Circumscribed Method, 129
  Constant Render Mode, 399
  By Edge Method, 130
  Inscribed Method, 129
  Place Block tool, 127-28
  Place Regular Polygon dialog box, 129-30
  Polygons tool box, 126
Polygons tool box, 126
Popup Calculator, 248
Positional Unit (PU), 29-31
positive right-hand coordinate system, 362
PowerSelector tool, 82-85, 329
  methods, 83-84
  modes, 85
precision input, 34, 228. *See also* AccuDraw
preferences, 34-36
Preview tool, 422-24
Preview Reference, 341
Preview Refresh tool, 98
Preview setting, 341
Primary class, 57
Primary Tools tool box, 8
Print Setup dialog box, 99
printer controls, 99
printing, 94-108
  Drawing Composition, 516
  drawings, 52
  Entity setting, 95-97
  fence boundary, 101
  margins, 100
  multiple copies, 98
  orientation, 99
  Page Setup dialog box, 99
  paper size, 99
  plot border, 101
  Plot dialog box, 94-95
  Plot Layout dialog box, 100
  Plot Options dialog box, 101-2
  Plot tool, 98
  plotter drivers, 103
  Preview Refresh tool, 98
  Print Setup dialog box, 99
  printer controls, 99
  printer name, 99
  reference files, 351
  scale, 100
  to scale, 104-7
  to scale, text size for, 152-53
  steps for, 108
ProjectBank, 538-40
  database functionality, 538-40
  history information, 540
  server-based usage, 538
Project setting, 13
prompts, 23
PU (Positional Unit), 29-31
pull-down menu bars, 7
pull-down menus, 10
  main pull-down menu, 10

R

radiosity, 399, 406
radius
  base radius, 441
  measuring, 163
  size, 119
Ray Trace Render Mode, 399
recalling an ACS, 393
Record tool, 425
rectangles
  Place Block tool, 127-28
  Place Regular Polygon dialog box, 129-30
rectangular arrays, 219-20
rectangular coordinates, 33
Redo tool, 26
Reference File dialog box, 353
reference files, 338
  accessing tools, 339-40
  active design referencing itself, 353-54
  Attach Reference File tool, 341-43
  capabilities, 338
  copying, 352
  Detach Reference File tool, 350
  Drawing Composition, 353-54
  Fit View tool, 351
  limitations, 338
  manipulation tools, 347
  Move Reference File tool, 347
  printing, 352
  Reference File dialog box, 353
  Reference Files dialog box, 341
  Reference Files tool box, 339
  Reference Levels dialog box, 345
  Reload Reference File tool, 351
  Scale Reference File tool, 348
  settings, 343-46
    Display, 344
    Locate, 344-45
    Snap, 344
  snapping, 344, 346
Reference Files dialog box, 340
Reference Files tool box, 339
Reference Levels dialog box, 345
reflection, 399
refraction, 399
refreshing, 59, 98
regions, 247
relative coordinates, 29, 33, 363
Reload Reference File tool, 351
Remove Face and Heal tool, 478
Render Modes, 398-99
  Constant, 399
  Filled Hidden Line, 398
  Hidden Line, 398
  Phong, 399

Ray Trace, 399
Smooth, 399
Wiremesh, 398
Render tool box, 397-99
rendering, 396
  Define Light tool, 404-8
  example design files, 402
  Format options, 401
  lighting
    Ambient Global Lighting, 406
    Area Source Lighting, 408
    Distant Source lighting, 408
    Edit Light Source tool, 410-11
    Flashbulb Global Lighting, 406
    Global Lighting, 404-7
    Place Light Source tool, 408
    Point Source Lighting, 408
    shadowing, 399, 408
    Solar and Sky Global Lighting, 406
    Source Lighting, 407-8
    Spot Source Lighting, 408
  materials, 411-12
    Apply Material tool, 412-15
    assigning, 414
    definitions, 414
    transparent materials, 415-16
  radiosity, 399, 406
  Render Modes, 398-99
    Constant, 399
    Filled Hidden Line, 398
    Hidden Line, 398
    Phong, 399
    Ray Trace, 399
    Smooth, 399
    Wiremesh, 398
  Render tool box, 397-99
  Rendering Setup dialog box, 417
  Rendering View Attributes dialog box, 416
  Save Image dialog box, 400-402
  saving rendered images, 399-402
  shading types, 399
    antialias, 399
    stereo, 399
  shadowing, 399, 408
  Target, 397
Rendering Setups dialog box, 417
Rendering Tools tool box, 403
Rendering View Attributes dialog box, 416
  Shadows setting, 408
Replace Cells tool, 266-69
  methods of, 266-67
  modes of, 268-69
reset/enter, 5
resizing windows, 6
resolution, 29-30
  changing, 31
Rotate ACS, 392
Rotate tool, 212-13
  Active Angle Method, 212
Rotate View tool, 377-78
rotating elements, 212-13
  with fence, 213
rotating 3D views, 377-78, 390-91, 392

S

Save Image dialog box, 400-401
Saved Views utility, 499-500
saving
  ACS, 388
  automatic, 36
  design files, 36
  rendered images, 399-401
  settings, 28
scale, 100, 104-107
  text size, for printing to scale, 152-53
Scale Reference File tool, 348
Scale tool, 210-11
  Active Scale Method, 211
scaling elements, 210-11
  Active Scale Method, 211
Screen coordinates, 364
scroll bars, 6
Section by View tool, 518-22
section lining. *See* patterning
sectioned views. *See* sectioning
sectioning
  Drawing Composition, 516
  Generate Section dialog box, 517
  horizontal, 518
  Section by View tool, 518-22
  section generation uses, 516
  sheet view, 522-27
  vertical, 521
seed files, 21-22
  global origin, 31-32
  MicroStation, 365, 430
  three-dimensional, 365, 430
  3D coordinate system, 363-64
seed3d.dgn file, 365
Select and Place Cell tool, 270
Set Active Depth tool, 373-74
Set Display Depth tool, 371-73
Set View Display Mode dialog box, 379
settings
  AccuDraw, 233-34
  changing, 28
  design file, 28
  dimensioning, 304-17
  Drawing Composition, 514
  Interface settings, 19
  reference files, 343-46
  saving, 28
  SmartSolids, 433-35
  Style, 19
shading, 399
shadowing, 399, 408
shape fence, 87
shared cells, 271
sheet view, 494-95
  drawing in, 508
  editing the model, 508
  example layout, 503-8
  modifying layout, 511-14
  sectioning, 522-27
Shell Solid, 473
Show Active Depth tool, 375

Show Display Depth tool, 375
Show Pattern Attributes tool, 295
slabs, 436-38, 445
Smart Tools, 228
  chamfered vertex, 242
  definitions, 238
  mixing vertex types, 242
  modifying elements, 244
  Place SmartLine tool, 238-44
  rounded vertex, 239-41
  sharp vertex, 239
  SmartSolids, 430-57
  straight lines, 239
SmartLine Modification Settings, 244
SmartMatch, 176
SmartSolids solid modeling
  common settings for, 433-435
  Construct Difference, 451
  Construct Intersection, 452
  Construct Union, 449-50
  Export, 453-57
  locating solids, 431
  Place Cone, 441, 450
  Place Cylinder, 439-40
  Place Slab, 436-38
  Place Sphere, 438-39
  Place Torus, 442
  Place Wedge, 443
  solids for Boolean operations example, 445-48
  3D Main tool box, 432
  3D Modify tool box, 444
  3D Primitives, 433
  3D seed file, 430
Smooth Render Mode, 399
snaps, 70-74
  accessing, 70-74
  ACS Plane Snap, 385
  Active Snap, 74
  Intersection, 72
  Keypoint, 71
  locking, 78
  overriding Snap Mode, 74
  Parallel, 74
  Perpendicular, 73-74
  reference files, 344, 346
  Snap Mode, 74
  Tangent, 72-73
Solar and Sky Global Lighting, 406
Solid Area type, 282
solids
  for Boolean operations, 445-48
  Construct Difference, 451
  Construct Intersection, 452
  Construct Union, 449-50
Source Lighting, 407-8
spheres, 438-39
spot light, 407
Spot Source Lighting, 407
Standard tool box, 25
standardized drawings. *See* cells
status, dropping, 246
status bar, 23
stereo shading, 399

Style attributes, 56
Style setting, 19
SU (Sub Unit), 29-31
symmetric objects, 214-15

T

Tangent snap, 72-73
Target, for rendering, 397
tearing off tool boxes, 8
templates. *See* seed files
tentative, 70-74
Terminator button, 263
Terminators category, 310-12
text
  adding to drawings, 50-51
  data field, 149-51
  dimensioning, 313
  Drawing Composition, 515-16
  Edit Text tool, 145-46
  Place Text dialog box, 143
  Place Text tool, 143-45
  placing with AccuDraw, 235
  size, for printing to scale, 152-53
text attributes, 138-41
  accessing settings, 138
  Change Text Attributes tool, 148
  fonts, 139
  height, 140
  justification, 140
  line spacing, 140
  Match Text Attributes tool, 147
  underlining, 141
  width, 140
text boxes, 7
Text category, 313
Text dialog box, 138
  View button, 139
Text Editor dialog box, 143
text fields, 7
text height units, 305
Text Node, 141
Text tool box, 141
Thicken to Solid, 474
three-dimensional space, 362. *See also* Drawing Composition; SmartSolids solid modeling; rendering
  AccuDraw in, 390-91
  ACS Plane Lock, 385
  ACS Plane Snap, 385
  active vs. display depth, 370
  Change View Perspective tool, 369
  Change View Rotation tool, 376
  coordinate system, 363-64
  depth tools, 370-75
  drawing coordinates, 363
  Fit View tool, 373
  positive right-hand coordinate system, 362
  Rotate View tool, 377-78
  rotating views, 376-78
  screen coordinates, 364
  seed files, 3D, 365
  Set Active Depth tool, 373-74
  Set Display Depth tool, 371-73

Show Active Depth tool, 375
Show Display Depth tool, 375
3D View Control tool box, 366-67
view controls, 366
View Rotation tool, 376
Zoom In/Out tool, 368
3dm.dgn file, 365
3D Modify tool box, 444
3D Primitives tool, 433
3D View Control tool box, 366-67
TIFF format, 402
title bars, 6
Tolerance category, 314
tool bars. *See* tool boxes
tool boxes, 8-9
  ACS, 384
  Arcs, 122
  Cells, 260
  Change Attributes, 174
  Dimension, 318-21
  docking, 8
  dragging, 8
  Ellipses, 118
  Fence, 86
  grouping, 10
  Groups, 245
  Linear Elements, 113
  Manipulate, 204
  Measure, 160
  Modify, 177
  Patterns, 283
  Polygons, 126
  Reference Files, 339-40
  Render, 397-99
  Rendering Tools, 403
  Standard, 25
  tearing off, 8
  Text, 142
  3D Construct, 464
  3D Modify, 444, 475
  3D View Control, 365-67
tool buttons, 6, 8-9
tool tips, 8
torus, 442
tracking Help, 10
transparent materials, 415
Trim tool, 187-95
  verb/noun order, 188
trimming, 188-95
Type setting, SmartSolids solid modeling, 433

U

underlining text, 141
Undo tool, 26
Union Method
  area measurement, 166
  Crosshatch Area tool, 291
Unit Format category, 317
Unit Roundoff setting, 234
Units category, 315-16
Update Dimension tool, 329
Update View view control, 59

V

verb/noun order, 188
vertices
  chamfered, 242
  dimensioning, 318
  mixing types, 243
  rounded, 240-42
  sharp, 239
view attributes, 63
  for rendering, 416
View Attributes dialog box, 63
View button, Text dialog box, 139
view controls, 58-62
  Fit View, 62
  Pan View, 62
  three-dimensional, 366
  Update View, 59
  Window Area, 61
  Zoom In, 59-60
  Zoom Out, 60
view coordinates, 364
view levels, 64-67
View Levels dialog box, 64
View Parameters, 497
View Rotation tool, 376
View Windows category, 35
views
  Attach Folded View, 502-3, 511-12
  Attach Saved View, 499-502
  Attach Standard View, 502, 505-8
  attaching, 497-99
  detaching, 509, 511
  Modify Hidden Line, 510, 514
  modifying existing attached views, 509-10
  moving, 509, 513
  sectioned. *See* sectioning
Void-Clip fence mode, 89

W

wedges, 443
Weight attribute, 57
width of text, 140
Window Area view control, 61
windows, 6-7
  background color, 34
  borders, 6
  closing, 6
  dragging, 6
  maximizing, 6
  minimizing, 6
  resizing, 6
  view windows, 6
Windows BMP format, 402
Wiremesh Render Mode, 398
Working Area, 30
working units, 29-31
  changing names and resolution, 31
Workspace, 14-19
  interface, 19
  project, 14-16
  setting, 14
  style, 19

user, 16-19
Write to ACS, 393

Z
Zoom In view control, 59-60
Zoom In/Out tool, 368
Zoom Out view control, 60
zooming, effect on the grid, 76